Horsekeeping on a Small Acreage

Horsekeeping on a Small Acreage

Designing and Managing Your Equine Facilities

Second Edition

Cherry Hill

Illustrations by
Richard Klimesh

Photographs
by Cherry Hill and
Richard Klimesh
(unless otherwise noted)

Storey Publishing

Edited by Deborah Burns

Art direction and cover design by Kent Lew

Front cover and spine photograph taken by Blake Gardner at O'Connor Farm, Rhinebeck, NY; back cover photographs by Richard Klimesh; photographs on page ii, vii, and 157 taken by Blake Gardner at Kildare Stables LLC, Millbrook, NY; timothy on page 223 © Peter Lilja/age fotostock

Text design and production by Karin Stack

Indexed by Susan Olason, Indexes & Knowledge Maps

The mission of Storey Publishing is to serve our customers by publishing practical information that encourages personal independence in harmony with the environment.

Printed in China by Elegance Printing
10 9 8 7 6 5 4 3 2 1

Library of Congress Cataloging-in-Publication Data

Hill, Cherry, 1947–
Horsekeeping on a small acreage : designing and managing your equine facilities / Cherry Hill.— 2nd ed.
p. cm.
Includes index.
ISBN-13: 978-1-58017-603-3; ISBN-10: 1-58017-603-8 (alk. paper)
ISBN-13: 978-1-58017-535-7; ISBN-10: 1-58017-535-X (pbk. : alk. paper)
1. Horses. 2. Horses—Housing. 3. Farms, Small. I. Title.

SF285.H55 2005
636.1'083—dc22

2004025198

To Richard, for making my horsekeeping dreams come true.

And to Pat Storer, for her enthusiasm and love for animals and for her desire to share with others.

Contents

PART TWO

Designing Your Acreage

PART THREE

Management

Preface to the Second Edition

Since I wrote the first edition of this book almost 15 years ago, I have come to appreciate the privilege of owning horses more with every day. As our planet continues to host larger populations of people, its resources are being spread dangerously thin. More than ever, we horse owners must be diligent caretakers of the land and environment so our children will be able to know the joy of owning, caring for, and riding horses.

Since 1990, many farms and ranches on the outskirts of towns and cities have been subdivided into small acreages, often of 1 acre or less. Some subdivisions are specifically designed for horses and include communal arenas, trails, and other facilities. The sense of horse community is evident in other ways, too, including the rallying of assistance during times of fire, flood, and drought. So although the earth seems to be shrinking as we live in closer proximity to each other, we horse owners are coming closer together in purpose.

I want to be a good steward of the land while providing my horses with the best care possible. If you share these goals, I hope you will find some practical help in the pages of this book. Without further ado, let's get to work! Happy horsekeeping.

Preface to the First Edition

It was a hot, sticky July day in northeast Iowa. The auctioneer had moved through most of the household goods and furniture and all of the shop and farm tools. The crowd was thinning, and those left were congregating under the huge shade trees on the side lawn. My husband and I had purchased a rake and a stepladder and stood leaning against them while we waited for the final item to be sold — the house and 10 acres.

The house was modest, and the well might have been an early experiment in hand-dug wells. The garage was Model T size, but at least there was no old barn that had to be torn down. The 10 acres were as flat as a pancake and covered in shoulder-high ironweed. There wasn't a single fence post or rail in sight, but there was rich Midwest soil and subirrigated fields. The road out front was lazy and peaceful, yet it was only 10 miles to a fair-sized town. It was far from ideal, yet if it went for an affordable price, it might make a nice little horse farm.

Finally, just a dozen folks were left, and only two parties were bidding. When the gavel sounded for the last time, my husband looked at me with a smile and said, "I guess I'll put the ladder and the rake back in the garage!" We were ready, once again, to set up horsekeeping!

I simply cannot imagine life without horses. It's not that horses are more important than or a substitute for people and other activities, but horses do have a special way of making life's big picture complete. When my schedule or the weather does not allow me to ride, the day feels as if a piece is missing. But riding is just one part of the horse experience. Conscientiously caring for animals brings a wonderful sense of satisfaction. There is nothing quite so fulfilling as a job well done, and the satisfaction of owning a healthy, fit, well-trained, and happy horse is great. Just imagine being able to see the fruits of your efforts as you glance out the window to check on your broodmare and foal, as you stroll through your well-manicured pasture, or as you open the door to your tidy tack room to prepare for a morning ride.

Even though owning a horse and boarding it away from home is better than not having one at all, keeping a horse at home offers many advantages. It allows you to be involved in and to attend to every single detail of horse care. And because keeping a horse at home makes it more convenient to undertake routine handling and training, you will find that you are able to spend more quality time with your horses.

The suggestions and information offered in this book are based on recommendations from Extension agents all over North America as well as on my experiences owning and managing horses in Alaska, Arizona, Colorado, Idaho, Illinois, Iowa, Michigan, and Alberta, Canada. The herds have ranged from two to more than one hundred horses, from newborns to geriatrics, including "idle" horses and those in all phases of development and training. I have been involved with breeding operations and training businesses for both English and Western riding. The size of the facilities has varied from a single acre to 160 acres. Some farms were relatively complete on my arrival, others required remodeling or repair, and some were mere tracts of bare land. From these experiences I will share with you what I believe contributes to a horse's well-being.

I hope that this book will prove to be a valuable reference for you and will inspire you to formulate some ideas of your own. I have approached the subject of horsekeeping by providing information on the behavior and needs of horses before outlining the design of the various facilities and the development of a management scheme. The more thoroughly you understand horses, the more appropriate your plans will be and, likewise, the more successful your horsekeeping venture. Because the emphasis of this book is facilities and management, I mention health and nutrition topics only briefly. (Consult the recommended reading list for other helpful books on these topics.)

My suggestions about facilities are not meant to be the final word on such things as horse barns and fences. There are simply too many options to discuss them all, and the choices are constantly changing. By including photos of various products, I do not mean to imply that they are in any way better than other options.

I hope that you will begin to develop a horsekeeper's consciousness as you read about various time-honored methods and promising new options in horse facilities and management. As you read this book and observe existing horse farms, you will begin to formulate lists of questions to ask and characteristics to look for when planning the various aspects of your farm.

Expensive facilities do not guarantee good horsekeeping, and, similarly, simple facilities do not, in and of themselves, indicate poor care. The well-thought-out and conscientiously applied management plan is the tie that binds a venture together and makes it successful. Good management requires knowledge, dedication, and a sincere interest in the well-being of horses.

For a helpful resource guide and current information from Cherry Hill's Horse Information Roundup, visit Horsekeeping Books and Videos at www.horsekeeping.com.

Acknowledgments

Thank you to the following people and companies for their help with photos and information for this book:

Barnmaster, Inc., Lakeside, California

Ray and Diane Bawol, Rhinebeck, New York

Best Friend Equine Supply, Saint Joseph, Michigan

Cheval Publications, Houston, Texas

Lisa Clark, photo shoot coordinator, front cover

Classic Cover-Ups, Oxford, Pennsylvania

Co-Line Welding, Sulley, Iowa

Cover-All Building Systems, Saskatoon, Saskatchewan, Canada

Mattie and Thomas Duggan, Kildare Stables LLC, Millbrook, New York

Randy Dunn, Bath Brothers Ranches, Laramie, Wyoming

Horse Pal Fly Trap, Omro, Wisconsin

Joyce and Jerry Hubka

KESA Quarter Horses, Fort Collins, Colorado

Dr. Tony Knight, Colorado State University, Fort Collins

Debbie and Gary R. Little, The Sandman, Boise, Idaho

Livermore Fire Protection District, Livermore, Colorado

Ron Lonneman, Ron's Equipment Company, Inc., Fort Collins, Colorado

Mac Mountain Tack Repair, Windham, Maine

Robert Malmgren, Fort Collins, Colorado

Ernie Marx, Larimer County Extension, Fort Collins, Colorado

Tom McChesney, Wellington, Colorado

Don McMillen, Fort Collins, Colorado

Nelson Manufacturing, Cedar Rapids, Iowa

Premier Fence Systems, Washington, Iowa

Priefert Manufacturing, Mt. Pleasant, Texas

RAMMfence, Bryan, Ohio

The Russell Meerdink Company, Ltd., Neenah, Wisconsin

Smith Irrigation Equipment, Kensington, Kansas

Spalding Laboratories, Arroyo Grande, California

Triple Crown Fence, Milford, Indiana

A special thanks to Kathleen Grayson, owner of O'Connor Farm, Rhinebeck, New York, for her help with the front cover photograph.

A big thanks to my dear husband and to my horses, who you will see in the photos throughout this book. With them, life is complete. Without them, this book would not have been possible.

Richard Klimesh: artist, blacksmith, farrier, photographer, videographer, facilities manager, best friend

Zinger: 1975 American Quarter Horse Association (AQHA) mare "Miss Debbie Hill"

Sassy: 1976 AQHA mare "Sassy Eclipse"

Zipper: 1984 AQHA gelding "Doctor Zip" (son of Zinger)

Dickens: 1990 Quarter Horse–Selle Français gelding (son of Sassy)

Seeker: 1994 Quarter Horse–Trakehner mare (daughter of Zinger)

Aria: 1994 Quarter Horse–Trakehner mare (daughter of Sassy)

Sherlock: 2000 Quarter Horse–Akhal Teke gelding (son of Sassy)

Knowing Horses

PART ONE

1 Ownership

Did you ever wonder why so many young girls collect statues of horses and decorate their bedroom walls with photos of horses? Why some, when they have had an opportunity to stroke a real, live horse are reluctant to wash their hands so they can keep that heavenly scent around for as long as possible?

How about the middle-aged woman who secures a position with a trainer or instructor as a working student, trading labor (cleaning horses, tack, and stalls) for riding lessons? And why do professionals from so many fields find that weekend rides help them make it through their workweeks, no matter how hectic?

What makes a horse owner postpone his own medical checkups but religiously schedule routine veterinary appointments for his horse? Why would a person consistently skip his breakfast or vitamins and grab fast food, yet never dream of shortcutting rations for his horse? What makes a horse owner forgo a new coat, yet not bat an eye when slapping down several hundred dollars for a new winter horse blanket?

These and other "horse-crazy" behaviors demonstrate the effect a horse can have on a person. There is something noble about horses that makes us want to treat them well. When we treat horses with the respect they deserve, they provide us with many unique opportunities to find a type of nobility in ourselves as well.

Animals are a reflection of their care and handling, and in no case is this more evident than with the home-raised, home-trained, home-kept horse. The relationship between horses and their people should be a partnership. Both have certain obligations to each other, and when those are met consistently on both sides, the partnership is solid.

The Benefits of Horse Ownership

The relationship between a person and a horse can be simple and fulfilling and without all the complications that can occur in the human world. A horse doesn't talk back but does tell you, using body language and other nonverbal communication, how he interprets your actions. A horse will reveal your true character — your confidence, the shortness of your temper, how consistent you are — and working with

A horse reflects the commitment of his owner. Good care and communication are evident in a healthy, content horse.

Zipper provides the ultimate reward: a spring trail ride in the beautiful Colorado mountains.

horses can give you the opportunity to become a better person. Caring for and interacting with horses has made many people more reliable, thorough, trustworthy, honest, and consistent. People who have difficulty working with other people often learn the meaning of teamwork with a horse.

Horses often become affectionate companions. Once the ground rules have been established and a horse feels secure in his role, he can become a friend as well as a partner. A soft nicker when you approach your horse's stall or pen says welcome. If you head toward the pasture to catch your horse and he meets you at the gate, he has given you the highest compliment. As R. S. Surtees said, "There is no secret so close as that between a rider and his horse."

A trustworthy horse can provide invaluable therapy for someone caught up in a hectic pace. Riding is an engaging activity, so it helps stop the mental conversations that contribute to stress. Few experiences equal a trail ride in the fresh air, especially if there is gorgeous scenery. Riding down a road or in an arena is also enjoyable and beneficial for both horse and rider. There is nothing quite like a rein-swinging walk to get back into a natural rhythm; nothing like a brisk trot with its metronome-like quality to physically invigorate; nothing like a rollicking canter cross-country to rekindle a sensation of freedom.

The exercise associated with the care and riding of horses can also add to your fitness. Grooming, cleaning tack, hauling feed, cleaning stalls, and riding involve many muscle groups and types of activities; the composite exercise is well balanced and definitely not monotonous.

Another physical benefit of owning animals is to satisfy the human need for contact, the desire to touch and be in close proximity to a warm, responsive being. You contact your horse from head to tail when you groom, and your horse can become an extension of your own body when you ride. It doesn't get much better than that! Horses are beautiful to watch as they rest, graze, play, and move with energy and grace. They continually provide valuable lessons in animal behavior. Their reactions and interactions are fascinating and provide material for stories and learning.

Horsekeeping has so many different aspects that over a lifetime you can learn new things about behavior, breeding, selection and use of tack, the use and care of land, exercise and conditioning, nutrition, health care, various styles of riding, training, and much, much more.

Do you view horsekeeping as a task that must be done or as an adventure and opportunity? Taking care of a horse's needs can be a great gift; it can help establish good habits and routines and bring order to a chaotic life.

Finally, being involved with horses offers social benefits. Many local, regional, and national organizations provide opportunities for individual and family participation. Groups are available for all types of horse involvement: trail riding, lessons and clinics, competitions of all types and levels, and groups for "backyard horsemen" of varying interests. In addition to providing a great place to share experiences, horse groups are good for exchanging ideas, forming friendships, and creating a network for group purchases and business transactions.

The Responsibilities of Horse Ownership

Although there is something almost magical about working with horses, the "wild and free" aspect is often romanticized in stories while the realities of horse training and care are skipped or glossed over. Horse ownership is a huge responsibility that requires hard work, dedication, a substantial investment of money and time, legal obligations, and a commitment to the environment.

Your domestic horse depends on you because he can't take care of himself as a wild horse would. Horses need care when they are idle as well as when they are actively being trained or ridden. Their needs do not diminish if your interest does. During winter, when you might be least likely to ride your horse regularly, he actually needs the most care. Here are some of the realities of horse ownership.

HARD WORK

Many parts of horse ownership involve hard physical labor, not only the energy-expending kind but the backbreaking kind as well. Shoveling manure, toting bales, carrying water, giving a vigorous grooming, and instituting a conscientious exercise program for your horse all go more smoothly if you are physically fit.

DEDICATION

When you own a horse, you must give a part of your life to the horse. There will be occasions when you must give up other things you like — such as sleep, warmth, and comfort — to ensure that your horse receives proper care. Horses pick inconvenient times to have foals, become ill, or get injured. It's not unusual for these things to happen in the middle of the night, just as you are leaving for an important meeting in your three-piece suit, during the worst blizzard your area has seen in more than 15 years, or moments before the kickoff of the championship football game. Even routine horse care will sometimes seem to intrude on other plans. For example,

Hard work is a daily requirement for conscientious horsekeeping. Chores contribute to your physical fitness, and with a positive attitude can be fun!

your veterinarian may be able to make it to your farm only on the morning of your best friend's wedding shower; your horse may be seriously injured the day before you plan to leave on vacation; the person you had lined up to do chores may become unavailable; and so on. The horse comes first.

FINANCIAL COMMITMENT

The initial purchase price of a horse is just the first of the costs associated with horse ownership. You can certainly cut costs by being diligent and innovative, but horsekeeping still requires weekly expenditures and budgeting.

Sample budget per horse per year*

ITEM	COST
FEED	
Hay (20 lbs. per day × 365 days = 7300 lbs., or 3.65 tons × $120 per ton)	$438
Grain (4 lbs. per day × 365 days = 1460 lbs. × $0.20 per lb.)	$292
Salt and minerals	$40
Bedding (2 bags/wk × $5/bag)	$520
VETERINARY SUPPLIES AND CARE	
Immunizations	$60
Deworming (6 times at $10 per)	$60
Dental and miscellaneous	$100
Farm-call charge	$75
FARRIER	
Shoeing (6 times at $80 per)	$480
Trimming (3 times at $25 per)	$75
TOTAL	**$2140**

*Costs are for example only; substitute prices from your area.
Note: This estimate is for basic care only. It does not include specialized or therapeutic shoeing, neutraceutical supplements or medications, emergency medical care, blankets or other tack, breeding fees, specialized feed for breeding animals or young horses, costs related to facilities, or other incidentals.

TIME

You must be willing and able to spend time attending to your horse's needs at least twice a day every day. You will have to tend to feeding, cleaning, grooming, and exercise every day, as well as associated chores such as buying feed and repairing tack and facilities.

LEGAL OBLIGATIONS

Horse owners have legal obligations to their horses, neighbors, other horse owners in the area, and to pedestrians and motorists passing by the property. Check the liability laws that apply to your specific location; they may be described by a phrase such as "ordinary care and diligence," which can be open to a wide range of interpretations. When farms were larger, "Good fences make good neighbors" was about all you had to worry about. Now, with horse properties becoming smaller and neighbors getting closer, perhaps "Good senses make good neighbors" would be more appropriate. Be aware of how your horse operation looks, smells, and sounds to your neighbors.

RESPONSIBILITIES TO THE ENVIRONMENT

Horses can be hard on land. If overgrazed, a pasture can become a dirt lot in a hurry. Once bare, land either blows away, washes away, or is taken over by weeds. Although it is ideal to afford a horse as much turnout on pasture as possible, we must learn and use techniques that will allow us to balance our use of the land with our care of the land. In addition, every horse owner must deal with the reality of manure management and pest control, and implement environmentally responsible practices when using pesticides, herbicides, and other potentially toxic substances.

2 Behavior

To make wise decisions when designing facilities and devising a management plan, first learn all you can about how and why horses behave the way they do. You can't significantly change intrinsic behaviors that have been part of the horse for more than sixty million years, so it's best to design facilities specifically suited to horses and their habits.

Horses are not humans, nor are they puppy dogs or glass ponies. Horses are horses and should be treated as such. Even though horses can elicit emotions similar to those we feel for our family or friends, dealing with horses as if they are humans is a dangerous anthropomorphic trap, and is unfair to them.

Horses are not pets. They can be partners, but we humans need to be the leader of the team. Horses are quite content to know this, and if they are treated consistently and fairly, they bond closely to humans, just as they would to a dominant horse in a herd.

Although horses can do some very cute and charming things, they are large and potentially dangerous animals. Horses are works of art, whether peacefully grazing or in breathtaking motion, but they aren't collectibles like porcelain statues or framed oil paintings. They are living, breathing creatures with deeply ingrained reflexes, routines, and needs.

The horse is a gregarious nomad with keen senses and instincts, highly developed reflexes, a good memory, and a strong biological clock. In the animal world, the horse is a prey animal, one that is hunted for food by predators such as wild felines, canines, and humans. That's why a horse is inherently wary. The more you understand the nature and characteristics of the horse, the more likely it is you'll be able to help your horses adapt to domestication, confinement, and training.

Characteristics

It is common for horse owners to say a horse is misbehaving when the horse is merely behaving according to his inherited instincts. While a horse's natural behavior patterns need to be altered somewhat to make him safer and more useful, it's best to work with, not against, existing instincts and reflexes to minimize stress and ensure long-lasting results. Take time to observe horses in herds and in various styles of domestic confinement so you can develop insights that will help you make good handling and management decisions.

Whether or not there is action, a horse can be exhibiting behavior. A sullen horse, rigid and unyielding, is exhibiting a behavior; a wildly bucking horse is exhibiting a behavior. Repeated behaviors, even if not part of a formal lesson, become habits. Horses are constantly learning as a result of their handling and the environment.

Even though the modern horse is relatively safe from predators, his long history of struggle for survival has resulted in a deeply embedded suspicion of anything unfamiliar. Because of this, the horse is one of the few domestic animals that still retain the instincts necessary to revert to a wild state. These instincts can make a horse awe inspiring and challenging at the same time.

GREGARIOUS

Gregarious animals are sociable and prefer to live and move in groups. If the domestic horse has a choice, he will stand not alone but in close proximity to another horse, finding safety and comfort in numbers. An entire band panicking from an imaginary beast, a group huddling tightly against the wind or snow, and buddies participating in social rituals — all of these are examples of a horse's social behavior. Horses seek the companionship of other horses and are most content when they are with other horses, near other horses, or at least can see other horses. For this reason, bands of horses turned out on large pastures will often choose to congregate over the fence that separates them.

Dickens and Sassy are free to roam their 20-acre pasture, yet they prefer to graze near each other.

The Herd

Because horses are most secure when they are with other horses, it is understandable that an unhandled horse could be restless or even panic if separated abruptly from the herd. If a horse has not had sufficient handling, socialization, and bonding with a human handler — that is, to the point that he feels as safe with a human as he does with the herd — then the horse might desperately attempt to stay near or communicate with the herd, a preferred companion, or even the barn. This insecurity is often referred to as *herd bound, buddy bound,* or *barn sour.* The insecure horse links food, comfort, companionship, safety, and security with the herd or barn. A horse that is separated from other horses might pace back and forth along a fence line or stall wall, paw or weave, or scream shrilly in an attempt to maintain contact. If the horse is being handled or ridden, the horse might wheel, bolt, rear, or buck to try to rejoin the other horses.

What might begin as a temporary insecurity may evolve into a longstanding and dangerous habit. In order to prevent such a bad habit from forming, from a very early age horses should be handled individually so they develop confidence, and they should never be put in a position of panic. The distance and time away from the other horses should be gradually increased. All training and housing facilities should be strong and safe. A horse that is quiet and attentive to his handler can be rewarded by feeding or grooming away from his companions. It also helps if the horses in the group (herd or barn) are content, so they don't call to or answer a horse that is learning to be separated.

Top: Even when separated from Sassy, Dickens is content because he can see her. _Middle:_ This gelding is herd bound. Although he can see his herd mates, he paces fervidly along the fence and screams shrilly. _Bottom:_ Even though Sassy's band and Zinger's band both have access to large pastures, they choose to congregate over the fence. Consider this when planning safe fencing.

Mutual Grooming

When two horses have a strong bond, one way they show their admiration is mutual grooming. You may have seen two horses standing head to tail, nibbling each other along their back, neck, and mane. This is a natural social ritual, but it can ruin a beautiful mane in a single session. Horses that are aggressive mutual groomers may need to be separated for turnout. Capitalize on the potential for bonding by finding your horse's favorite spot and grooming him there frequently.

Top: **Dickens and Sherlock enjoy a mutual grooming session.** *Bottom*: **I've found one of Sassy's favorite spots and am giving her a good rub.**

Pecking Order

Just because horses desperately want to be with other members of the band doesn't mean all horses get along well. Particularly when there is limited food or space, personality conflicts will appear. Battles may be fought with teeth and hooves or merely with threatening gestures. Once the clash is over, a pecking order or dominance hierarchy emerges. This social rank makes future aggression unnecessary unless a particular horse is not thoroughly convinced of his status and continually tests the horses immediately above him. The most assertive horse generally earns his choice of feed, water, and personal space.

In planning facilities, therefore, you should avoid introducing acute angles and tight spaces, especially around feeders and waterers, and places where horses could get cornered and hurt. Assume that horses will fight at feeding time, and plan either to feed every horse separately or to feed groups of horses in a large space using more hay piles or feeders than there are horses.

Because of the potentially violent behavior associated with the establishment of status in a pecking order, new horses must be carefully added to an already established group. It is best if the new horse can spend a few days in close proximity to but not in direct contact with the band. Putting a new horse across the fence from an established group is not a good way for them to become acquainted; it will almost certainly result in injury. If possible, allow the new horse to settle in for a few days in a pen near the other horses but where they can't reach each other. After a few days, turn out the new horse with a few of the neutral members of the herd; each day add a few more horses until the group is complete.

Humans occupy a spot in the pecking order too, and various horses will test you to see just where you stand. You must convince each horse that you are kind but firm and, yes, that you are the top gun. Not only will this encourage your horse to respect you, but it will also give him a great sense of confidence, because horses are basically followers. If you are a good leader, your horse will be content to do as you ask.

Above: **A meeting of the board of directors: stallions at Randy Dunn's Laramie, Wyoming, ranch establish pecking order.**
Right: **Seeker, named for her appetite, waits before I allow her to approach the hay I've delivered.**

Until a horse knows that you are above him in the pecking order, at feeding time he may come toward you aggressively, perhaps with laid-back ears and threatening body language. The worst thing you can do at this point is to reinforce the horse's aggressive behavior by dumping the feed and leaving the pen or stall; you will have rewarded him for his pushy behavior. Instead, make the horse back off and wait until you give a clear signal that he can approach the feed.

With a very aggressive horse, in some instances it might be necessary to halter the horse, tie him to the hitch rail, put his feed out, and then return him to his pen or stall. This way you develop a positive association and establish in-hand control first. In other cases you might just have to stomp a foot toward the horse to get his attention. Or you may have to slap the horse across the chest with the end of a lead rope to keep him from crowding you. You should issue a firm voice command such as a stern *Wait!* and require the horse to remain attentive until you leave the pen. Then use a command such as *OK* to indicate the horse may now approach the feeder. If a horse continues to be pushy or intimidate you or make you fearful, seek help from a professional trainer or find another horse.

NOMADIC

If horses had a theme song, it would be "Don't Fence Me In"; they are born wanderers. But we do have to confine our horses, especially when horsekeeping on a small acreage. Their nomadic tendency can lead to confinement behaviors such as pawing, weaving, and pacing. These vices are a response to inactivity, lack of exercise, overfeeding, and insufficient handling. Regular exercise is essential for the horse's physical and mental well-being. Adequate turnout space and exercise time can prevent the development of these vices.

Horses that are kept in box stalls or small pens need to be turned out and allowed to be horses. Otherwise, they may become either very bored with their existence or extremely hyperactive. An introverted horse that has "tuned out" is just plain dull: lazy, unresponsive, and balky. The overly energetic horse is "wired": anticipatory, nervous, irritable, and possibly unsafe.

SENSES

Keen senses allow horses to pick up very slight changes in the environment. More sensitive to subtle movements, far-off sounds, smells, and possibly barometric pressure than humans, horses are frequently alerted to potential danger while we notice nothing out of the ordinary. Horses are capable of feeling vibrations through their hooves warning them of approaching predators or other horses.

Horses have a very discerning tactile sense. Their lips, skin, and hairs accumulate information that we normally gather with our hands. Horses are dexterous with their lips and can open gates with intricate

Top: **Dickens is on high alert because a mysterious woman snuck into his pasture to take his picture.** ***Bottom:*** **Sassy Eclipse, also known as Lips Eclipse, demonstrates her desire to explore with her lips and their dexterity.**

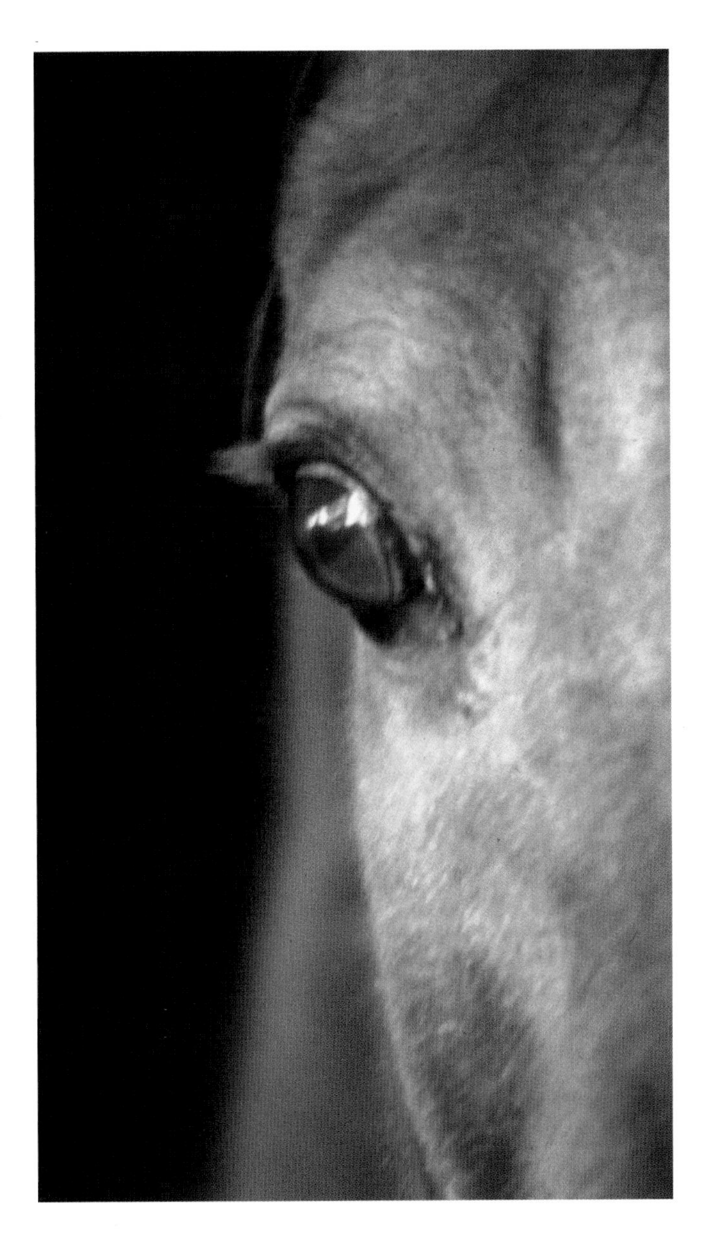

Because a horse's vision is both monocular (seeing a separate image with each eye) and binocular (seeing a combined image with both eyes), his reactions to objects and motion are often quite different from ours.

latches. They can also determine whether an electric fence is operating by checking it with the hairs on their lips.

With all of these keen senses ready to put him on red alert, it seems unfortunate that a horse's vision doesn't provide much help in resolving his apprehensions. First of all, a horse has blind spots — the areas directly in front of his face, below his head and neck, on his back, and directly behind him.

He also has a lesser ability than humans do to focus both near and far, which is demonstrated by a horse's wide range of head and neck positions when trying to see something — from craning and straining to lowering and peering.

The horse's eye is slower than ours to adapt to light changes, which explains why a horse must take a few more seconds to get his bearings when stepping out of a dark barn into the bright light or from the bright light into a dark trailer. And finally, because horses see with both monocular and binocular vision, at the junction of the two fields of vision images might jump or be blurred, causing visual distortions and concerns for the horse.

Horses generally have an avid sense of curiosity. They are not content just to look at an object. They must inspect, fiddle, meddle, smell, nuzzle, paw, knock over, and in general fool around with almost anything they can get to. A horse's curiosity should never be discouraged because it is a valuable key to training, but it should be taken into consideration when building and managing facilities.

REFLEXES

Horses can assume thundering speeds from a standstill. They can rise from a recumbent sleeping position and instantly run. They can strike or kick in the blink of an eye. These lightning reflexes helped the horse survive for more than sixty million years. The same automatic responses allow today's horse to perform in a vast array of spectacular performance events, but they can also prove dangerous for humans.

Much of training is designed to work with and/or systematically override a horse's natural reflexes. An example is the withdrawal reflex. This is the natural reaction of a horse to pick up his leg when something touches it. In order to be able to wash, clip, and bandage a horse's legs, you must override this reflex so he will keep his hoof on the ground as you touch his leg. But you must keep in mind that you will also want to pick up a horse's hoof to clean it, so you will want to make a discernable difference in the way in which you request each behavior.

FLIGHT

When the horse is convinced that danger is imminent, he almost always chooses to flee rather than to fight. It is the rare horse that chooses to stick around and reassess the situation in the event he might be imagining things. Horses can be taught to trust their handlers' good sense, however. The horse out on pasture, left to its own devices, would probably avoid the "black hole" that in reality is only an 8-inch-deep spot in the creek. When the trainer (who treats her horse fairly) assures the horse by voice and body language that it is safe to step into the water, the skeptical but trusting animal will reconsider. As long as you make wise decisions and never ask your horse to negotiate something unsafe, your horse's instinctual fears can be overridden by his confidence in you.

Zinger knows our well-behaved Vizsla, yet still exhibits the flight response because a deeply ingrained instinct tells her dogs are predators and horses are prey.

MEMORY

If a horse lacks confidence or has received poor handling, he can behave very unpredictably and spook with the slightest provocation. Because a horse has an excellent memory, he will remember remote experiences, especially if they relate to his imagined safety. Horses are believed to never quite forget these fears. All a handler can hope for is to bury the bad experiences beneath layer upon layer of good ones.

For example, suppose a horse is turned out on a new pasture for the first time. As he trots around snorting, with head high, inspecting the boundaries, a couple of dogs pop out of a wooded area at the edge of the pasture and begin chasing him. In his panic to escape his modern-day predators, he mindlessly heaves his body at the wire fence and manages to stretch and break enough wires to allow him to return to the barn. The stray dogs quit the chase, as they are leery of the humans usually around the barn; your horse stands quivering and bleeding alongside one of his buddies.

What do you think will cross your horse's mind the next time you turn him out on that pasture? Even if no dogs are present, do you think he might avoid the wooded area altogether? Will every moving leaf in the woods make him suspect that killer dogs will emerge? Will he go through the fence again? Unfortunately, your horse will be suspicious of that pasture, and especially the woods, for a long time. Similarly, a horse that reaches into his water tub for a drink and receives a shock from a tank heater with an electrical short will very likely refuse to drink even if his body is in a life-threatening state of dehydration.

The best plan is to prevent such things from occurring in the first place. Once something traumatic does happen, however, you must allay your horse's apprehension by systematically planning good experiences to replace the bad ones.

BIOLOGICAL CLOCK

Horses perform daily routines in response to needs and a strong biological clock. A horse is a creature of habit, following his natural rhythms where possible and being most content when his management has a predictable pattern. Many routines are socially oriented: small groups graze in tight-knit bands on huge ranges, participate in contagious pawing and rolling sessions, or engage in running and bucking games. At regular times of the day, individuals in stalls or groups on pasture can be observed to eat, drink, roll, play, and perform mutual grooming. The desire to participate in these rituals is not diminished, and in fact is probably intensified, for the horse in confinement. In spite of bathing, clipping, and blanketing, most horses love a good roll in the mud, much to the chagrin of their human grooms!

Once a horse has established a routine of urinating in his stall, he will often, to the stall cleaner's dismay, "hold it" all day while out on the pasture only to flood the stall the instant he is returned to it. And the behavior inspiring the old adage "You can lead a horse to water, but you can't make him drink" is based on a horse's firmly implanted habits and his strong biological clock (although not drinking can be influenced by many other factors, including a horse's keen sense of smell). The horse's biological clock is especially evident near feeding time, when whinnies will get you out of bed if you happen to oversleep. It is best to feed a horse the same ration at the same time of day each day.

Vices

Vices are undesirable behaviors that horses might originally develop for legitimate reasons but once formed often become habits that appear even if the original cause has been removed. Vices, such as cribbing, pawing, and weaving, tend to be performed in confinement — in a horse's stall or pen — whether or not humans are around.

Bad habits differ from vices in that they are undesirable behaviors that horses learn in response to handling and training, and they almost always occur during in-hand work or riding. Examples of bad habits are rearing, bolting (running away), and biting the handler. Bad habits can be avoided with proper training.

Vices are related to horsekeeping and thus are the emphasis here. Vices are almost always caused by confinement, lack of exercise, overfeeding, and stress. The more natural the horsekeeping, the more content the horse. The more content the horse, the less likely he will be to form vices.

Keeping a horse in a stall or small pen is contrary to his desire to roam and have regular exercise. Prolonged confinement is one of the leading causes of vices such as pawing, pacing, weaving, and stall walking. Feeding a horse a high grain ration is contrary to his natural diet of grass and can lead to excess energy and wood chewing.

Vices can be prevented with proper management. Understanding common stable vices can help you identify early signs, take appropriate steps, and modify the horse's behavior.

WOOD CHEWING

The beaverlike gnawing of wood rails, planks, buildings, and feeders is costly and unnecessary and can be dangerous to the health of a horse. Wood chewing can afflict a horse of any age and can result in colic from wood ingestion or damage to the gums and lips from splinters, to say nothing of the damage to facilities.

Housed in a pen with wood rails, this horse has developed the vice of wood chewing. Cold, wet weather, boredom, teething, or lack of roughage might have been the initial cause.

Young horses may begin nibbling out of boredom, curiosity, or perhaps to relieve an itching of the gums during teething. Serious wood chewing can initially be caused by low fiber intake in relation to a horse's needs, especially during cold and/or wet weather. Horses appear to be relaxed and comforted when they are able to spend a good deal of time chewing long hay. Horses deprived of this natural satiation may be seeking oral gratification and an increase in fiber intake from the wood. Weather-related wood chewing is thought to be a result of the frustration and anxiety a horse feels when he is uncomfortable. Precipitation softens the wood, making it more palatable and aromatic.

Be sure your horse's diet is well balanced and adequate in fiber in the form of long-stem hay. Provide ample exercise. To cure the chronic wood chewer and to prevent others from acquiring the vice, coat all wooden surfaces with an effective, safe wood antichew product. Cover all wooden edges in stalls with heavy metal corner trim. Run electric fence wire along wooden fence rails.

When a horse is a cribber, he is obsessed with grabbing on to an object with his incisors and gulping air. He'd rather crib than eat.

CRIBBING

Cribbing, or wind sucking, is a debilitating vice. The cribbing horse grabs the edge of a partition, the top of a door or post, a feeder, or other solid object with his incisors, arches his neck, and swallows air in labored gulps. Although, at first glance, you may think a cribber is a wood chewer or vice versa, these are two very different behaviors. The horse that started chewing wood and later became a cribber probably would have become a cribber even if he was kept in an all-metal stall. Cribbers are often nervous, neurotic individuals that find comfort in the ritual. Research suggests that cribbing releases endorphins (opiates) from the brain, giving the animal a natural, habit-forming high, so it is easy to understand why cribbing tends to be a permanent vice. Horses that crib are often thin, have abnormal dental wear, and sometimes suffer colic from swallowed air.

Because cribbers are often more interested in cribbing than in eating, they waste a lot of time and energy pursuing the vice and tend to be hard keepers. Cribbing is usually managed, but not cured, by the use of an anticribbing collar. Drugs and surgery are other options you can discuss with your veterinarian.

PAWING

Pawing is initially a signal that a horse wants or needs something, but once a horse has been allowed to perform such behavior, it may become a habit that no longer has any specific cause. The wild horse or the pastured horse uses pawing for many practical purposes such as uncovering feed under snow, opening up a water hole, digging up roots during a dry season, inspecting an unfamiliar object, and softening the soil before rolling. Pawing can also indicate pain and restlessness, as with a colicky horse or a mare that is foaling or expelling a placenta.

Lack of training, lack of sufficient exercise, boredom, and confinement can lead to pawing. Pawing is an expression of a horse's restlessness or desire to wander. It is damaging to the horse and the facilities, so should be prevented by proper management and training.

The pawing instincts of the stalled horse are especially evident around feeding time: many horses paw to indicate their grain or water pail is empty, and some just paw in anticipation of being fed. Unfortunately, feeding a pawing horse is a form of reward and encourages him to repeat his behavior, often more intensified, in the future. Therefore, it makes more sense to do something like this: halter the pawing horse, take him for a short walk, tie him, put his feed out, and then return him to his stall.

Horses that have not been made to accept the confinement of cross ties or a hitch rail often paw out of impatience or nervousness. Other horses paw in response to confinement, lack of exercise, and overfeeding. To release excess energy, the underexercised horse might paw in a variety of ways: making flat slapping sounds on the ground, making repeated swipes through the air, or bearing down and scooping up earth. The latter type of pawing results in damage to stall floors, hooves and joints, and shoeing. Bare hooves can be worn horribly out of balance in one short pawing session. A shod horse that paws can catch the shoe on a fence and pull it off or can loosen the clinches from the repeated pounding.

To prevent pawing, be sure that the horse receives conscientious handling and adequate exercise and turnout and is not being overfed or inadvertently rewarded for pawing at feeding time. These are the best safeguards against the unwanted habit. If the pawing appears to be caused by boredom, a companion animal or diversionary stall toy could help.

WEAVING

The horse that weaves stands with his head and neck over a stall door or fence and sways his body from side to side. The rhythmic, lulling movement appears to be soothing to a nervous horse or to one that has insufficient exercise. It does use a lot of energy, so often a weaver is a horse in poor condition. This obsessive, repetitive movement can wear unshod hooves unevenly and even distort the growth of the hoof wall.

A horse confined an excessive amount of time (for a week or so) may try this behavior to fill his need for exercise. If he is then properly exercised, he may still retain the habit, even when pastured.

TAIL, MANE, AND BODY RUBBING

Rubbing, like weaving, is often a continuous, rhythmic, swaying motion. Like other vices, initially it may have had a legitimate cause, but even when the cause has been removed, the rubbing habit often remains. A horse may start rubbing during shedding, when wearing a dirty blanket, when there isn't sufficient

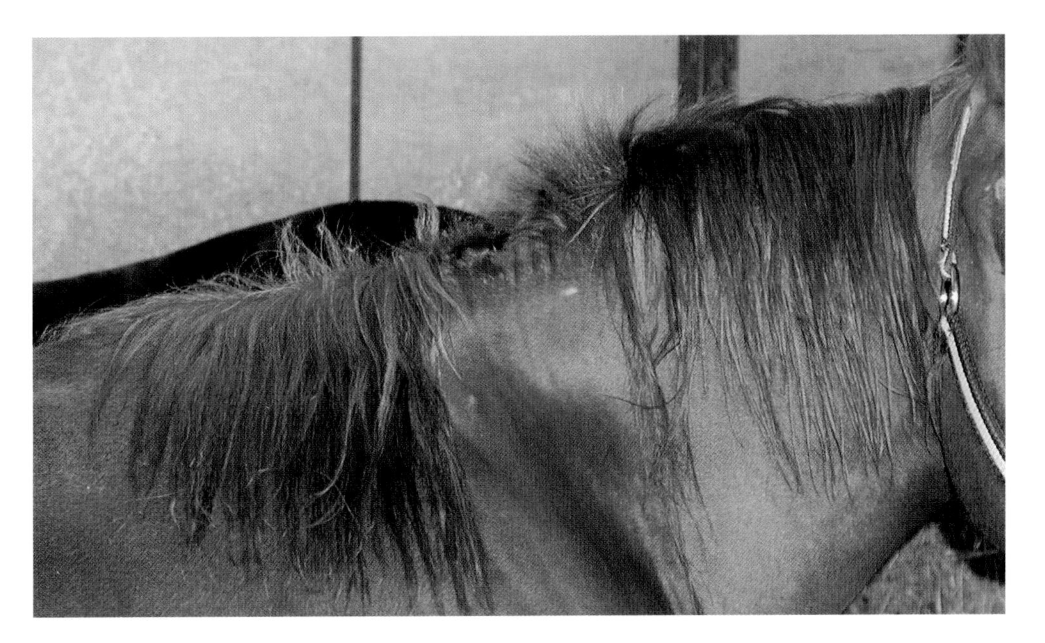

This broodmare might have had lice or simply an itch from shedding, but she did not stop rubbing until she removed the entire middle portion of her mane.

room to roll, or when the mane, tail, anus, udder (if a mare), or sheath is itchy from poor sanitation, lice, or ticks.

The chronic rubber is hard on facilities, actually knocking rails down, stretching wire fences, breaking branches and small trees, and damaging stall walls. The horse that persistently rubs his tail or mane often ends up with bald spots. Cleanliness is frequently the key to preventing rubbing.

STALL KICKING

Few vices can be as destructive to both the horse and the facilities as stall kicking. In some cases the horse stands with his hindquarters near a wall and rhythmically thumps the wall with one hind foot while his head bobs in a reciprocating motion. Other horses let fly with both hinds at once in an explosive burst. This type of kicker can wipe out a stall wall in a single kicking bout, to say nothing of the damage that can be done to his lower hind legs. Capped hocks and curbs are often associated with chronic stall kickers.

Insufficient exercise, excess feed, and unsuitable or constantly changing neighbors can cause stall kicking. The vice can be contagious, but it is not always an act of aggression toward another horse. It can be a response to training pressures or confinement, or it can be a game. Neighboring horses might be caught up in the game, or they might interpret the action as threatening and respond with a defensive kick.

Some horses have learned that stall kicking is a great way to get attention and feed. The kicking noise brings a human to the stall, often with a diversionary flake of hay. This gives the horse what he wants and rewards him for the kicking behavior.

If stall kicking is due to boredom or confinement, additional work or turnout time usually helps. If the kicking is due to a particular neighbor, simply shifting the horse's location in the barn might do the trick.

If management has been evaluated and modified and the vice still remains, you could try butt boards. These are horizontal boards set on edge around the inside of the stall at rump height to prevent the horse from getting close enough to the wall to kick, or set at hock height so the horse will punish himself if he does try to kick.

Before implementing any remedial methods to cure a vice, first be sure that a horse receives adequate work and exercise and appropriate feed and has reasonable neighbors.

Needs

To be healthy and content, a horse's needs must be met. The most basic needs of the domestic horse are the same as those of the wild horse: food, water, shelter, companionship, exercise, and rest. Because we confine our horses and want to provide optimum care, the domestic horse also needs regular veterinary and farrier care.

Feed

Feeding horses is an art and science unto itself, and entire volumes are devoted to the subject. Learn how to select feed and balance a ration for horses of all ages and activity levels (see recommended reading for helpful books on the topic).

Understanding certain principles ahead of time will help you make appropriate plans for facilities and management.

- Because horses evolved as grazers, their digestive systems are adapted to many small meals each day. That is why confined horses should be fed at least two times each day. Three times a day at 7-hour intervals is ideal.
- Horses have an extremely strong biological clock, especially when it comes to feeding. Feeding late or inconsistently can result in colic and other digestive upsets. They do best when fed the same amounts at the same time every day.
- The horse's digestive system is adapted to a high amount of bulk and a low amount of concentrate. High-quality hay should be the mainstay of your horse's diet. Do not feed too much grain.

Estimating horse weight

The approximate weight of your horse can be determined by using a livestock scale, weight tape, or a heart girth table, or by calculation.

To calculate an estimate of weight, measure the heart girth as described at right and the body length. Body length is measured in a straight line from the point of the shoulder to the point of buttocks.

GIRTH

LENGTH

HEART GIRTH × HEART GIRTH × BODY LENGTH ÷ 330 = BODY WEIGHT

To use a heart girth table, measure heart girth, which is the circumference of the horse's body just behind the withers and elbows. With the horse standing square, place a tape measure around the horse's body just behind the withers and about 4 inches behind the front legs. Pull it tight enough to slightly depress the flesh.

HEART GIRTH in. (cm)	WEIGHT lbs. (kg)
30 (76)	100 (45.5)
40 (102)	200 (91)
45.5 (116)	300 (136.5)
50.5 (128)	400 (182)
55 (140)	500 (227)
58.5 (148)	600 (273)
61.5 (156)	700 (318)
64.5 (164)	800 (364)
67.5 (171)	900 (409)
70.5 (178)	1000 (455)
73 (185)	1100 (500)
75.5 (192)	1200 (545)
77.5 (197)	1300 (591)

• Feed each horse individually according to his specific needs. This avoids competition, fighting, and some horses gulping and getting too much while others get too little.

• Know exactly what you are feeding. Read and understand the feed tags of commercially prepared feed. Have your year's supply of hay tested for nutrient content if possible.

• Know how much your horse weighs.

• Know exactly how much you are feeding. Feed by weight, not by volume. Feeding by volume contributes to overfeeding and wasted money. Feed hay at an approximate rate of 1.50 to 1.75 pounds per 100 pounds of body weight. This means that a 1000-pound horse will require about 15 to 17.5 pounds of hay per day. It is best if you weigh hay at each feeding, or you can weigh several flakes of the hay you are feeding to determine the average weight of a flake. Flakes from standard hay bales (also called *fleks, leaves, slabs,* or *slices*) can vary from 2 to 7 pounds, depending on the type of hay, moisture content, how tightly the hay was baled, and the adjustment on the baler for flake thickness.

• Feed grain to young, growing horses, horses in hard work, and lactating broodmares. Because grain should be fed by weight, not volume, don't rely on a scoop to measure unless you've determined beforehand the weight of grain the scoop holds. Oats are much lighter than corn, for example, so a quart of oats will weigh far less than a quart of corn (see box).

The energy values of grains vary greatly too. A pound of corn contains nearly a third more energy than a pound of oats. Before adding grain, determine how much additional energy the horse needs beyond what he receives in hay.

• Make all changes in feed gradually. Whether it is a change in the type of feed or in the amount being fed, make the changes in small increments and hold the amount at the new level for several feedings. If you are feeding 2 pounds of grain per day and want the horse to have 4 pounds per day, increase to 2½ pounds and feed that for at least 2 days. Then increase to 3 pounds for 2 days, and so on. If you are making a change in hays, feed one part new hay and three parts previous hay, hold for 2 days, and then feed half and half for several days, and so on.

Weight per quart of common grains

GRAIN	WEIGHT lbs. (kg)
Bran	½ (0.23)
Oats	1 (0.45)
Barley	1½ (0.68)
Corn	1¾ (0.79)

• Be aware that a pasture- or grain-fed horse that is brought suddenly into work can suffer *azoturia*, or tying up. This usually afflicts a horse that is vigorously exercised after a period of inactivity (several days or more), during which the feed was not decreased. When the idle horse is forced to exercise, excess lactic acid accumulation in the muscles results in tenseness and soreness, often preventing the horse from moving at all. To avoid such a situation, decrease your horse's grain ration if he will not be exercised for 2 days or more. When you resume work, be sure the horse is given a thorough warm-up and cool-down. And when you start him back on his regular grain ration, do so gradually.

• When turning a horse out to pasture for the first time, do so when the pasture grasses are mature. Each horse responds to pasture differently, but follow this plan as a guide. First, let him fill up on grass hay before you turn him out. Limit his grazing to 30 minutes per day for the first 2 days, then 1 hour per day for 2 days, then 2 hours per day for 2 days, then 2 hours twice per day for 2 days, then turn the horse out for half a day for 2 days, then for the full day for 2 days, and then you can turn the horse out on pasture full time. Keep a close watch on horses that are on pasture, as they can quickly become overweight or suffer the devastating condition *laminitis* (founder) from too much rich or young green feed. If a horse has been off pasture for a week or more, reintroduce him to the green feed gradually.

• Be sure it is impossible for a horse to get to the feed in your storage areas. Horses do not know when to stop eating and can literally eat themselves sick. An excess amount of grain can cause colic or laminitis, both of which can be life threatening.

• Horses have sensitive digestive systems. Do not feed a horse immediately after hard work, and do not work the animal until at least 1 hour after a full feed.

• Feed the highest-quality hay you can find. Take the time to shop around and become familiar with the characteristics that constitute excellent hay. (See chapters 13, Land, and 18, Routines, for more information on hay.)

• Be sure feeders are clean and safe. Do not let feed accumulate in the bottom of feeders. Moldy or spoiled feed can create problems for your horse and large veterinary bills for you. Routinely check all feeders for sharp edges, broken parts, loose wires or nails, and any other hazard.

• Do not feed horses on the ground where they might ingest sand or decomposed granite along with their feed. This can cause sand colic, a dangerous type of impaction. Feeding on clean concrete pads,

Above: **Dickens enjoys his first taste of summer pasture. Waiting until pasture grasses are mature is often safest for the horse and easier on the land.** ***Right:*** **Feeding hay to Zinger on pasture is best when the land is covered in snow. This eliminates the risk of her ingesting sand with the feed.**

rubber mats, or snow-covered pastures can be helpful in preventing sand colic.

• To take the edge off an overeager horse's appetite, consider feeding hay first and following it 10 to 20 minutes later with grain. Horses that gulp or bolt their grain can suffer choke, colic, or poor feed utilization. To encourage a horse to eat his grain more slowly, mix large hay wafers, cubes, or "cakes" in with his grain ration or leave several baseball-sized smooth rocks in his grain feeder. A large, shallow grain feeder will cause a horse to eat more slowly than will a narrow, deep grain feeder such as a bucket, which invites gulping.

• Balance your horse's ration by providing free-choice trace mineralized salt. This contains sodium, chloride, and usually iodine, zinc, iron, manganese, copper, and cobalt.

• Depending on the horse's age and type of feed, determine whether calcium and phosphorus need to be supplemented and in what ratio. If calcium is deficient, limestone can be added to the grain. If phosphorus is low, monosodium phosphate can be added. If both calcium and phosphorus are low, dicalcium phosphate can be used.

Below: Free-choice trace mineral salt (red) and plain salt (white) lets a horse decide what he wants, when he wants it. _Right:_ Zinger enjoys a deep drink from a cool, naturally aerated Rocky Mountain creek.

Water

Horses require between 4 and 20 gallons of drinking water a day. Water should always be available, clean, and of good quality. (See chapter 14, Water, for more information on water quality.) A horse's water intake will increase with environmental heat, exertion, lactation, increased hay ingestion, some illnesses, and increased salt intake. Horses drink less water in extremely cold weather and during some illnesses.

If a horse doesn't get the water he needs on a regular basis, he could suffer impaction colic, in which the contents of his intestines aren't moist enough to move properly through his digestive system. To determine if a horse is dehydrated, perform the pinch test. Pick up a fold of skin on the horse's neck between your thumb and index finger. Release the skin. It should return to its normal, flat position in 1 second. If a ridge remains, the horse is slightly dehydrated. If the skin remains peaked, called a "standing tent," the horse is dehydrated and could require immediate veterinary attention.

In very cold weather, providing freshly drawn water might encourage a horse to drink more water than if he were forced to get his needed moisture by eating snow or drinking from an icy pond or trough. However, many horses will not drink artificially warmed or too-hot water. If you are in a cold climate and do not have heated watering devices, the best bet is to draw fresh buckets from the tap or hydrant several times a day to offer each horse. Horses seem to prefer freshly aerated water at 35 to 40°F.

A horse drinks by closing his lips and creating suction with his tongue, so it can take quite a bit of time for a horse to get his fill of water. Horses drink about ⅓ of a pint per swallow or 1 gallon in about 30 seconds, generally coming up for air after about ten swallows.

Because horses have such keen senses of smell and taste, they often refuse to drink foreign water when they are away from home, even though the water may be perfectly safe to drink. To prevent a horse from going "off water" when traveling or when you move to a new location, you can flavor the horse's home water for a while before leaving. You'll want to flavor the new water with the same substance so it will smell and taste the same as the flavored water the horse was used to at home. You can try a few drops of oil of peppermint, oil of wintergreen, molasses, apple juice, or soda. Use the additive sparingly and test each horse ahead of time so you know what will work before you move him.

The trouble with snow

Snow is approximately 5 to 10 percent water. If a horse needs 10 gallons of water per day and is forced to eat snow, he would have to eat and melt between 100 and 200 gallons of snow per day.

Above: **Large trees provide ideal shelter from sun and insects. However, the land around a tree is soon bare, and if the trunk is unprotected, horses can quickly strip the bark by rubbing and chewing. *Opposite, left:* Three-sided run-in sheds allow horses to take shelter during bad weather. *Opposite, right:* Some horses, like Dickens, prefer to stand outside and weather the storm.**

Shelter

It is not necessary to have an airtight, heated barn for horses; in fact, that is one of the unhealthiest environments in which a horse can live. (See chapters 7, Barn, and 8, Interior, for help designing your barn.) A horse's shelter requirements are pretty basic: a place to get out of the cold wind and hot sun and a way to stay dry during cold, wet weather. Trees, bushes, large rocks, hills, and other terrain features can provide natural shelter. Man-made shelter can be in the form of sheds or barns from the simple to the sublime. Blankets can provide additional protection. But no matter what type of shelter you provide,

your horse may prefer to stand out in the weather. (See chapter 4, Program, to learn more about shelter and choose a management style.)

Exercise

Exercise is essential for the health of every horse and for the proper development of young horses. It maintains a balance between feed ingested and bodily waste and is essential for bodybuilding and repair. Exercise, in contrast to the progressive training effects of conditioning, is often referred to as maintenance. The term *idle,* used when formulating rations, does not indicate that the horse is not allowed or encouraged to exercise but that the horse is not being used for regular, strenuous work at that particular time. All horses of all ages need exercise every day — a ride, longeing, or a minimum turnout of 2 hours in a large pen or pasture.

A regular exercise program invigorates the appetite, tones muscles, increases lung and heart capacity, and helps develop reflexes and coordination. Exercise increases circulation, which increases the activity of the skin and lungs, which in turn helps remove body heat and the waste products (especially lactic acid) of exercise. Exercise aids in the development and repair of tissue and improves the quality and strength of bones, tendons, ligaments, and hooves. Regular stress creates dense, stress-resistant bone. Exercise also conditions and stretches muscles and tendons, resulting in less chance of injury and lameness. Allowing horses to play in moderately soft footing can help develop elasticity in tendons.

Horses that are allowed ample exercise rarely develop vices such as pawing, stall kicking, and wood chewing, which are often results of boredom.

Adult horses take the largest portion of their exercise at the walk, but young horses, testing their physical limits, exhibit explosive outbursts. Because foals and yearlings are characteristically insecure, vulnerable, excitable, and unpredictable, it is essential to provide them with an extra-safe place to exercise. And

Top left: **Turnout areas for foals like Sherlock, who tends to exercise at full throttle, should be safe but with varied terrain to promote development.** *Top right:* **Some horses seem to prefer barn life and when turned out hang around the pasture gate waiting to be brought back in.**

since a horse's vision is less than perfect, it is important that any exercise area is safely fenced and free from hazardous objects. Footing should be soft but not excessively deep. Hyperextension of the fetlock in deep sand can do permanent damage to tendons.

Riding is the obvious exercise choice because that is the reason most of us have horses! But for the days on which you cannot ride, free exercise (turnout) is the least labor-intensive and a natural way of allowing your horse to exercise. However, many horses turned out on pasture use the opportunity to eat rather than exercise, so turnout is counterproductive. Others that prefer barn life stand at the gate waiting to come back in.

Here are other exercise options for your horse.

Ponying, or leading one horse while riding another, is a good choice, especially for young horses. Ponying can start in an arena but can be expanded to include work in open spaces on varied terrain. Ponying a young horse on the surface that he will be worked on when he is an adult provides an opportunity for specialized adaptation of tissues. And the variety in scenery and experiences during ponying is good for any horse.

Longeing, working a horse around you in a 66-foot-diameter (20 m) circle, is an option for horses over 2 years of age. Due to the uneven loading of the legs associated with repetitious work in a circle, longeing a horse younger than 2 years or any horse in a pen smaller than 60 feet in diameter may result in strain. Free longeing can be conducted without a longe line in a round pen. If you don't have a round pen, you can longe a horse on a 30- to 35-foot longe line in an arena.

Long lining is ground driving the horse through various exercises and patterns much as you would longe a horse except that you are holding two long "reins" (that is why it is also called *long reining*). This gives not only your horse good exercise but you as well, as you will be walking briskly to perform many of the maneuvers.

In-hand work is a practical way to introduce your horse to various new areas on or near your property. It is a great warm-up and manners review and good exercise for you, too.

Electric horse walkers can be useful for occasional sessions but should not be viewed as the mainstay of a horse's exercise program. Thirty minutes on a walker once or twice a week might be a good alternative on busy days. Depending entirely on a walker for exercise could result in a stiff carriage, resistance, laziness, and boredom.

Free-run exercisers are a recent innovation and consist of pens that rotate in a circle in a concentric

Sherlock demonstrates that longeing can be a great exercise option and a convenient way to work on balance and form; 2 to 3 inches of footing is ideal.

pathway around the central motor hub. Unlike with a conventional walker, the horse is not tied and can move his head and neck freely.

Treadmills can also be used for an occasional workout, providing that the horse is gradually conditioned to the work and carefully monitored for signs of stress. A continuous climb at the 5- to 7-degree slope characteristic of most treadmills can be fatiguing. A workout using a treadmill is accomplished in about half the time required for most other forms of exercise. If a young horse is asked to perform on the treadmill for even a few minutes beyond his physical capabilities, he might become injured or sour toward work. Treadmills are used successfully for muscle development, particularly of the forearm, chest, stifle, and gaskin.

RAMM Fence

A free-run exerciser has pens that rotate around the perimeter of a circle.

Swimming allows horses to receive a good deal of exercise without traumatizing the joints. Horses are naturally good swimmers. An oval pool with a walkway along the edge allows a handler to walk along the perimeter while holding the lead rope of the swimming horse.

Fresh Air

To safeguard a horse's respiratory health, the air he breathes needs to be clean, with no mold or dust particulates in suspension. When working in arenas, manage the footing to reduce dust. In barns, use dust-free bedding and feed and minimize the amount of hay handling that takes place there. The healthiest place for a horse to live and be ridden is usually outdoors.

Top: Zinger, resting in the standing position, has locked three legs and is resting a hind. *Middle:* Zinger, in the sternal recumbent position, could rise in an instant. *Bottom:* This Wyoming foal feels content and safe with his dam nearby and so rests in the lateral recumbent position.

Rest

Horses usually rest in one of three positions: standing, sternal recumbent, and lateral recumbent. When standing, a unique stay apparatus and a system of check ligaments allow a horse to lock his legs and sleep on his feet. And that is how most horses rest. In order for a horse to have quality rest while dozing on his feet, he simply needs a comfortable place to stand, one that is relatively level and free from weather extremes, noise, light, insects, and anything threatening.

On a sunny day or after a particularly hard workout, your horse may lie down to rest. First, perhaps, he will try the sternal recumbent position by simply kneeling, then tucking his hind legs under his body and lying on his belly. Horses often take a snooze while tucked in such a cozy little ball. If suddenly startled, however, most horses can rise from this position in an instant because the hind legs are under the body, ready to push it up.

If a horse is very relaxed and unthreatened, he may roll over from the sternal recumbent position and lie flat on one side, extend all four legs, and lay his head and neck on the ground. It takes more time to get into a "red alert" position from this lateral recumbent position than from the other two, but a horse can still get up more quickly than you might imagine. That's why you should approach a recumbent horse cautiously and be ready to move quickly yourself.

Most horses love a good roll, which is a form of self-grooming, and rolling often precedes or follows lying down. When a horse lies down, whether for rolling or sleeping, he first bends his fetlocks and knees and lets his weight fall toward the ground on his forehand. Usually at the same time, he is folding his hind legs under to fully lie down. However, old horses or any horse lying down on hard ground might pause in the kneeling position and swivel around, looking for the best place to lie down, or the horse might get up and down from his knees several times before he finally lies down. If the ground is hard or coarse, a horse can develop chronic abrasions on the front surfaces of his fetlocks and knees.

Zipper enjoys a good roll in soft sand while wearing a fly sheet. Be sure that blankets are well fitted and straps are properly adjusted and checked often.

That's why it is good to provide a soft, comfortable place for a horse to lie down. Pastures usually provide ideal cushion. In or near barns you can use bedding, smooth pea gravel, or sand to provide your horse with a comfortable place to lie.

It is perfectly natural for a healthy, fit, sound horse to lie down occasionally, but if your horse spends more than 2 or 3 hours a day lying down, he may have hoof or leg problems; get professional advice and help.

Companionship

Since a horse is a social creature, a single horse can often be lonely if he can't interact with others of his species. Horses do not need actual physical contact with one another, but if they can be near enough to see, smell, and hear each other, they will often be content. The more you interact with your horse, the more you will provide him with some of the aspects of companionship that he would normally get from other horses. This point should not be taken too far, however, because you are not a horse and he is not a human. If you keep things in perspective, you can develop a healthy partnership with your horse and you will both be better off for it.

Sometimes a companion animal such as a goat or a burro will help alleviate a single horse's loneliness. Horses have been successfully pastured and housed with many types of animals, including cattle, sheep, and goats.

Left: **As yearlings, Aria and Seeker were very close.** ***Above:*** **If you have just one horse, a goat or burro can often make a good companion.**

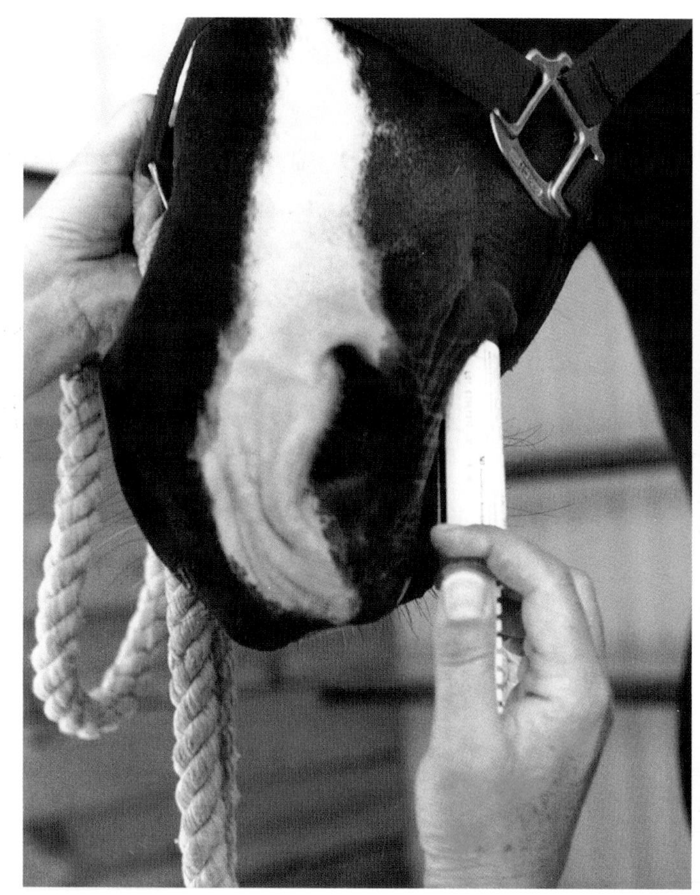

Above left: **A horse that receives routine veterinary care usually has fewer urgent medical needs.** ***Above right:*** **Richard rasps Sassy's hind hoof in preparation for shoeing. Maintaining a regular hoof-care schedule is important for your horse's comfort and safety.**

Veterinary and Farrier Care

If you follow good management practices, your veterinary and farrier bills will be routine and minimal. When you find a good veterinarian, discuss your overall management plan and determine which health care tasks you will perform as a horse owner and which your veterinarian will provide. For example, you might feel comfortable deworming your own horses, but you'd rather your veterinarian give your horses their yearly vaccinations, dental care, and checkup. (For more information, see Hill, *Horse Health Care* [Storey, 1997].)

Look for a competent farrier and ask for his or her recommendations as to which of your horses should be shod and which should be left barefoot and trimmed and how often. Then stick to the schedule. You'll probably need to schedule a farrier visit every 6 to 8 weeks. Earn the respect of your veterinarian and farrier by being the very best horseman you can, and they will be there to help you when you need them the most. Provide these valuable professionals with a safe, comfortable place to work (see chapters 7, Barn, and 8, Interior, for more on work areas). Don't try to cut corners on hoof or health care. Your horse's well-being is at stake and you can't afford to risk it. (For more information, see Hill and Klimesh, *Maximum Hoof Power* [Trafalgar, 2000].)

For books dedicated to these important subjects, see the recommended readings on page 298.

Program

4

There is no one right way to keep horses. But horsekeeping will go more smoothly if you design a program, or management routine, that fits your lifestyle, facilities, and locale. If you have ample pasture but little time for daily horse care routines and plan to ride only on weekends, then keeping your horses on pasture full time might be the best choice. If, on the other hand, you have limited space but have the time and interest to do barn chores and can provide daily exercise for your horse, then stabling could work. Another popular way to keep horses is in a partially sheltered pen or run. No matter which management style you choose, there may be times when your horse will benefit from a blanket; this chapter's final section will help you choose appropriate horse clothing.

To tailor your own horse management plan, consider the pros and cons of the three most common methods of keeping horses: stall, pasture, and pen.

Keeping a Horse in a Stall

The smaller your acreage and the closer you live to an urban area, the more likely your horse will spend part of his time in a stall. Although it is a space-efficient way to keep a horse, it demands a large investment of capital and time. Keeping a horse in a stall requires that you have a well-designed barn and that you feed at least twice a day, clean the stall at least once a day, and exercise the horse every day by riding, longeing, driving, ponying, or providing active turnout. Even with all that, stall life doesn't suit every horse. For the best chance of success, start with a good stall.

A horse can be happy and healthy living in a stall if the stall is well designed and well maintained.

STALL ESSENTIALS

A good stall environment begins with a minimum space of 12 by 12 feet with an 8-foot by 4-foot door. Many horses over 1100 pounds or 15 hands are much neater and more content in a 12-foot by 14-foot or 12-foot by 16-foot stall.

For hoof and respiratory health, the barn should be located on a well-draining site. The base of the stall floor should be porous material such as 10 to 12 inches of gravel. The flooring, which goes on top of the base, should be comfortable and safe, such as rubber mats. The bedding must be nontoxic, clean, dust-free, comfortable, and something the horse won't eat.

The stall walls and doors should discourage rubbing, be able to withstand damage from kicking or chewing, allow ventilation to flow through the stall, and allow the horse to look out of the stall. There should be a clean place for the horse to eat hay (preferably at ground level), a grain feeder, and a large water pail or automatic waterer. The stall should be located where there is not a lot of noise or bright lights. The barn environment overall should be healthy — plenty of ventilation (windows, doors, vents, or fans) that keeps the temperature in the 30 to 80°F range and humidity in the 35 to 60 percent range.

PROS

Stalls are convenient and allow you to give each horse individual attention.

Space. If you have limited acreage but still want to keep a horse or two at home, stalls can make this possible provided you have the time for daily care and exercise.

Convenience. When you have limited time for daily riding, keeping a horse in a stall usually ensures that he is clean and close at hand so you can quickly tack up and get to work.

Hair coat. A stall-kept horse usually has a better hair coat, mane, and tail than a horse kept outdoors. If you want to delay or minimize the growth of a winter coat or speed up shedding in the spring, keeping a horse in a stall will allow you to use lights on

Stalled horses must receive daily exercise. Riding a stalled horse is convenient because the horse is clean and ready the moment you arrive.

timers to affect his seasonal biorhythms and hair growth.

Feeding. Individual stalls allow you to feed each horse separately, so you know each horse gets his entire ration and no more.

Safety. Individual stalls prevent injury from normal herd behaviors such as biting, kicking, chewing manes and tails, and fighting over feed.

CONS

Stalls are restrictive, and confinement is contrary to a horse's natural behavior. If his needs aren't met, things can go wrong in a hurry.

Vices. Excess feed, a small living space, too little exercise, and not enough socialization can lead to the development of stable vices. Because a vice can render an otherwise excellent horse into a very undesirable one, vices should be prevented.

Respiratory health. Ammonia that is released from decomposing manure and bedding and dust that is associated with some hay and bedding often lead to respiratory problems. Humid air encourages mold growth, which can further complicate respiratory disease.

Stocking up. Some horses, even if regularly exercised, will "stock up," or carry excess fluid in their lower legs, when they stand in a small space for long periods of time. Such horses may need their legs to be rubbed and wrapped daily to manage the swelling.

Safety. When a stalled horse rolls, he can become cast — getting stuck against the wall and unable to get up. Some cast horses panic and fight, leading to injury; others suffer colic while waiting for rescue.

Fire. Loss of horses by fire is more common for stalled horses than for those kept in pastures or pens.

Expense. A well-designed barn with stalls and storage areas is a considerable investment. Bedding is a substantial regular cost.

Labor. Stalls are labor intensive. You must commit to cleaning, bedding and manure hauling, daily exercise, late-night checks, and water maintenance.

Manure management. Stall refuse requires a more intensive manure management plan.

Maintenance. A barn and stalls can require considerable maintenance (structural, electricity, plumbing, flooring, antichew).

Body rollers

A body roller has projections on the top, something like the handholds on a vaulting surcingle, that discourage a horse from rolling over on his back and getting "cast" in his stall. If a horse rolls over completely near a stall wall, he may get his legs trapped in a curled-up position between his body and the stall wall, preventing him from getting up. It can be dangerous for a horse's digestive system if he is trapped on his side for longer than an hour or so; therefore, it is a good idea to frequently check a horse that is prone to roll in his stall.

This is an ideal field for horses; it is a well-drained, grass-mix pasture.

Keeping a Horse on Pasture

Part of the dream of having a horse is the visual satisfaction of seeing a horse peacefully grazing on a well-maintained pasture at your home. Pasturing a horse might be the most natural way to keep a horse, but unfortunately, it is out of reach for many and can be far from ideal from a horse's viewpoint. For the best chance for success, start with a good pasture.

PASTURE ESSENTIALS

A good pasture has a stand of plants suitable for horses. The best kind of horse pasture is a well-drained grass mix with few weeds and no poisonous weeds, trees, or shrubs. If there is a good grass stand established, you have decent rainfall or access to irrigation, and you mow, harrow, and reseed as necessary, you should be able to keep one horse on 2 acres of pasture during the growing season. However, arid ranchland with minimal browse plants can require 20 acres or more to support a single horse. To get a better idea of the specific stocking rate for your property, contact your county Extension agent.

A pasture needs to be enclosed with safe fencing and gates. Pasture fences and gates should be at least 5 feet tall and well maintained to increase the horses' safety and decrease the liability of loose horses on public or private property. Using electric fencing in

Above: Farm equipment in a horse pasture presents potential for injury. *Top right:* Stagnant water, such as in this irrigation ditch, can be hazardous to your horse's health. *Bottom right:* Allowing horses access to wetlands is bad for hooves and destroys the wetlands, which perform important functions.

conjunction with conventional fencing decreases the wear and tear on fences and adds to security as long as the electric fence is checked daily to be sure it is working.

There should be no old dumps or farm equipment in a pasture; horses can easily get hurt on items hidden by tall grass.

There should be easy and safe access to free-choice, good-quality water. Natural sources should be running, not stagnant. Know the source of the water your horse drinks. If it contains agricultural runoff, it could be high in nitrates (see chapter 14, Water). A trough or automatic waterer should be kept clean and situated to minimize mud and to prevent a horse from being crowded into a corner or against a fence.

Pastures should be well drained with no bogs or stagnant water. Ideally, the soil will not be sandy.

The pasture should provide shelter — either natural (trees, rocks, or terrain) or man-made (shed or windbreak) to ward off sun, wind, cold precipitation, and insects.

There should be free-choice salt and mineral blocks at all times.

A good field shelter

A good field shelter has the following characteristics:

- ☐ Three sides with the fourth side completely open
- ☐ Back to the prevailing weather
- ☐ Front entry height at least 12 feet
- ☐ Made of weatherproof, safe, durable materials that don't invite chewing
- ☐ Safe, comfortable footing
- ☐ Enough room for all horses (approximately 150 square feet per horse)
- ☐ Easy access for cleaning

Left: Pasture is an ideal place for Sherlock to rest.
Right: Sherlock, on the move, gains valuable benefits from free-choice, low-level exercise.

PROS

Pasturing can be an ideal way to keep horses.

Contentment. Many horses are most content when they are at pasture because they are allowed "to be horses." Pastured horses rarely develop vices.

Rest. Pastures usually provide a comfortable place for horses to lie down.

Socialization. Living with other horses provides companionship and the chance for valuable herd interaction so a horse can learn limits of behavior such as biting, kicking, and crowding.

Soundness. Pastured horses exercise freely, so tend to stay "legged up" — tendons, ligaments, bones, and hooves receive moderate, continuous stress; thus, they become tough yet resilient and are less prone to lameness from a misstep or slip. As long as the pasture is not excessively wet, muddy or boggy, or very rocky, many bare hooves improve with pasture turnout.

Respiratory health. Horses that live in fresh air and sunshine tend to stay healthier with fewer of the respiratory diseases that are seen with stalled horses.

Exercise. Pasture turnout is an ideal choice for broodmares and foals and for growing young horses, as it allows them to satisfy their need for exercise while adapting to natural terrain.

Fitness. A horse that is an energetic self-exerciser will retain a higher level of fitness on a pasture than in a stall, even if ridden the same amount. Horses in training or work that are kept on pasture rarely suffer from *azoturia* (tying up).

Nutrition. A well-kept pasture can offer excellent nutrition, especially minerals and vitamins A, D, and E.

Recreation. Pasture turnout is often good for a horse that needs a break from training or performance or for a horse that is recuperating from an illness or injury.

Labor. For day-to-day tasks, labor is decreased because the horse is on "autopilot" for eating, drinking, exercising, and self-grooming. There are no stalls or runs to clean. Manure management is less intense, especially on large pastures.

Cost. If you already own well-maintained pastures, you will incur low hay and grain costs and will likely have no bedding costs.

CONS

In spite of a seemingly idyllic lifestyle, pasturing horses presents a substantial list of concerns.

Discomfort. Although many horses are happy on pasture, not all are. Barn horses sometimes prefer stall or pen life with daily human care and individual hay and grain rations. Some thin-skinned horses find pasture life too irritable with insects, brush, sun, and extreme weather. Very old horses might enjoy short pasture turnout but tend to do better when fed individually and provided with a comfortable, quiet place to rest.

Inconvenience. Pastured horses are less handy to catch and ride than those kept in a stall or pen.

Wild behavior. Horses on pasture that are not handled often can forget their training and revert to "wild" behaviors. Unlimited freedom and no structured interaction with humans can result in pushy, feisty, and headstrong horses.

Herd bound. Although socialization is desirable, many pastured horses become herd bound or buddy bound. They form strong attachments to certain individuals or to the herd and when separated can become uncontrollable.

Mane damage. That natural, social ritual of mutual grooming can destroy a lovely mane in a single session.

Fighting. Horse herds have a pecking order. Horses can violently compete for feed, a salt block, or a place at the water hole. Because of this and poor fencing, pastured horses generally get injured more often than horses that are stalled or penned individually.

Hair coat. In winter, pastured horses grow thick, long coats that make them sweat more easily and become more difficult to cool out. Grooming a long coat is a catch-22 — a thick coat requires more grooming than a short coat, but grooming a thick coat removes some of the waxy buildup on the skin that provides protection from moisture and wind. In summer, hair coats bleach and tails get thinner and shorter from wind, swatting flies, brush, and burrs.

Hoof damage. Generally, there tends to be a higher incidence of lost shoes on pasture. If the terrain is hard or rocky, bare hooves can be bruised, broken, or worn excessively.

Overeating. Certain horses on pasture tend to overeat, be overweight, get out of condition, and become lazy. If the pasture is very lush, some horses become soft and fat, have a difficult time breathing when worked, and sweat heavily.

Laminitis. Laminitis is a severe inflammation of the hooves caused by a digestive overload. Grass founder is a common occurrence in horses grazing on lush, early-growth pastures.

Sand colic. If the pasture soil is sandy or if a horse drinks from a water hole with a sandy bottom, sand can accumulate in the horse's gut and cause sand colic, a life-threatening obstruction of the horse's bowels.

Zipper maintains a good level of fitness for riding by exercising vigorously when he is turned out in his mountain pasture.

Top left: **Although Seeker likes wooded areas for shade and shelter, they can be a source of ticks and injuries.** ***Bottom left:*** **During hunting season, a blaze orange sheet identifies Zinger as a horse, not an elk — we hope.** ***Above:*** **Power poles and guy wires are hazards. Cover guy wires with visible plastic tubes; these are often available from the local power company.**

Parasite reinfestation. Horses shed parasite eggs in their feces, the eggs hatch, and the larvae crawl up grass stems and are ingested by the horse, resulting in constant reinfestation.

Undesirable plants. Many pasture plants, trees, and shrubs are poisonous to horses. Others can trigger a photosensitive reaction that causes unpigmented areas to become red, thickened, and scaly.

Insects. Pastures usually have more insects than do buildings, and the pests are more difficult to control because breeding grounds are near water, in tall grass, and in wooded areas. Bloodsucking flies, ticks, gnats, and mosquitoes can carry diseases, including West Nile virus (mosquitoes) and Lyme disease (ticks).

Handler safety. It can be difficult to safely go into a group of horses on pasture and bring one out or to safely turn a horse into a group of horses on pasture.

Wild animals. Other animals that live in or cross through pastures can cause problems for horses. Poisonous snakes, bears, coyotes, mountain lions, loose domestic dogs, and the holes of gophers, prairie dogs, and marmots can cause injury.

Wooded areas. Heavily wooded areas are a home to ticks and are a source of leg and eye injuries.

Hunting season. In many rural areas where there are open hunting seasons on deer and elk, horses are in danger of being mistaken for game.

Safety. Utility poles often pass through pastures and pose a potential risk.

Security. Horses are generally more secure when kept near homes and buildings. Horses pastured in remote fields can more easily be stolen.

Labor. Pastured horses need an up-close-and-personal check once or twice a day; on large pastures, this can take some time.

Maintenance. Pastures require daily, weekly, and seasonal maintenance: daily fence and pasture patrol, fence maintenance, manure management (either remove and compost or rotate pastures and harrow), mowing, spraying for weed control, irrigation, and reseeding.

Cost of land. Good pastureland, especially close to urban areas, is getting scarce and more expensive. Unfortunately, it often makes more economic sense to the owner of the land to use it for something else, like more houses or a strip mall.

Cost of fencing. Safe, secure fencing is essential and initially is a considerable expense.

Not year-round. No matter how good a pasture is, it will need to rest for part of the year, and you will need to provide other living arrangements for your horses during that time.

Damage. Horses can be very hard on pastures, trees, and fences. A horse selectively grazes 16 out of 24 hours each day, eating 5 to 6 pounds of grass per hour. If a pasture is carrying too many horses, overgrazing will quickly ruin it. Overgrazing destroys plant and root structure and results in soil erosion and takeover by weeds. Pawing before rolling and to expose more succulent grass damages plants. Because horses select certain areas to graze, to defecate, and to congregate, some patches are constantly grazed short, some are ignored because of fecal contamination, and others turn into bare dirt or mud holes. Horses that chew wood are destructive to fencing and can quickly kill trees.

Supplemental feed. Pasture horses will have to be fed hay and possibly grain during the late-fall, winter, and early-spring months. How you choose to do this will depend on how many horses you are feeding. Remember, horses will fight at feeding time, so if you have personality conflicts within your herd, or great numbers of horses, you will need to devise a way of separating horses until each gets his fair share of feed. This usually requires you to put each horse in a separate stall or pen at feeding time, which defeats some of the laborsaving bonus of pasture management.

If you are feeding just hay, spread it out in the pasture in an open space (provided it is not a windy spot). Make several more piles than the number of horses you are feeding and place the piles far apart. If you need to feed grain or more closely regulate the feeding of hay, you can construct several small pens in which you can separate horses.

Out of sight but not out of mind. Turning a horse out to pasture can be a natural, low-maintenance style of horsekeeping, but even if a horse is out of sight, he should not be out of mind. Pastured horses need special care, which starts with proper preparation of the pasture and the horse and continues with regular maintenance.

To learn more about pasture management, read chapter 13, Land. To learn more about caring for a horse on pasture, see chapter 18, Routines.

This poor tree has been girdled; that is, the bark has been completely eaten off by horses. What was once a nice shade tree will soon be dead and gone.

Keeping a Horse in a Pen or Run

When you want your horse to have some room to move around but you don't have access to a pasture, a good setup can be a group pen or individual run. These are usually located adjacent to a barn or other covered shelter and can vary in size from a bare minimum of a 16-foot by 60-foot individual run off a stall to a 60-foot by 100-foot or larger pen off the end of a barn or loafing shed for a group of horses.

PEN AND RUN ESSENTIALS

A good pen has safe, durable fencing and comfortable, well-draining footing. The pen should be located on high ground and be situated such that the horses can take shelter from cold wind, wet weather, hot sun, and insects as needed. There should be a clean place to feed and a comfortable place for horses to lie down. To prevent feed from blowing away, windscreens can be attached to the outside of the panels.

Top: This loafing shed has individual shelters and runs. Although bedding the shed provides a comfortable place for a horse to lie, it also encourages defecation and urination. *Bottom:* The rubber windscreen in Dickens's pen keeps the west wind from blowing away his hay.

Top: **Sherlock and Aria practice their pas de deux in the large sacrifice pen.** ***Bottom:*** **Zipper nibbles hay off mats in his sheltered eating area. The mats eliminate sand ingestion and feed waste.**

The land in pens and runs is considered a "sacrifice" because no vegetation is expected to survive the constant traffic. If the natural lay of the land doesn't slope away from the barn or shed, then excavation should remedy this so that the shelter under the building is high and dry and the pen or run gradually slopes about 2 degrees away from the building. There should be a buffer zone around the sacrifice pen, especially on the downsloping side. The buffer zone can be well-established grass, trees, or bushes that will act as a filter for pen runoff.

Depending on the native soil, footing can be added to provide cushion and minimize mud. Some choices are decomposed granite, road base, and pea gravel.

A sheltered feeding area with rubber mats allows a horse to eat off ground level without ingesting sand or wasting feed.

In the loafing area of the pen, bedding can be used to encourage a horse to lie down, but it usually invites a horse to defecate and urinate there also. This behavior can be minimized or eliminated by locking a horse out of the loafing or eating area except during specific times.

Pen fencing can be made from metal panels or continuous fencing. Panels don't require setting posts, so are more adaptable to changing pen size or shape. Whatever pen fencing is used, it needs to be tall enough (5 feet is OK; 6 feet is better) and strong enough to withstand roughhousing, rubbing, and

Top: Pens can be made with panels. *Middle:* Pea gravel makes good pen footing. *Bottom:* Zipper's water barrel is a 55-gallon drum that once contained vanilla and has had the top removed.

playing across the fence. Panel connections should be tight and safe. (See chapter 11, Fencing, for panel details.)

PROS

A well-designed individual pen can combine some of the best features of pasture and stall life. The horse will receive individual rations, privacy for resting and eating, shelter, fresh air, and the ability to take moderate exercise. Watering pen horses is usually less labor intensive than watering stall horses, as you can use a trough or barrel. He will be convenient to gather up for riding and can be outfitted with a turnout sheet to keep him clean. Hooves stay in good condition because of the managed footing.

Many senior horses fare better in spacious pens with regular turnout than they would on pasture full-time. Senior horses thrive when given a quiet, dark place with soft footing in which to rest when standing and a comfortable, safe place to lie down and roll. Locating the feed area, the water, and the loafing area a good distance from each other encourages movement.

CONS

A group pen can be more convenient from a management standpoint and will provide horses with social interaction and more room for exercise, but it does have some serious drawbacks.

Life in a group pen has many of the same challenges of pasture life without the benefit of the grass. Group pens present feeding problems, fighting, buddy bonding, mane chewing, and injury associated with groups. Fencing for group pens must be especially safe to protect each individual from "peer pressure," which is amplified in confinement.

Exercise. A pen horse's need for exercise can be easily overlooked since the horse is able to move around a little bit. Depending on the size of the pen, the horse will need turnout in a larger space and/or a certain number of formal exercise sessions per week to keep him content, fit, and healthy.

Fighting. Horses fight and play across pen fences, which can be hard on the facilities and the horses.

Horses often lie down next to panels and get trapped. That's why panels must be smooth and safe, as shown here.

Sore points. It is impractical to bed an entire pen, so when a horse rolls or lies in the dirt or gravel portion of the pen, he can develop sores on knees, hocks, fetlocks, and elbows and might need protective boots to help heal or prevent such injuries.

Holes. Pawing and traffic patterns can develop holes in pens that can turn into messy wallows during wet weather.

Safety. Horses can lie down or roll next to the pen panels and get injured or trapped.

Combination Management Styles

To find your ideal style of management, customize and combine until you find the right fit.

Night stall/day turnout. If you like the security and convenience of keeping your horse in a stall at night and you have access to pastures for daily turnout, you can devise a turnout schedule that will satisfy your needs and result in a happy horse.

Night pasture/day barn. Keeping your horse in a stall or sheltered pen at the barn during the day will make it handy when you want to groom, ride, or perform health care tasks. If you turn your horse out to pasture after dusk and bring him in before dawn, he will miss the thickest fly and mosquito times, yet will have 8 to 10 hours of grazing time during the night. Depending on the quality of your pastures, you might only have to offer him some "busy hay" during the day.

Run with stall. If you like to keep your horse in a stall when the weather is bad or when you want him to be clean for a competition, keeping him in a stall with an attached run is ideal. You can choose to keep him in or out at a moment's notice.

This specially designed "senior center" has three separate areas to which the horse has free access: a 12'x12' matted, covered eating area; a 36'x24' covered "sandbox" for comfortable loafing; and a 3000-square-foot exercise area with pea gravel footing.

Horse Clothing

Whether your horse lives in a pasture, stall, or pen, you might want to outfit him with some protective clothing. Knowing about the different types of horse clothing will help you choose the most appropriate sheet or blanket and save you time and money when you go shopping.

If you will be keeping your horse in a stall a good deal of the time, you will be choosing stable clothing that is comfortable and durable but not necessarily waterproof. If your horse will be living in a pen or on pasture, you'll choose turnout sheets and blankets that are tough and weatherproof.

STABLE CLOTHING

Sheets and blankets for stall use should be comfortable, breathable, and well fitted. They should be durable enough to withstand rubbing but because they won't get wet and muddy, they don't need to be weatherproof.

Average blanket sizes

SIZE (in.)*	CATEGORY OF HORSE
48–54	Small pony
56	Pony
60	Pony
64	Pony
66	Yearling
70	2-year-old
72	Small horse
74	Small Thoroughbred, Saddlebred
76	Thoroughbred, Saddlebred, small Quarter horse
78–80	Small Warmblood, Quarter horse, large Thoroughbred
82–84	Warmblood, Draft cross

*Measured in inches from the center of the chest to the center of the tail.

Day sheet. This lightweight sheet is designed to protect a horse from dust, flies, and drafts without adding substantial warmth. Day sheets might be made of linen or cotton, which will shrink, or synthetic fabrics, which won't shrink. Always check fiber content and the manufacturer's size recommendation so you can purchase the correct size. Some day sheets make suitable liners under winter blankets.

Scrim. This lightweight nylon or cotton net cooler is draped over a stalled horse's body to keep flies off without making him too hot. These are typically not the sturdiest of sheets, and they don't fit as securely as a sheet (fewer straps and less fitted), so are often relegated to day use only and to horses that don't do a lot of rolling or rubbing.

Antisweat sheet. An antisweat sheet is a hot-weather net or mesh sheet made of cotton or a synthetic wicking fabric. It is used to prevent a stalled horse from breaking into a sweat on very hot days. It can also be used as a hot-weather cooler after a workout or bath to enable a horse to dry.

Winter stable blanket. A winter stable blanket needs to be warm and comfortable and stay in place. Leg straps usually prevent shifting, a cut-back withers prevents mane rubbing, and a nylon lining polishes the coat. The blanket's exterior must be tough enough to withstand a horse rolling and rubbing, and although it doesn't need to be waterproof, it must be breathable (allow sweat to evaporate) to prevent overheating. The middle layer of insulation can be light, medium, or heavy for various temperatures. A heavyweight blanket is not necessarily heavy in weight but just provides maximum warmth. Generally, a stable blanket is used in temperatures below 40°F.

Hood. A hood is sometimes used in conjunction with a stable blanket on show horses that are kept indoors and are short coated all winter.

Stable wraps. In addition to blankets, you may wish to use leg wraps for your stalled horse. They are not usually necessary unless the horse is in heavy work. Stable wraps (or standing bandages), which usually extend from below the knee or hock down to the fetlock, can be made of wool, flannel, or fleece,

Cool Zinger on early-summer pasture in her fly sheet and mask. For gnats, use a mask with ears; for nose bots, use a mask with a muzzle guard.

with or without a cotton quilt underneath. They offer warmth and protection and might keep a horse in heavy exercise from "stocking up" (accumulating fluids in the legs) as he stands in his stall. Stable wraps are not like exercise wraps. The latter are designed for temporary support and are applied with a greater amount of tension. Stable wraps are often left on overnight (but should not be on for more than 12 consecutive hours), so they must not interfere with circulation in any way. Stable wraps are applied much like shipping wraps, in thick layers with moderate tension, except that stable wraps end at the fetlock and shipping wraps cover the bulbs of the heels.

Cooler. A cooler is a large square of absorbent or wicking material that is draped over the horse's body for drying after a bath or cooling out after work, especially in cold weather. Coolers are traditionally made of wool, but since those often require hand-washing in cold water, new materials, including synthetic fleece, are available. A cooler is often held in place only by straps at the brow band and tail and therefore is most suitable for a horse drying out at a hitch rail or in cross ties.

TURNOUT CLOTHING

Sheets and blankets for turnout should be well fitted, waterproof, and breathable. They need to be tough enough to resist rolling and chewing by pasture mates and be outfitted with safe straps that won't entangle a horse when he rolls.

Fly sheet. A fly sheet designed for turnout must be tough and should provide UV protection. There are several types; the most popular is the PVC-coated mesh fabric sheet. Because horses in summer turnout are active, the sheets must fit well and are often outfitted with elastic leg straps to keep them in position without the leg straps breaking.

Fly mask. Fly masks made of various types of mesh are available to cover eyes, ears, and the muzzle. It is important to choose eye masks of good design so that the mask does not irritate the horse's eyes or cause hair loss on his lower jaw from rubbing.

Turnout sheet. A turnout sheet must be safe, well fitted, and tough. Whether a turnout sheet needs to be waterproof, water-resistant, or just dirt-resistant depends on its intended use. If your climate is dry, a tough, UV-resistant, dirt-resistant sheet will suffice. If you live in a very wet climate, you'll want to choose

Blanket types

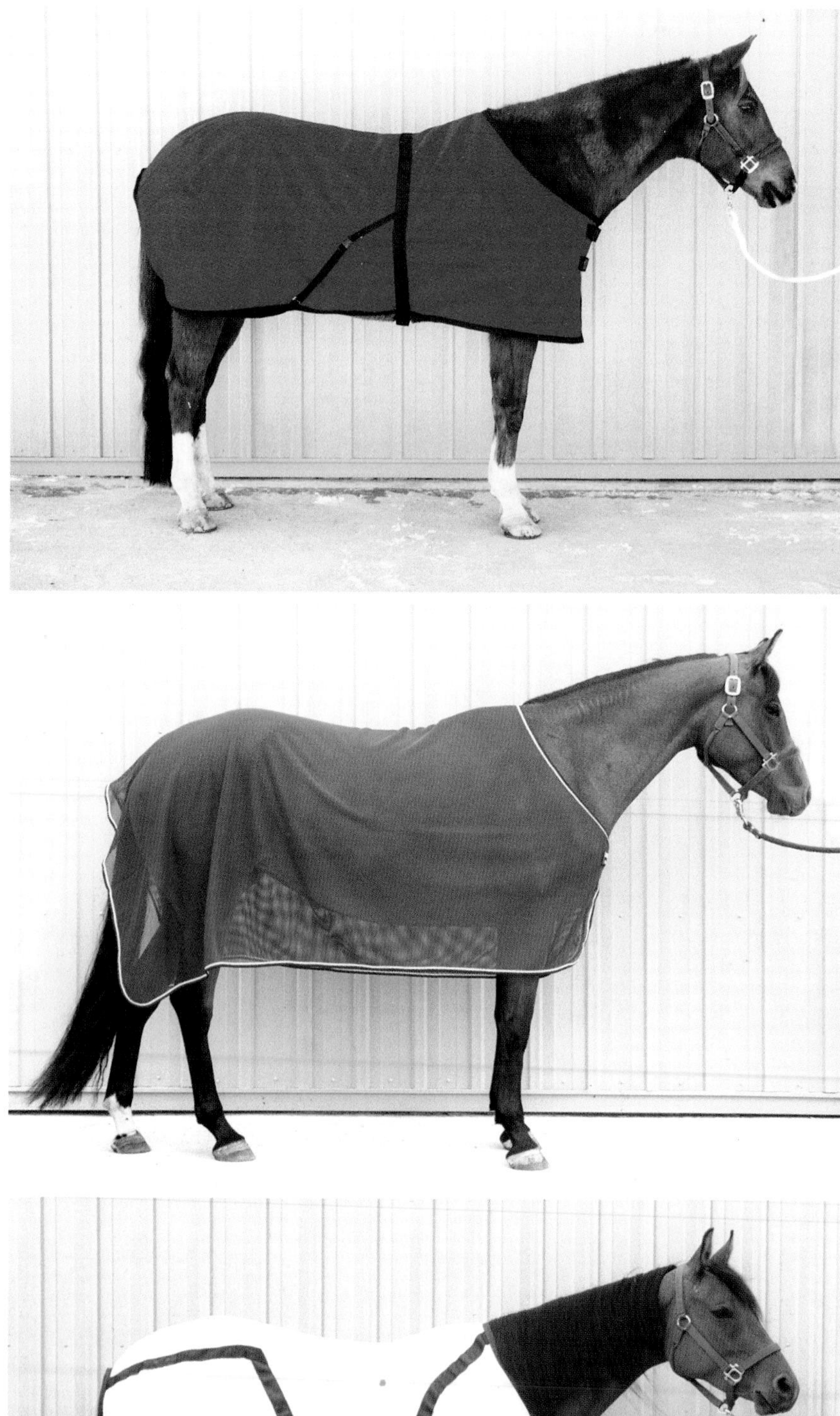

Dickens models a day sheet, which may be used for turnout or barn use depending on the material; a weatherproof material should be used for turnout.

Seeker's scrim is a loosely fitted, lightweight, mesh, cooler-style sheet for stall use only.

Seeker's fitted, antisweat sheet is made of rugged cotton mesh but is designed for stall use only, in very hot, humid climates.

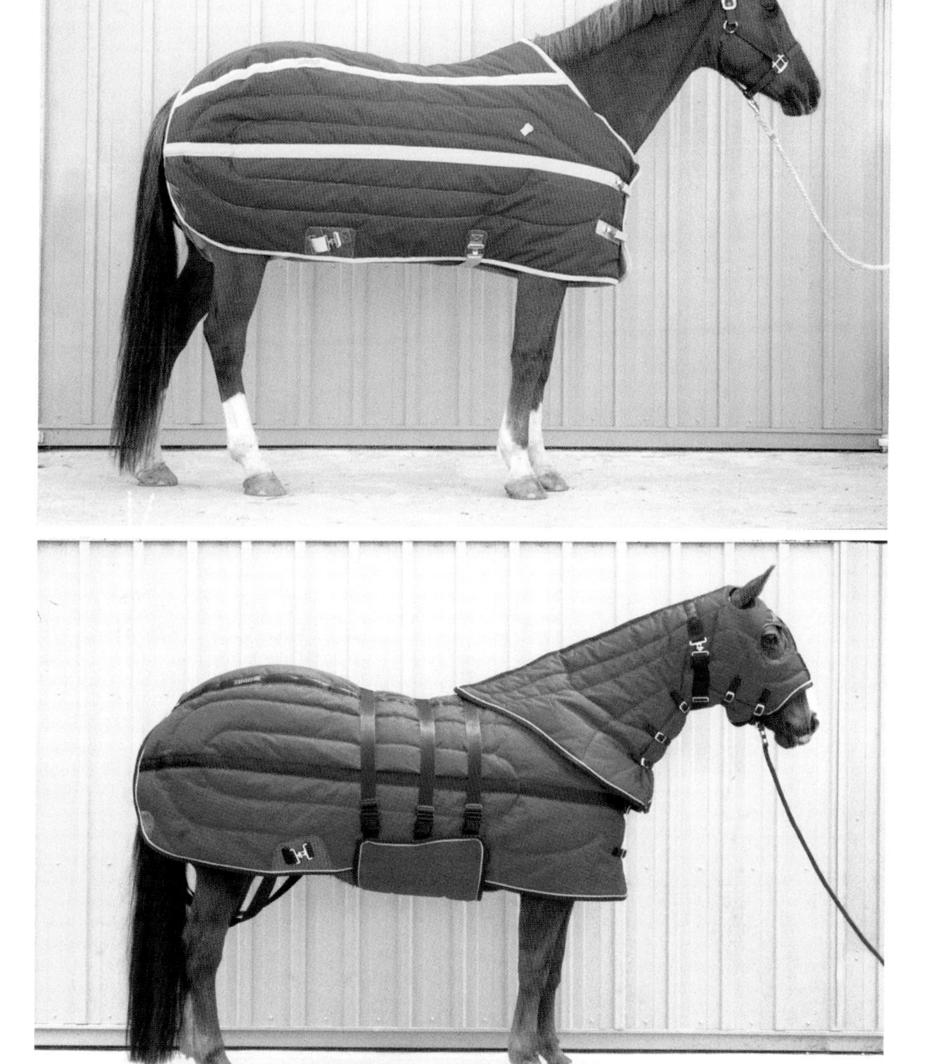

Dickens wears a quilted, insulated stable blanket; it's very sturdy and cozy but isn't weatherproof.

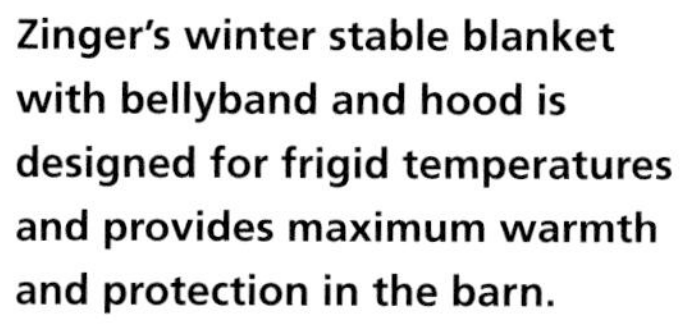

Zinger's winter stable blanket with bellyband and hood is designed for frigid temperatures and provides maximum warmth and protection in the barn.

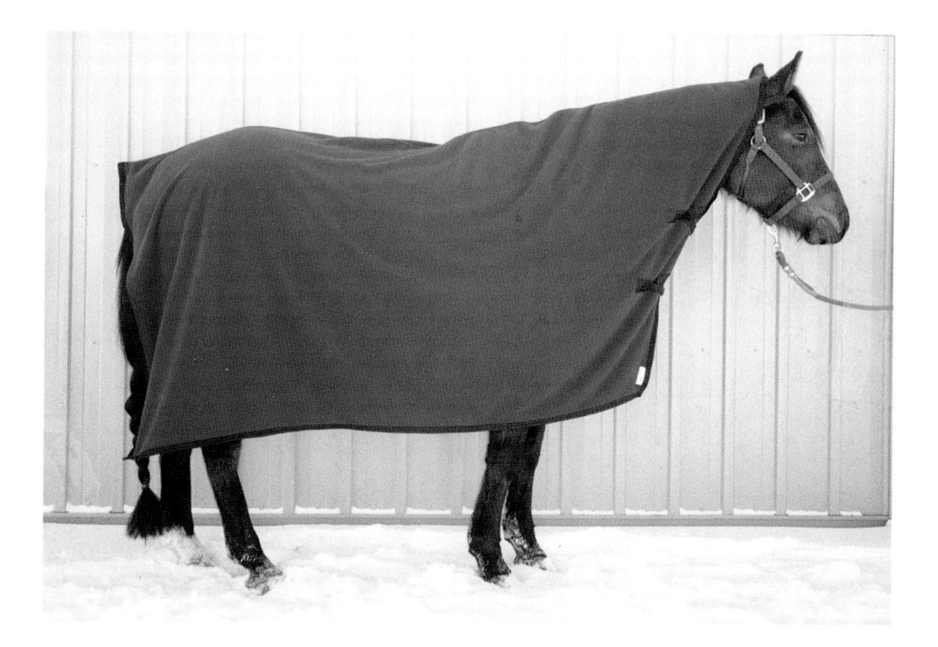

Seeker's blue wool cooler prevents a chill after a winter ride or a bath.

Aria's waterproof, breathable turnout sheet with tail flap proves its worth when a sudden cold fall rain catches her in a turnout pen.

a waterproof and breathable turnout sheet. If you turn your stalled horse out for exercise when it is not raining, a water-resistant sheet will work because even if he rolls on wet ground, he will likely stay clean. Since horses that are turned out are very active, the sheet must fit well and stay put. Sheets with properly adjusted elastic leg straps often are the best choice for turnout.

Winter turnout blanket. When a horse spends a lot of time out in winter weather, either you can let him grow a long winter coat or you can provide him with a winter turnout blanket. He will still grow a winter coat, but it won't be as thick and it will stay cleaner when blanketed.

Winter turnout blankets range from waxed cotton duck with wool lining (New Zealand Rug) to puffy, quilted ski jacket–type blankets. Generally, a turnout blanket has two surcingles (bellybands), front chest straps, hind leg straps, and possibly a tail strap and/or tail flap. Some have neck extensions that reach as far as the ears.

No matter what type you choose, the blanket must be waterproof and breathable to allow a horse to vent excess body heat and moisture from exercise or sudden environmental temperature increase. If a horse sweats and then chills in winter temperatures, it could set the stage for illness.

Winter turnout blankets usually require a very large front-load washing machine to clean. If you don't have one, you could field-wash the blanket, or you might find a self-service laundry that allows horse blankets, but many do not. In areas with large horse populations, you might be able to locate a horse blanket washing and repair service.

Hoods are sometimes used with turnout blankets.

BLANKETING TIPS

Proper blanket fit is paramount. Blankets that are too small or of the wrong cut can cause rub marks and sore spots on the withers, shoulder, chest, and hips. Too-large blankets have the reputation of slipping and twisting, possibly upside down, which can cause the horse to become dangerously tangled. Measure your horse from the center of his chest to the center of his tail to determine blanket size. To get an idea of what size blanket your horse will wear, see the table on page 44.

Blankets must be kept clean or they will cause discomfort to and possibly disease in your horse. A dirty blanket can cause a horse to rub and roll in an attempt to relieve an itch. Manure and mud on blankets fatigue the material and cause them to tear and disintegrate. Have any damaged blankets repaired and keep all leather straps and fittings well oiled.

Overheating can be a real problem with blanketed horses. Horses are often turned out to exercise in the same blanket that they wore all night. What is appropriate for low nighttime temperatures in a barn is not necessarily desirable for a sunny paddock, even though snow is on the ground. An unblanketed dark horse has the capacity to absorb much of the sun's energy and can actually feel hot to the touch on a cold, still, sunny day.

Waterproof blankets that are not breathable do not allow heat and moisture from normal body respiration to escape. Too many layers can cause the horse to sweat, then chill, which lowers the horse's resistance by sapping his energy. This is an open invitation for respiratory infections. Check for overheating by slipping a hand under the blanket from the shoulder toward the heart girth area. The horse should feel dry and cool to warm but never hot or damp.

Horses that have been body-clipped or trace-clipped must be blanketed. Clipping a horse's coat allows him to be more easily worked, cooled out, and groomed in the winter months. The first clip may occur in October and may need to be repeated several times throughout the winter and early spring, depending on the horse's work, blanketing, and housing. If you want to clip your horse's head and neck as well, you might want to use a hood made out of the same material as your blanket. Be sure the eyeholes and ear holes fit your horse properly, or it may cause him to rub. Your tack repair shop can modify hoods. To prevent damage to the mane and rub marks on the face, be sure a hood is lined with a smooth material such as nylon.

Designing Your Acreage

PART TWO

5 Site

When choosing country land, you have many factors to consider. Arrange the following items in an order that is appropriate for your location, financial situation, and goals. Be sure to seek competent, experienced advisers on matters with which you are unfamiliar. Contact your county Extension agent to learn about soil, water, climate, and plant growth in the area. Visit other horse farms to get an idea of how well various facility features work in your locale.

Location

Before you even begin looking at properties, decide how far away from a town you and your family can realistically live. How many miles and minutes are you willing to drive to your job(s) each day? Where is the closest school? Will you spend most of your time in the car ferrying your children into town for activities? Will you have convenient access to the services of a farrier, veterinarian, and feed store? Some of the most beautiful horse country and the best land bargains are a stiff commute from employment centers. Don't make the mistake of falling in love with a place that would be perfect for horses but tough for the realities of human existence.

Determine a reasonable distance you are willing to commute daily and, using a compass, draw an appropriately sized circle on a map to identify the area in which to begin looking. But don't automatically eliminate a property slightly outside of the circle, because its other advantages may result in a quality of living far beyond what is available inside the radius. If you are considering a remote location, make test drives to the property and see how much time traveling usurps. There is little sense in living in a wonderful spot if you can see it only occasionally in the daylight!

Climate

Although climate is somewhat predictable, be aware of small pockets of unusual weather. Such microclimates can make one property ideal and another nearby windblown, flooded, gloomy, or dry. Consider temperature, precipitation, wind, humidity, and length of growing and riding seasons. Become thoroughly familiar with an area before you buy. I have seen one property sell about ten times in 20 years because every spring the majority of the land, including the homesite, stands underwater for several months. Do your research.

Water

The availability and quality of water are of utmost importance to you, your family, and the health of your horses (see chapter 14, Water). If a property does not have a well, determine if there are legal restrictions that prevent drilling one. If the property will require irrigation, see if water rights are included or available for purchase that will meet the needs of the land. Find out the depth of the groundwater table. This will determine whether you will be able to do any underground construction and may affect the installation of a septic system.

A beautiful horse property located a reasonable driving distance from work, school, and shopping is convenient but usually expensive.

Rolling terrain, trees, and rocks provide natural windbreaks and shelter for Sassy and Dickens.

Topography

Note which way the slopes face and what kind of natural or man-made structures may block the sun, be a natural windbreak, or affect drainage. In temperate climates, north-facing slopes or areas obstructed from light may mean increased heating costs, icy driveways and barnyards, increased wind chill and snow drifting, and a "later morning" in the winter. But the same sun-blocking features may be advantageous in reducing heat on a warm-climate acreage.

The area you are researching may be noted for cold winter winds, but rocks, trees, and terrain can substantially block the wind and provide pockets of shelter. Conversely, if you are looking in a desert valley where hot, stale air tends to settle and stall, you might want to make note of natural topography that could interfere with the flow of desirable cooling breezes.

Moderately rolling grassy hills are desirable, as they usually drain well and provide a good environment for the exercise and development of horses. A slope of 2 to 6 percent (2 to 6 feet of rise or fall per 100 feet) is ideal. A much greater slope than this could result in erosion problems and would probably require extensive excavation costs for any improvements you might add to the land. A slope much lower than 2 percent would offer a poor chance for drainage and result in wet, marshy, boggy areas. Such areas, which are mosquito breeding

A word about wetlands

Wetlands provide vital filtering functions and are an important part of our ecosystem, but they are not desirable for horse pastures. If you have wetlands on your property, manage them appropriately (see chapters 13, Land, and 15, Sanitation, for more information).

grounds, contribute to the possible spread of equine infectious anemia, West Nile virus, and other diseases. In addition, wetlands encourage the growth of disease-causing organisms and are very damaging to a horse's hooves.

Soil

Along with the slope, the type of topsoil will greatly affect surface drainage. A sandy loam is ideal. The soil should not pack or become excessively muddy for long periods; nor should it be extremely sandy, or nutrients will be quickly leached out of the soil by rain and melting snow. In addition, sandy soil results in more cases of sand colic. Clay and adobe soils pack hard, cup and retain the uneven surface, and are slippery when wet. Gravelly soils lack nutrients and result in a greater number of hoof and leg injuries from the abrasive surface. The topsoil and subsurface soil can be tested and evaluated. Saline soils are not desirable for pastures or hay fields and are difficult to correct, so if the results show a high salt content, look for other land. The soil profile will also help you plan your planting, fertilization, and irrigation needs as well as deciding where to locate buildings and facilities. (See chapter 13, Land, for more information on soil testing.)

If you want to be able to ride without having to trailer your horse, look for property that adjoins public lands with bridle trails.

Other Natural Amenities

Since it takes so very long to establish a good stand of trees, and since it is impossible to transplant mountains, streams, bluffs, or ponds, each of these natural extras should be considered when tallying a property's assets. Besides adding beauty, these features can satisfy a horse's needs for shelter and water. An added bonus to any land is its close proximity to public lands that permit riding.

Utilities and Services

An undeveloped tract of land can be a real bargain, but before you put down a deposit, check on the availability of utilities and the cost to get them to your building site. These include electricity, telephone, municipal water or a well, natural gas lines, and irrigation. If you wish to heat with propane or wood, check on its availability and on any regulations that may prohibit its use. Find out whether the property is connected to a municipal sewage system and what the monthly charges are. If the property has its own septic system instead, be sure that the subsoil is permeable so percolation will take place, that the drainage field is sufficiently large, and that the tank is located where it can be cleaned easily. Inquire about the location of the nearest landfill and the availability of trash disposal services. Find out who maintains the roads and removes snow.

Determine who provides police service, fire protection, and ambulance service to the area. Will there be adequate water on the property for fire protection needs? Some of the answers may affect the fire insurance rates on your homeowner's policy.

Horsekeeping in the suburbs

As urban corridors are subdivided into small acreages, you may find yourself setting up horsekeeping in a residential neighborhood. Such a location involves some special considerations to maintain good neighborly relations and satisfy legal requirements.

- **MANURE MANAGEMENT.** Your sanitation plan must be impeccable. Manure piles must be concealed and well managed to control odor and flies (see chapter 15, Sanitation, for details).
- **NOISE CONTROL.** Whinnying horses, a blasting barn radio, a megaphone used during lessons — all of these seem perfectly normal to a horse owner but may not be to a neighbor who cherishes peace and quiet.
- **AESTHETICS.** Where you see a beautiful pasture, your neighbor might see a field of weeds and manure. Appearances matter when neighbors are close, so mow, paint, and repair as needed.
- **SECURITY.** It is imperative that your horses stay on your property and not be able to access your neighbors' garden, lawn, trees, or buildings by leaning, reaching, or escaping. If your neighbors also have horses, consider instituting a 20-foot buffer zone (double fence) between properties so the horses can't contact each other.
- **COMMUNICATION.** Stay in touch with neighbors so you can work out small problems before they become big ones. Stay informed. Be active in your local government so you can defend challenges to your rights as a land- and horse owner.
- **BOUNDARIES.** Buildings and other facilities must be placed a certain distance from property boundaries, street rights-of-way, and your home.
- **TRAFFIC.** The coming and going of truck and trailer, manure removal services, feed trucks, and the like may be a source of consternation for neighbors.
- **DUST.** Some properties are subject to a permit and fees for dust raised ("fugitive dust"). Any bare or disturbed surfaces, such as arenas, round pen, pens, runs, and driveways, could qualify.

Zoning and Other Legalities

Various city, county, state, and federal regulations may affect what you can or cannot do on your future property. Restrictions may include but are not limited to where you locate your buildings and well, how many horses you are allowed to have, sanitation requirements, regulations related to clearing and excavating, use of water, management of natural waterways and wetlands, endangered species protection, and even the management of stallions. In advance, research the zoning classification and regulations that affect the land you are considering. The classification determines the allowed uses of the land and whether you need to obtain prior approval and a permit for your particular needs. Places to seek out information are the local planning department and the health department.

There may be further restrictions on your use of the land if the tract is a part of a homeowners association in which membership is mandatory.

While a homeowners association could be restrictive, if it is made up predominantly of horse owners, it can offer benefits and protection for your horse interests as well. Sometimes a group of landowners

Barnmaster, Inc.

Homeowners associations and local government agencies may have regulations that dictate the construction and appearance of your barn.

Trail Ride Acres

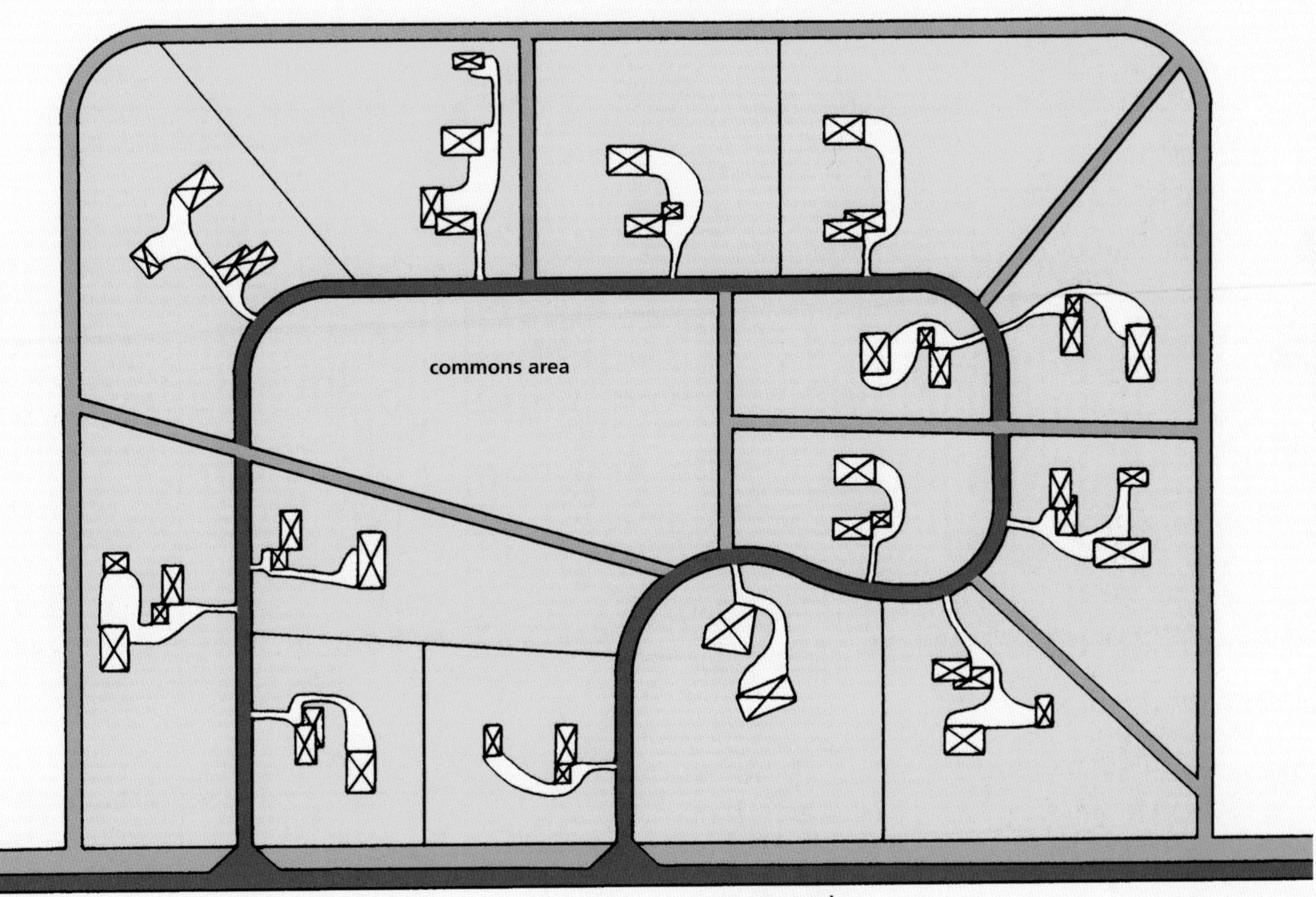

ACREAGES: 32 total acres. Parcels range from 1.5 to 2.5 acres. Each parcel is fenced for horses and has a house, garage, and barn, and some have other buildings.

COMMONS: There is a 4-acre fenced commons area in the center for riding or turnout.

TRAILS: A network of trails totals 1.94 miles of riding. This includes a 0.9-mile perimeter trail, the central access road, and the right-of-way along the county road.

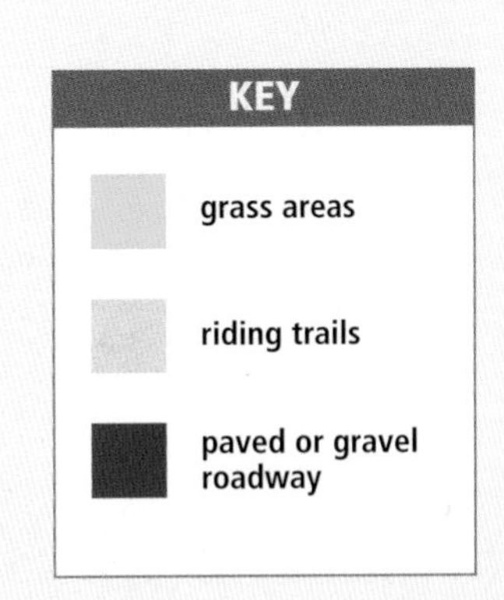

Canter Club Estates

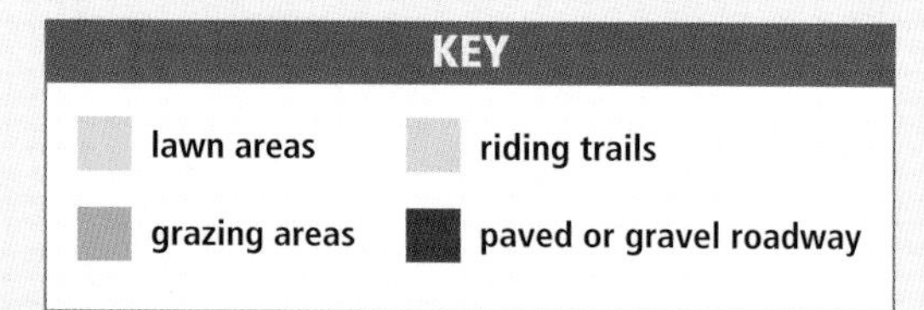

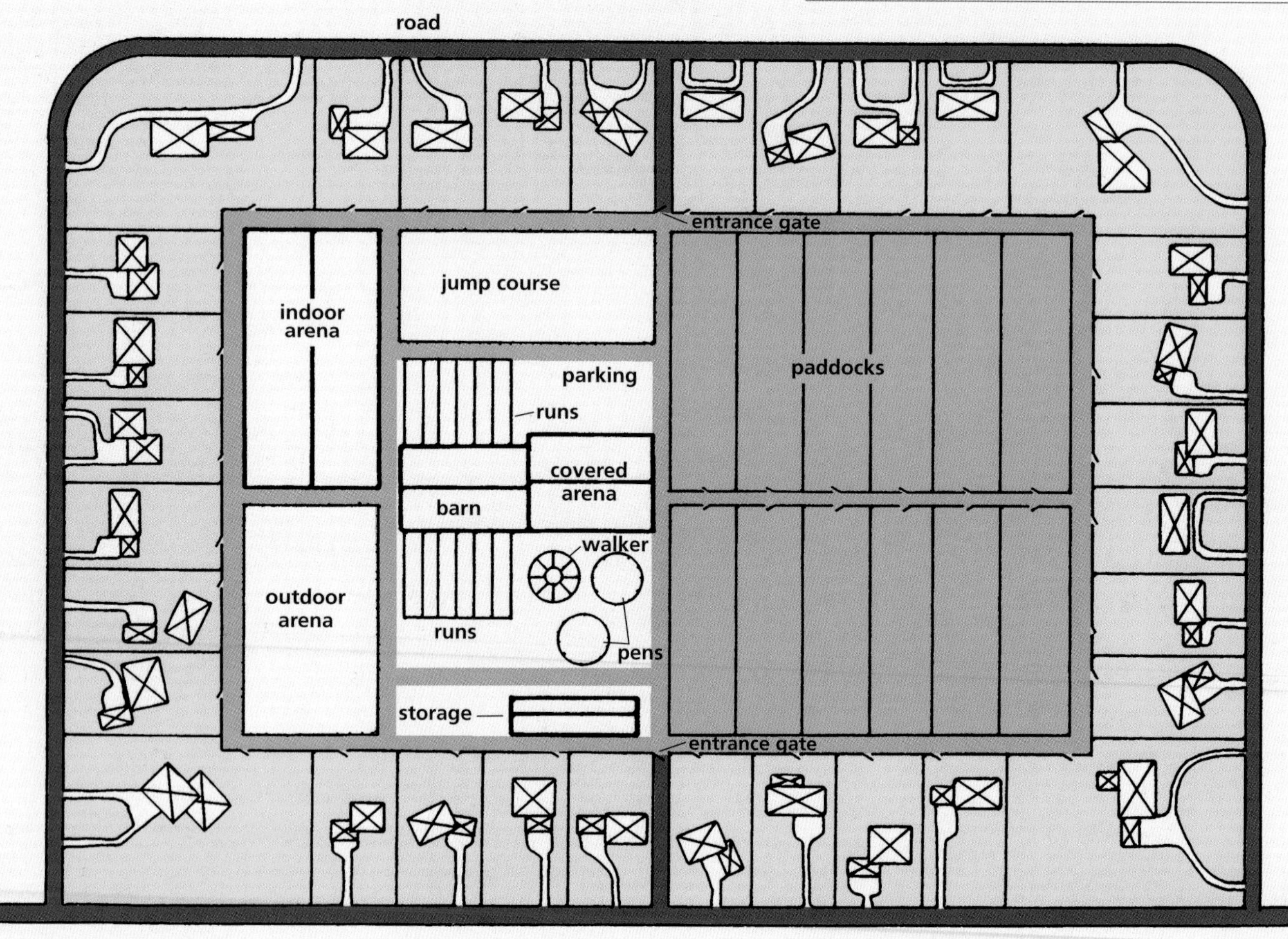

ACREAGES: 35 acres total. 17 acres for horse facilities; 18 acres for residences: thirty-two parcels at approximately 0.4 acre each, and four corner parcels at approximately 1.2 acres each. Each parcel contains a house and garage. Parcels are not fenced for keeping horses.

FACILITIES: Central facilities provide boarding for residents' horses, plus training and exercise facilities and riding trails. Facilities include:

- **BARN** (150'x100') with stalls for forty horses, tack and feed rooms, grooming areas. Some stalls open into 20'x100' runs off the barn.
- **COVERED ARENA** (150'x130') with open walls for training, riding, or free exercise.
- **INDOOR ARENA** (160'x300') for all-weather riding and for horse shows and clinics.
- **OUTDOOR ARENA** (160'x260') for riding, horse shows, and clinics; this arena could be divided into two or three smaller arenas.
- **JUMP COURSE** (130'x300') for riding, shows, and clinics.
- Two **TRAINING PENS** (66' diameter).
- A round **MECHANICAL WALKER** for exercising horses.
- Twelve **PADDOCKS** (80'x260') for daily turnout.
- **STORAGE BUILDING** (40'x160') for hay and equipment.
- **PARKING AREA** for cars, trucks, and trailers. During shows and clinics, the jump course or the outdoor arena can be opened for additional parking.

BRIDLE PATHS: The perimeter path is 0.6 mile long and 20' wide — enough for several horses and pedestrians to pass safely. Connecting paths throughout the facilities make a total of 1.1 miles.

GATES: Each parcel has a back fence with a gate that opens onto the perimeter bridle path. Residents are within easy walking distance of the horse facilities, and driving there is unnecessary. Two access driveways have gates that are kept closed — horses on the bridle path have the right-of-way.

Homeowners associations made up of horse owners often have communal trails, barns, and other facilities.

may initially unite for something simple, such as the maintenance of roads that are not cared for with public funds. Some associations expand to adopt covenants that specifically define building styles and land uses of the members. In any event, it is a good idea to get a feel for your neighbors and the people who live in the area you are considering.

Buildings added to any property must usually comply with county and/or state building codes that outline required building specifications.

Check to see whether the property has ever been surveyed and if the boundaries can be easily located.

Roads and Access

A horse operation of any size usually requires trailering and trucking. Be sure the property has all-weather access for the delivery of hay and grain, building materials, and gravel and cement. It will also be necessary to get a horse trailer in and out during all seasons in the event of an emergency.

If applicable, find out who owns the access road to the property. If the road crosses someone else's property, determine if there is a permanent easement for it. If there is not, before you purchase you should negotiate for a permanent legal right-of-way. Also, find out what easements are attached to your prospective property, such as roads, trails, utility lines, and irrigation ditches. Determine what rights these easements give other parties. This information will be found on the property deed or in the county recorder's office. While you are checking, see if there are rights that go along with the property, such as mineral and water, and see if there are any recorded rights that others have on the property, such as hunting and fishing.

Pollution

Air and water vary greatly in quality. Even if a parcel is far from a large population center, it may have one or more of the following hazards: smoke from factories, sawmills, or power plants; dust from mining or gravel operations; odors from agricultural operations, landfills, or sewage treatment facilities; noise from industry, highways, or airports; radioactivity from ore sites or mine tailings; or toxic waste from chemical dumps. Do your research!

Be sure to check the source of the local waters. Water may contain lead or aluminum or, in a highly farmed area, agricultural runoff, which could be high in nitrates.

Number of Acres

If you can afford it, buy more acres than you think you will need. First of all, buying a 5-acre parcel thinking that you'll have a 5-acre pasture is a mistake. The house and yard and the barn and the barnyard easily take up an acre or two. Second, although you may start out with one or two horses, you may soon find yourself expanding the herd. You may want to raise a foal or buy extra horses so everyone in the family can ride together or so there is an extra horse for visitors.

When planning your needs, read the town, city, county, or homeowners association code dictating minimum parcel size and number of horses permitted per acre. Determine whether you need to obtain a permit to keep a horse and what the requirements are. Ask your local Extension agent what the carrying capacity for pastureland is in the area. You may plan for 2 acres per horse with improved, well-irrigated pasture but may need 20 acres or more per horse of dry rangeland.

Acreage sizes

1 acre = 43,560 square feet

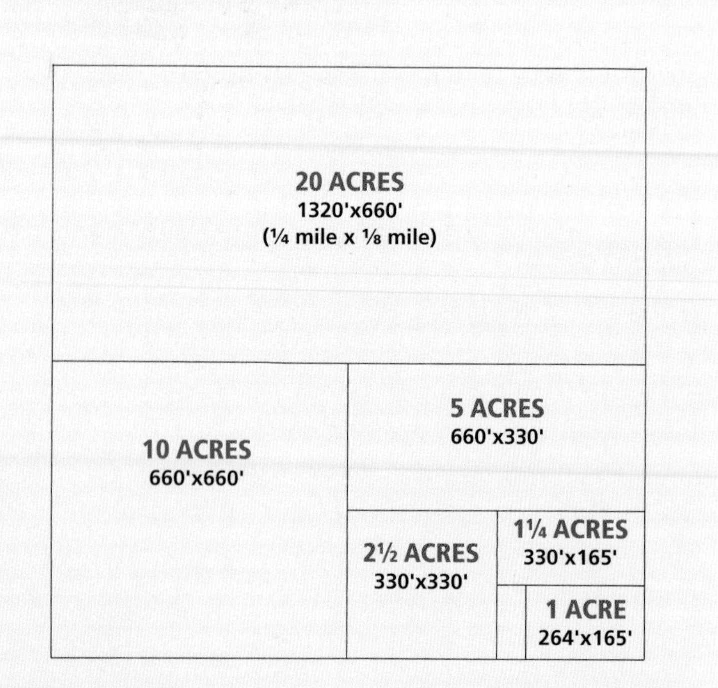

Pests

Some locations aren't very inviting for humans or horses. Areas that are heavily populated with flies, ticks, fire ants, killer bees, other bugs, poisonous snakes, or scorpions will provide constant management problems and in some cases disaster. That's not to say that you can't learn to live with these creatures, but do some research ahead of time so you won't be disappointed.

The Future

Look around the prospective property for several miles and try to determine where future construction or other changes may take place that could affect the quality of water, air, or the aesthetics of the place. Your local (county) planning commission can tell you of future plans and variances that have been issued that may affect your decision. Find out what the laws are regarding subdividing property or adding rental structures such as mobile homes.

While you are thinking ahead, consider the resale history of the property. A parcel that has been on the market for quite some time before you buy it will likely take a similar amount of time if you need to sell it. It may be a good value but suitable for few potential buyers.

Price

When shopping for property, price is often the first priority, out of necessity, but don't forget to consider the factors that affect the price so that you're evaluating it in a proper perspective. Compare the price of similar land that has recently sold. This is public information and can be found in the county recorder's office. When comparing two parcels, one may seem like a better deal when in fact its final cost

Price is driven by location, demand, recent sales, and aesthetics.

will be much higher once equivalent improvements have been made. Calculate the cost of financing, utilities, and essential improvements. Find out the yearly taxes and whether the property is located in any special taxing/service districts. If you are planning to pursue your horse involvement as a profit-making business, find out if the land is eligible for agricultural status and what you must do to qualify. And be sure to have enough cash remaining after the purchase for the horse facilities!

Buying an established farm

An existing farm may or may not be suitable for horses. To find out:

- Make a list of question to ask the seller.
- Hire a building inspector to evaluate the condition of the buildings, noting whether or not they are up to code.
- If the buildings aren't up to code, get an estimate from a builder for cost of renovations.
- Check to be sure all improvements comply with zoning and electrical, plumbing, and building codes.
- Hire an equine expert to walk through the property with you, noting the advantages and disadvantages.
- Test the water for quality and bacterial contamination.
- Take notes and bring this book with you so it will be handy for reference.

If you are considering a farm that was used for livestock other than horses, pay special attention to the following.

- In buildings, headroom (height of ceilings) in doorways and inside may be too low if they were built for cattle or other livestock.
- Aisles and handling areas are often too narrow. This is often a result of the type of livestock previously raised or the style of building. Some buildings have supporting walls or posts throughout, making it difficult to remodel to the open spaces necessary for horse buildings.
- Flooring in dairy barns is often concrete. While this is OK for aisles in a horse barn, it is not OK for stalls.

6 Layout

Now comes a very enjoyable aspect of your endeavor. Here's where you can test all of your dreams and ideas on paper, where you can build an ideal horse farm, or where you can improvise and remodel existing facilities. Country folks make an observation regarding the activities of newcomers to the area, as they note the progress of property improvements. If the animal facilities show a marked priority over the home, the person is a true farmer or rancher at heart!

Know your property. Either live on the acreage or visit it regularly for several months before you start permanent construction. This will give you valuable insights so that building location, orientation, and spatial arrangement can be tailored specifically to your land.

Take plenty of time at this planning stage, because you are going to have to live with your decisions. It's far cheaper to make mistakes on graph paper than with building materials! The larger the scope of your endeavor, the more I would urge you to consider hiring a professional planner, one who is experienced with horse facilities.

First, identify your goals. Keep in mind that your needs are very likely to expand. Try to determine exactly how many horses you want to keep with what style of management, and what kind of training facilities you desire. Make a timeline and list first things first. For example, trenches for utilities come before the building site is finished and before driveways are installed. Next, do some comparative shopping for building materials or for bids on the work. Finally, confer with your budget and make compromises if necessary.

No matter whether you are starting with a bare tract of land or a functional farmette, make your plans with natural principles and conservation practices in mind. Look at the lay of the land, the soil, the weather patterns, the wildlife, the plant life. Consider the natural forces as you make your plans. Don't fight the runoff from a slope by locating your pens at the base of it. Don't subject your roofing to the effects of a wind tunnel by poor building placement.

Enter landownership with the idea of improving it, not squeezing all you can out of it. Pasture improvement will be discussed in chapter 13, Land, but there are many other ways by which you can improve your land. Getting involved in water conservation, soil protection and erosion control, conscientious manure management, brush and tree management practices, and plant and wildlife protection will improve your land and add to your quality of living. You may even be eligible for cost-share assistance on a variety of programs administered by local, state, and federal agencies.

Take advantage of the natural lay of the land as you design your facilities.

If you are starting with a piece of bare land (or one with just a house), you have a big job ahead of you, but you can design things exactly as you want. Plan your facilities in relation to the improvements already there, such as residence, utilities, and fences. If the land has no source of water, the very first step is to drill a well in a location that will be convenient to the residence and to the horse facilities. Take the natural features — the trees, rocks, streams, and hills — into account so that you don't lose what they have to offer but can instead incorporate them into your scheme.

If you have an operating small farm or a partially developed acreage, you must evaluate the existing facilities. Are they suitable as they are or do they need modifications? Are they usable temporarily as transition facilities? Do they have salvageable parts but require extensive renovation? Will the renovation cost more than a new building would? Would it be impractical to remodel the existing facilities, and should they be torn down instead?

Study as many existing horse facilities as you can, especially those in your locale, and take notes and/or photos of what you like and don't like. Make some initial sketches of what you want and show them to knowledgeable people in agricultural Extension and in the building and construction business, as well as to other horsemen. Take note of all comments so you can consider them when making decisions.

Common horse facility components

- Barn(s) with stalls
- Runs, pens, paddocks, pastures
- Storage for feed, bedding, machinery, tack, and other equipment
- Manure storage and handling areas
- Training areas: round pen, arena, track, hot walker or exerciser, treadmill, pool
- Work areas: grooming area, wash rack, shoeing and veterinary area, breeding shed, laboratory, office, tack room
- Driveways, walkways, parking areas, turnarounds
- Shelterbelts, windbreaks, wildlife areas, perimeter buffer zones (double fence)
- Water and other utilities

Facility Goals

When putting together a horse farm or ranch, keep the following goals in mind.

Safety. Facilities should be strong and well designed with horse behavior principles in mind.

Convenience. Everyday activities should flow efficiently. Buildings should be placed to conserve labor and time. Don't forget to plan ample space for access roads and turnarounds. Locate water within easy access to the places where it is needed.

Protection. Keep horse comfort in mind and provide protection from sun, wind, precipitation, cold, and insects.

Storage. Always plan for more storage space than you think you will need for feed, bedding, manure, tack, vehicles, and equipment.

Economy. Without sacrificing quality, consider alternative materials and plans. One place you don't want to skimp is the finished dimensions of buildings or access lanes. The layout often takes more space in reality than it looks like it will on paper.

Flexibility. Keep some degree of adaptability in mind as you plan. Always leave room for expansion on to your buildings. You may want to shift the emphasis of your acreage in the future, or you may need to sell it to someone without horses.

Locating Major Buildings

When locating major buildings, consider the following factors.

For most climates, plan for maximum sun and shelter in the winter and maximum shade and breeze in the summer. Check with the local weather authority to find out what the prevailing winds are during the various seasons. Go to the site itself during each season, especially winter, to determine

Above: **A natural spring revealed itself after the barn and pens were in place.** ***Right:*** **Choose a building site that is high and dry.**

which way the buildings should face. In the United States, prevailing winds usually come from the north, west, or northwest, so most farm buildings face the south or southeast. But local winds can be very different. You may live in an area that receives upslope storms and winds and receive more of your "weather" from the east. Or if you live along the coast, your weather might move from the coast directly inland. So do plenty of research before you orient your buildings.

Other buildings, trees, rocks, and slopes can also have an effect on your proposed building site. They can obstruct the light; change the flow of air, causing drafts, vacuums, or drifting snow; and contribute excess runoff to the new building site. Also consider proposed expansions to existing buildings.

Locate your buildings on dry ground, preferably high ground. Starting with as flat an area as possible will lessen excavation or fill costs. Ideally there should be a 2 to 6 percent slope away from the building in all directions for surface drainage. The building floor should be 8 to 12 inches above the outside ground level. If the building is located on a hill, you might need to dig a diversion ditch around the backside.

Ensure that there will be good subsurface drainage, especially for stall areas and runs, by having the subsoil evaluated. For a barn site, sandy or gravelly subsoil is preferred over clay or adobe soil. If necessary, have the critical areas of the site excavated. Refill the hole first with large rocks, then small rocks, and finally road base or limestone. Let the site settle for several months, or ideally a year, before beginning construction.

Be sure that key buildings have all-weather access for the delivery of building materials and eventually for hay, grain, bedding, and so on. Plan for ample space to turn large trucks and/or trailers around. Ensure that routine chores are possible without great hardship during all seasons.

Locate key buildings close enough to the house for security and convenience, yet far enough downwind so that flies and odor do not invade the residence. Formulate your fire plan as you plan your facilities. (See chapter 16, Disasters, for more about fire safety.)

Make the appearance as nice as possible without sacrificing functional aspects of your layout. Plan for safety, efficiency, and convenience.

Choosing a Builder

If you are not constructing your own facilities, you will need to either hire various builders and workmen for the different aspects of your plans or hire a general contractor to do this for you. If you hire a general contractor, be sure that this person is very familiar with horse needs and facilities.

When looking for professional tradesmen, ask for references from previous clients. In the case of a builder, take the time to look at his past work and talk to people who have hired him. Ask the builder what type of warranty comes with a building or project. It is important to be very comfortable with the person(s) you hire.

So that you have an equitable means of comparison among bids, you or your representative should ask to see detailed building plans and/or specifications from every contractor you are considering. This will assist you in determining the quality of materials and work you can expect. You will need to devise a detailed list of specifications you want in your barns, other buildings, fences, and any other facilities you will have built. When considering the construction of a pole barn, if you want an indication of joint strength, for example, you will need information on the system of bracing used and the location, number, and size of bolts or nails used to hold the braces together. Refer to the sample list of items to check on when getting an estimate for a simple pole barn. Similar checklists can be made for any other building project or improvement.

Above: A good builder has solid references and an experienced crew. _Right:_ This pole barn skeleton is ready for its roof, skin, and doors.

Once you have chosen a builder, you will probably sign a contract guaranteeing that the work will be done as promised and that you will pay the agreed-on price. Contract requirements vary from state to state, so be sure you are knowledgeable about the pertinent laws.

You may also want to get a lien release from all of the suppliers and subcontractors who work on your property. This will assure you that if your general contractor does not pay his steel supplier and the electrician he hired, they will not file a lien on your property in an attempt to get their money. If you have chosen your general contractor carefully, this type of situation will not occur, but some companies working with limited resources may not pay their bills on time and could put your property at risk.

Contracts

Contracts should be written in simple terms and should contain clauses that cover a list of the attached contract documents.

- Agreement: Brief statement of project and price.
- General conditions: Definition of responsibilities of all parties to the contract.
- The drawings: The plan or blueprints.
- Technical specifications: Specific definitions of number, type, quality, and/or brand of materials and products to be used.
- Supplement: Modifications to fit specific needs of custom project such as insurance and bonding.
- Statement of project, price, and responsibilities (such as who cleans up construction debris).
- An outline of how subcontracts and separate (but related) contracts are to be handled.
- Procedure of how plan changes are to be handled.
- Procedure of how disputes are to be handled.
- Statement of completion date.
- Outline of how and when payments are to be made, with interest charged for payments not made on time.
- Statement of insurance requirements and safety responsibilities.
- Procedure to accommodate price, material, or completion date changes.
- Outline guarantees and procedure for dealing with defective work.
- Procedure to terminate the contract.

Rules and Regulations

You have certain obligations as a landowner and a horse owner. As far as protecting other people, animals, and property from your horses, one of your main responsibilities is to install and maintain good fences. Look ahead to anticipate what potential problems are likely to occur, because you could be held liable for all results of your negligence. For example, a loose horse could kill or injure a human or animal, could cause a human or animal to injure himself, could infect other horses with disease, could become involved in an unwanted breeding, or could inflict damage to buildings, fields, plants, or yards. If you have a stallion, you must be absolutely sure the animal is under control at all times — a stallion is potentially too dangerous to risk him running loose.

Usually a state statute covers fence laws. Situations not specifically covered by law may be determined by court decisions or by conjecture. Often, state laws say that a person has the duty to fence animals in, and his neighbors do not have the responsibility to fence them out. Some open-range states require landowners to fence out free-roaming livestock.

If you share a division fence with a neighbor and the fence is put on the property boundary, state law will usually dictate whether you share the cost of installation. Adjoining owners usually cannot legally force one another to erect a fence of a particular height or materials. Imagine if you were not a horse-owning neighbor. Would you want to install a 6-foot horse fence? Once a boundary fence is built, depending on your laws, you might be required to maintain the half of the fence to your right as you stand on

Pole barn specifications worksheet

BUILDING DIMENSIONS

Length ____________ ft. (to outside of wall framing)

Width ____________ ft. (to outside of wall framing)

Height inside ____________ ft.

WALLS

POLES: Treated with ____________

________ in. x ________ in. or ________ in. diameter

Length ________ ft.; in ground ____________ ft.

Spacing ________ ft.

CONCRETE PAD:

Length ________ ft.; width ________ ft.; thickness ______ in.

SPLASH BOARDS: Treated with ____________

Dimensions ____________

(Or concrete foundation specs ____________)

WALL GIRT:

Dimensions ________ spaced ________ in. apart

WIND AND CORNER BRACING:

Type ____________

SIDING MATERIAL:

Steel: ________ Description: ____________

Wood: ________ Description: ____________

INSULATION:

Thickness ________ in.; type ____________

VAPOR BARRIER: ____________ mil thick

INSIDE SURFACE MATERIAL: ____________

ROOF

Designed and certified to withstand ____ psf? (per local code)

TRUSS: *Sketch truss design*

Length ________ ft.; spacing ________ ft. on center (OC)

Type ____________

CHORD:

Top ________; bottom ________; diagonals ________

ROOF GIRTS (Purlins):

Dimensions ________ spaced ________ ft. OC

ROOFING:

Description: ____________ e.g., steel, wood

Coating ____________

Dimensions ____________

Number of skylight panels ____________

INSULATION: Thickness ________ in.; type ________

VAPOR BARRIER: ____________ mil thick

CEILING MATERIAL: ____________

FLASHINGS: ____________

GUTTERS: ____________

DOWNSPOUTS: ____________

OVERHANGS: ____________

EAVES: Open or closed? ____________

SOFFIT: Material ____________

VENTILATORS: Number ____________

Type ________ ; size ____________

WINDOWS

Number: ____________

Sizes: ____________

Type: ____________

DOORS

Number: ____________

Sizes: ____________

Type: ____________

FLOOR

Flooring: ____________

Type ________ Thickness ________

Reinforcement ____________

BRACING

Sketch type with nail or bolt pattern.

Is brace included on each post? ____________

HARDWARE

Door: ____________

Type ________; track length ________ ft.

UTILITIES

ELECTRICAL CIRCUITS:

Number ________ Size ________

LIGHTS: Number ________ Type ________

Size ____________

SWITCHES: Number ______ Type ________

OUTLETS: Number ____________

WATER OUTLETS: Number ______ Type ________

DRAINS: Number ____________

HEATER: Number ______ Type ________

FANS: Number ______ Type ________

your property at the midpoint of the fence facing the division line, and your neighbor would be required to maintain the half of the fence to his right as he stands on his property at the midpoint of the fence facing the division line. If you can't come to an agreement with a neighbor on the type of fence or cost sharing, if you set the fence 2 to 6 inches inside the property line, it will become your fence and your responsibility. Note, however, that once so placed, in many states, the fence becomes, over time, the boundary between the properties. Although we all hope that good fences make good neighbors, it is always a wise idea to put any fence agreements with your neighbors in writing. This is especially important if you put up a perimeter buffer zone or double fence. To prevent problems in the future, draw up a simple memorandum with a drawing that indicates the property boundary in relation to the fence(s).

No matter who does the work on your land, buildings, or fences, it is ultimately your responsibility to comply with all legal codes and regulations. Check the local zoning regulations. These restrictions are designed to control the growth and development of communities by establishing particular areas for certain uses such as residential, commercial, industrial, agricultural. These laws define the type of building you can construct and the type of activity that can take place on your land. Zoning laws may also dictate building height and size, property size, legal distance from road or neighbors, and appearance of facilities. You may need to get approval from the zoning committee for your plans.

A safe, attractive fence adds to property value, increases security, and defines a property.

Typical snow and wind patterns

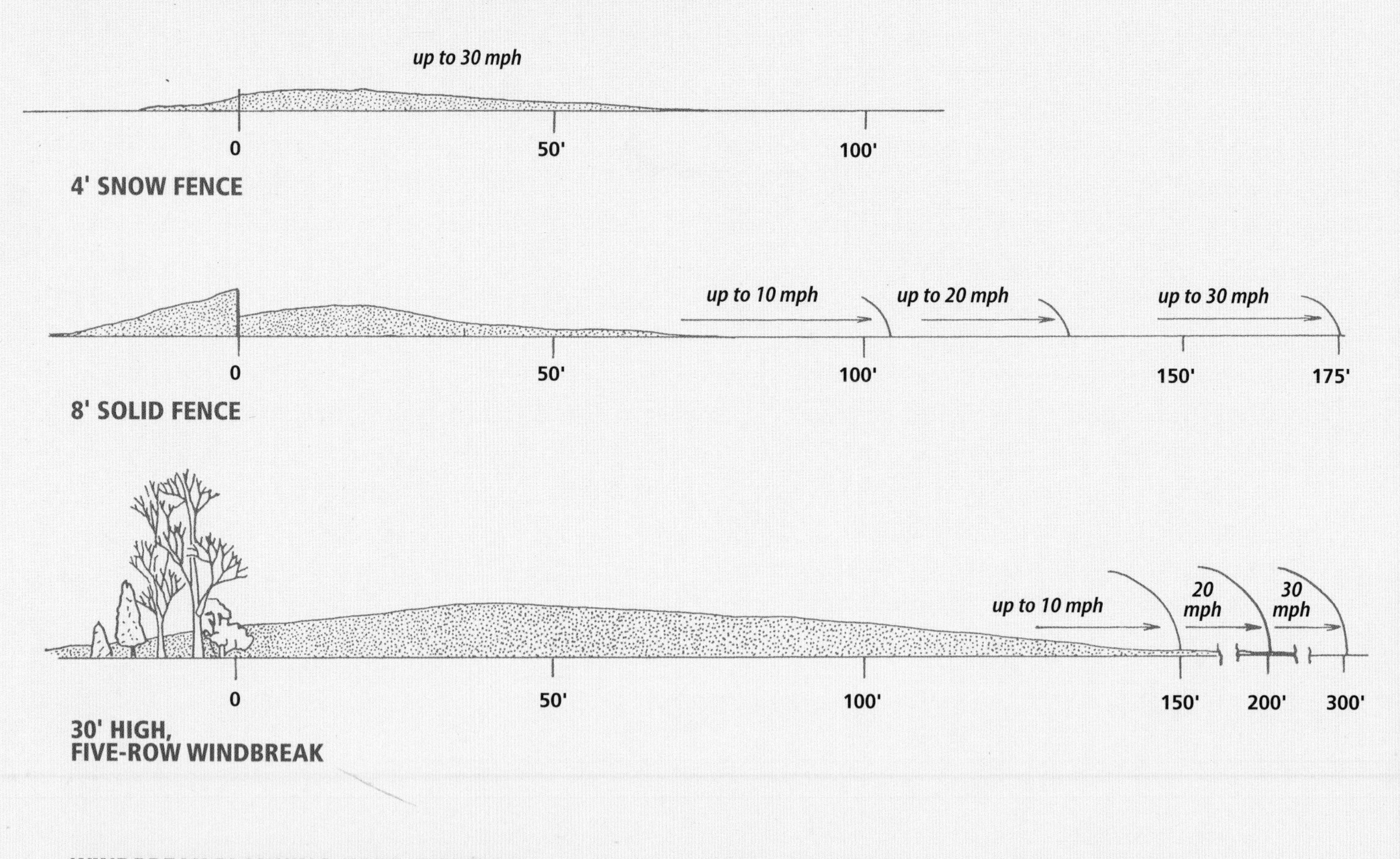

WINDBREAK PLANNING, WIND PATTERNS
With a 40 mph wind from the left, velocities will be reduced to about those shown. For other speeds, the reductions will be proportional. (From *Horse Housing and Equipment Handbook*, MWPS-15. Ames, IA: Midwest Plans Service, 1994.)

Your buildings must be constructed in accordance with the local building code, a group of regulations or construction standards concerning structural soundness, enacted by law. The regulations cover topics such as height and area restrictions, room size requirements, required method for design, minimum design loads for wind and snow, building materials, electrical requirements, plumbing and septic requirements, heating, ventilating, air conditioning, and sprinkler specifications. There are various building codes in the United States; the Uniform Building Code (UBC) is the most common. You can buy a copy of the UBC at bookstores or from your local building department. For information on local building codes, contact your county clerk or commission or your city building and zoning department.

Your local building official will advise you about which codes you must follow. You will need to submit plans for approval and get a building permit. The local building inspector will visit the construction site to ensure that the building is being constructed according to the approved plan and to applicable codes.

Finally, you must check with the health department and the Environmental Protection Agency to be sure you are designing your farm to comply with public health requirements, pollution criteria, and pest control standards.

Protecting open-front buildings

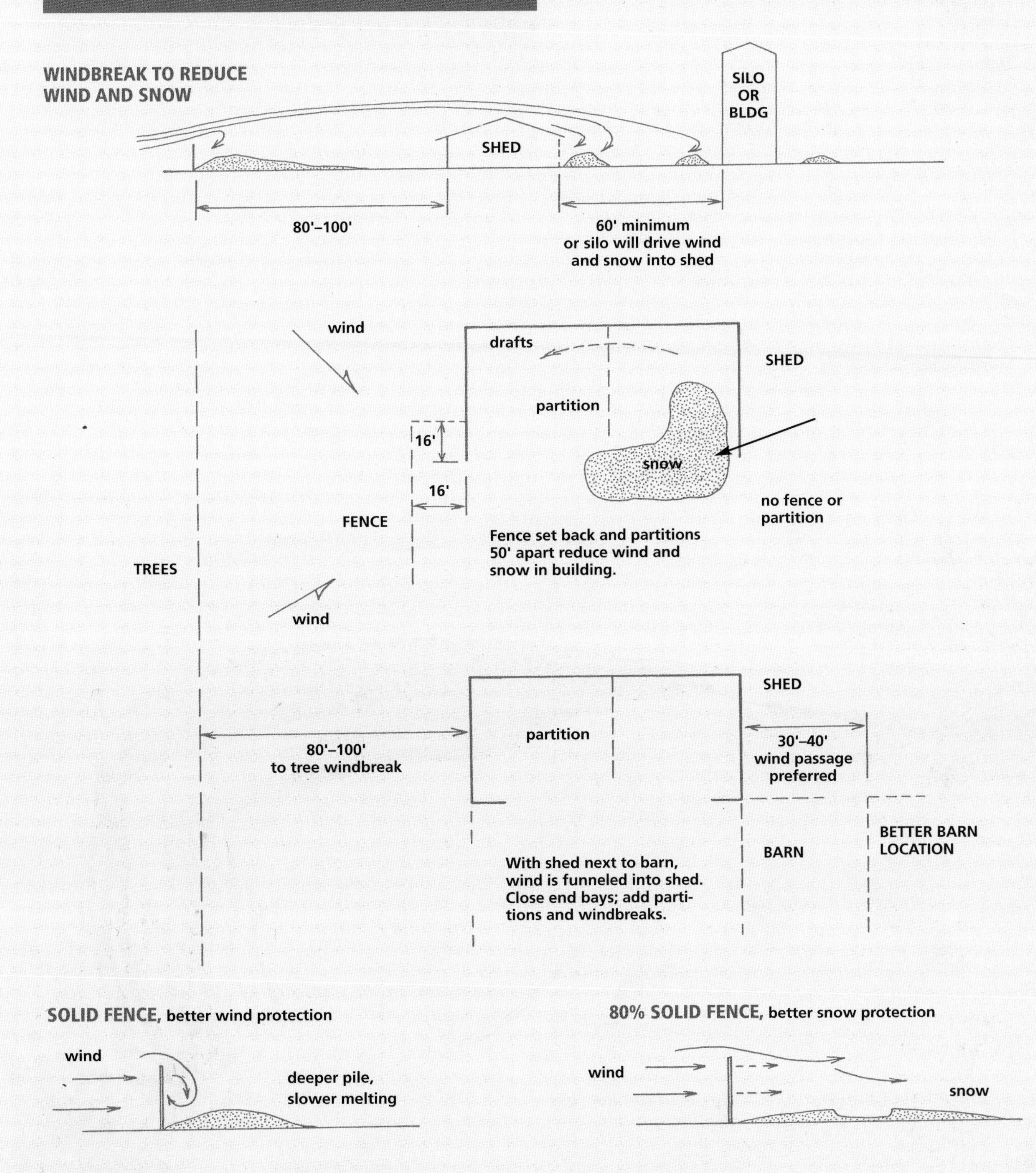

WINDBREAK PLANNING, ORIENTATION

Local experience is the best indicator of the distance facilities should be placed from shelterbelts. Generally, shelterbelts should be 100 to 300 feet away from protected areas. The shorter distance is suitable where snow accumulation is less severe. (From *Horse Housing and Equipment Handbook*, MWPS-15. Ames, IA: Midwest Plans Service, 1994.)

½-acre residential subdivision

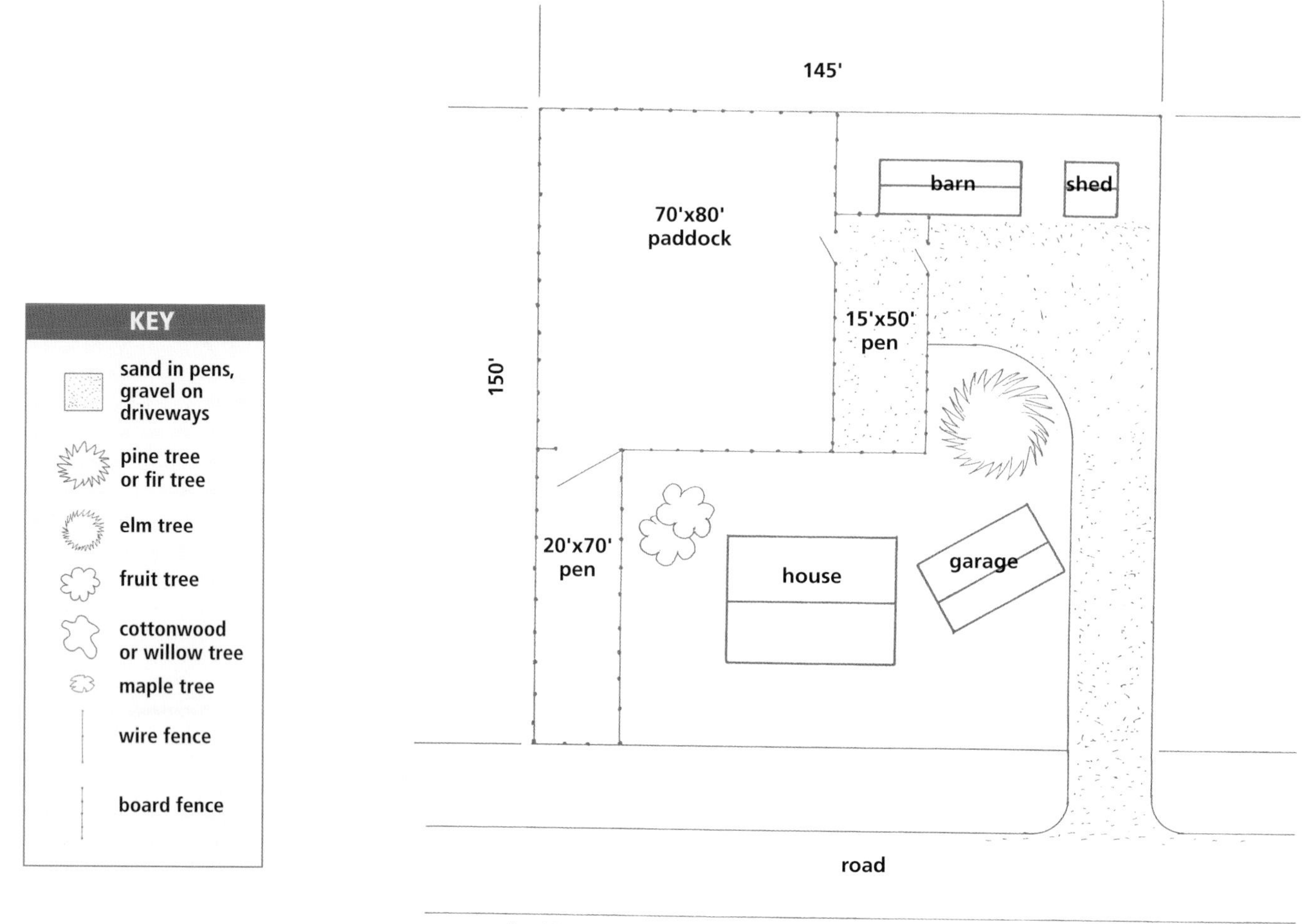

RESIDENTIAL SUBDIVISION: ½ acre. 150'x145' (road frontage), developed from bare land. Suitable for keeping one horse. No neighbors yet, but soon.

HOUSE: Single story, 30'x40', 1200 square feet with two-car (20'x30') garage. Barn and run and both pens are visible from the house.

BARN: 12'x34' modular with one stall, tack room, and central grooming area. One stall door opens into a pen and another opens into a grooming area in the barn.

FEED AND HAY SHED: 12'x12' storage shed located at the end of the driveway for feed, hay, and bedding.

PENS: 70'x80' paddock can be used for grazing or longeing; 20'x70' pen can provide some grazing if managed carefully; 15'x50' all-weather turnout pen has gravel or sand footing.

FENCE: Because of the close proximity of future neighbors, perimeter fence is V-mesh to keep out animals, children, and windblown debris. Electric fence wire can be installed on the top and middle of the fences later if horses are kept across the fence.

GATES: Gates between pens can be opened for larger exercise area and for total access for cleaning.

LANDSCAPING: Shade trees in yard. Acreage is too small to have shade trees in pens because horses would chew them, and fencing them off would take too much room.

1-acre residential subdivision

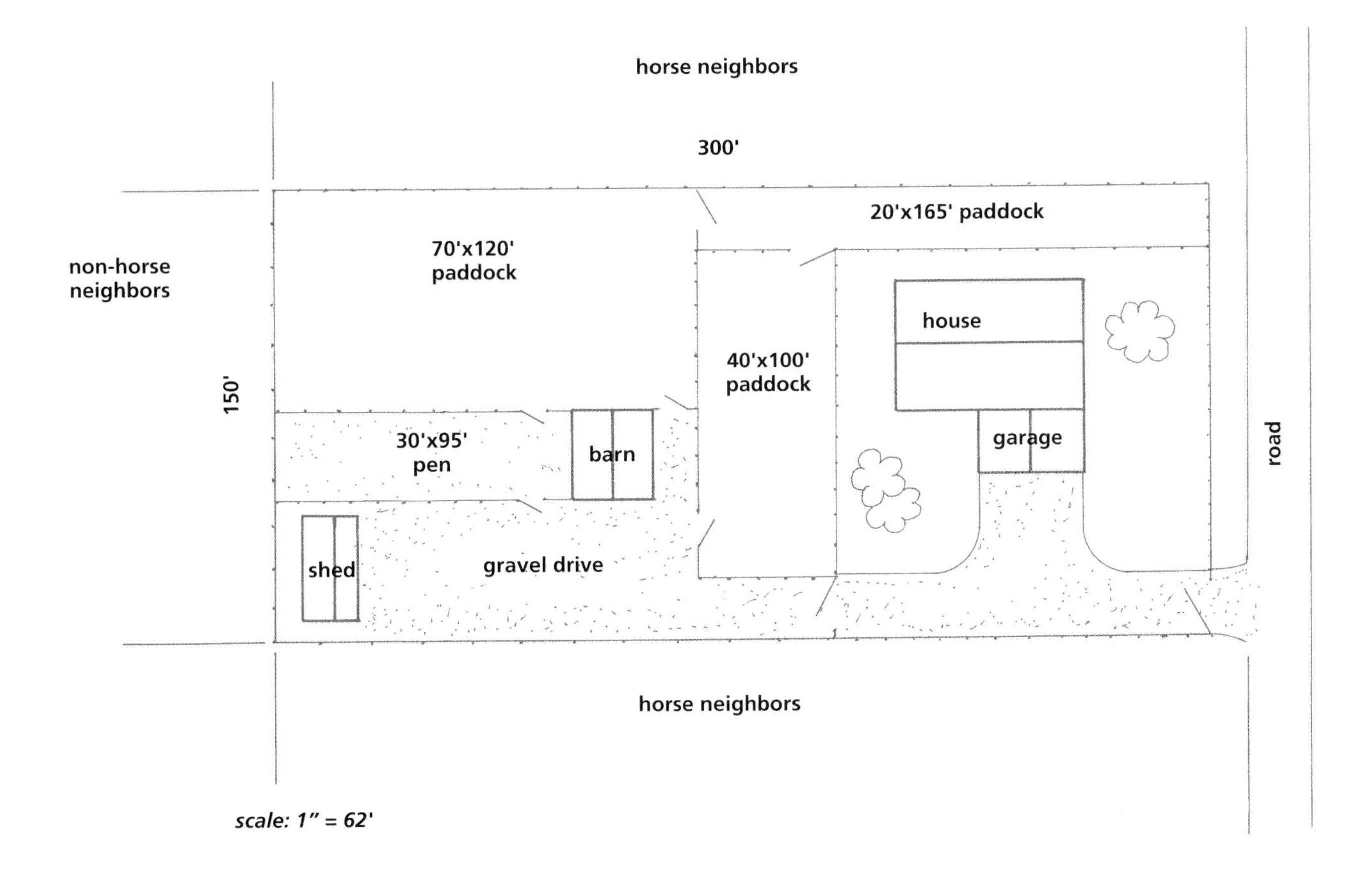

RESIDENTIAL SUBDIVISION: 1 acre. 150' (road frontage) x300', developed from bare land. Horse-owning neighbors on two sides, non-horse-owning neighbors on one side. Suitable for two horses maximum. Buildings are located 70' apart for fire safety.

HOUSE: Single story, 60'x40', 2400 square feet with two-car (20'x30') garage.

BARN: 24'x30' with two stalls, grooming area, and tack room. Each stall has two doors, one leading into the pen, the other into the barn grooming area. Barn is visible from the house but pen is not. Gravel drive in front of barn for maneuvering vehicles.

FEED AND HAY SHED: 15'x30' storage shed located at the end of the driveway for feed, hay, and bedding.

PENS: 70'x120' paddock is a good size to be used for grazing and longeing; 20'x165' and 40'x100' paddocks can provide some grazing if managed carefully. 30'x95' all-weather turnout pen has gravel or sand footing.

FENCE: Because of the close proximity of neighbors, perimeter fence is V-mesh or 2"x4" mesh to keep out animals, children, and windblown debris. Electric fence wire is used on the top and middle of the fences near horse-owning neighbors to keep horses from leaning over and from rubbing.

GATES: Entrance gate prevents loose horse from getting onto the road. Gate between driveway and gravel yard prevents loose horse from getting onto lawn. Gates between pens can be latched open to make larger exercise areas.

LANDSCAPING: A few shade trees in yard, but this acreage is too small to have shade trees in pens without horses damaging them.

2-acre residential subdivision

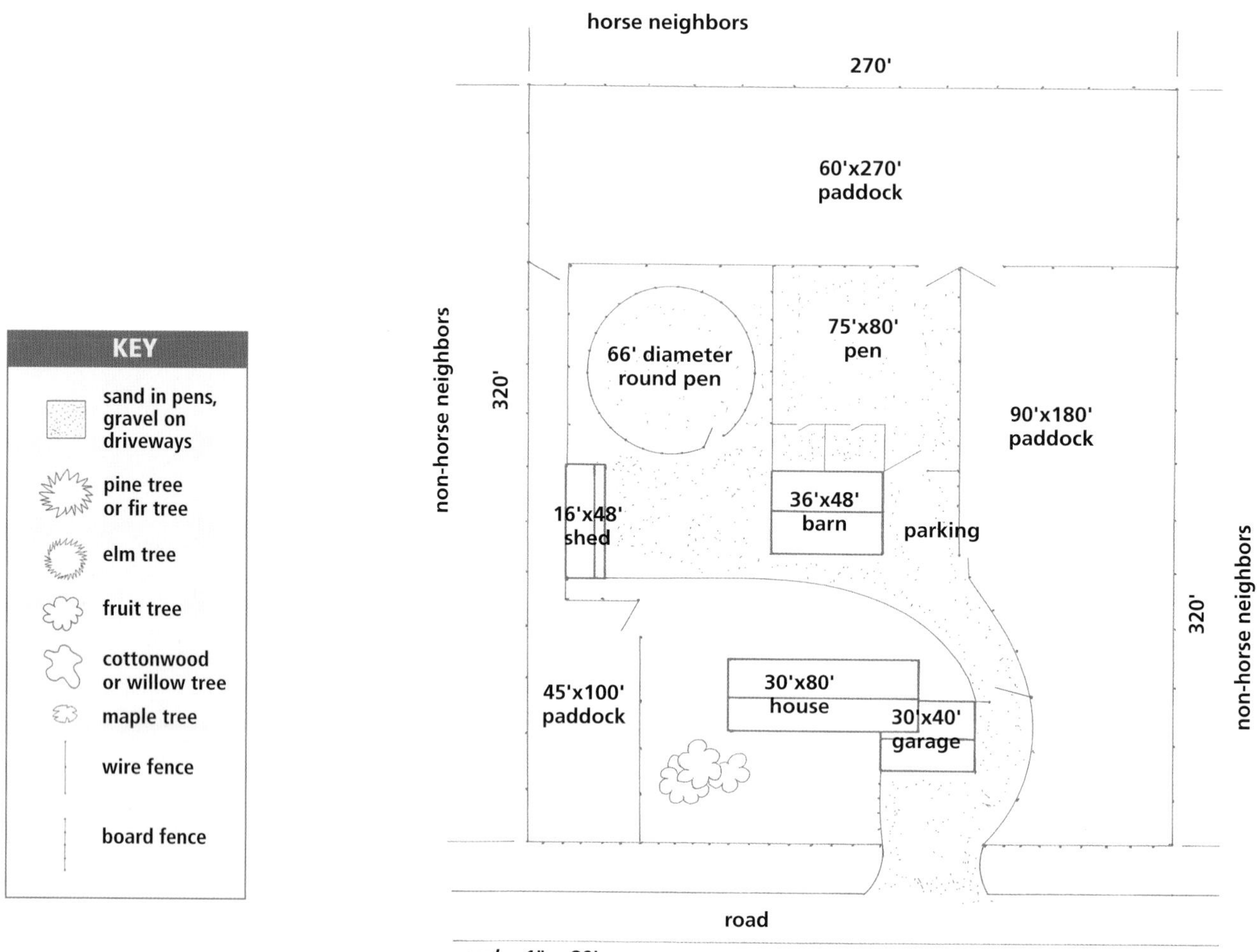

RESIDENTIAL SUBDIVISION: 2 acres. 270' (road frontage) x320', developed from bare land. Non-horse-owning neighbors on two sides, horse-owning neighbors on one side. Buildings are located 70' apart for fire safety.

HOUSE: Single story, 30'x80', 2400 square feet with two-car (30'x40') garage. Barn, round pen, and portions of all turnout areas are visible from the house.

BARN: 36'x48' with four stalls, tack room, feed room, and grooming area. One door of stall opens into pen, another opens into grooming area inside the barn.

FEED AND HAY SHED: 16'x48' storage shed located at the edge of the property for the bulk of hay and bedding.

PENS: 66' diameter round pen for longeing and training; 75'x80' "sacrifice" pen surfaced with gravel for daily and bad-weather turnout; three remaining paddocks can be used for turnout and limited grazing with judicious management; gates between the three paddocks can be opened for total access.

FENCING: The fence along the road is white PVC post and rail. The long borders with non-horse-owning neighbors can be fenced with wire rail or another type of attractive, secure fencing. The border with horse-owning neighbors should consist of a durable fence plus some type of electric fence to prevent horses from fighting and playing over and through the fence. The all-weather pen needs sturdy fencing such as metal panels or continuous pipe.

GATES: Gates between turnout areas provide access for equipment. Entrance gate is kept closed to prevent a loose horse from getting onto the road. Locating the gate past the garage means you don't have to open it for vehicles every time you leave and return home.

LANDSCAPING: Shade trees in yard. Acreage is large enough that shade trees could be added in the larger pastures, but they would need to be fenced off so horses wouldn't chew them.

5-acre residential subdivision

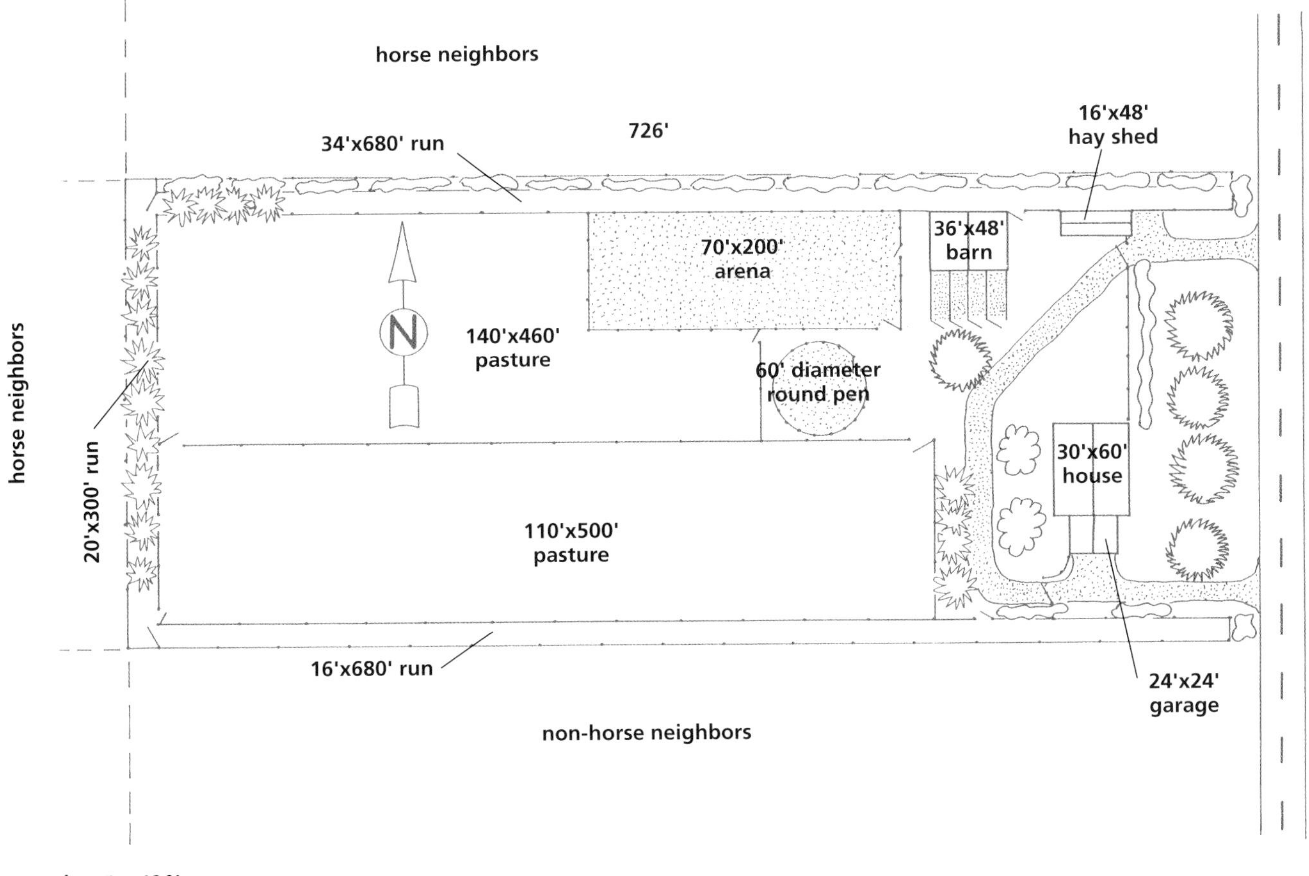

scale: 1" = 120'

RESIDENTIAL SUBDIVISION: 5 acres. 300' (road frontage) x726', developed from bare land.

HOUSE: Single story, 30'x60', 1800 square feet with two-car (24'x24') garage.

BARN: 36'x48' with four stalls, tack room, wash rack, and four 12'x 32' runs. Runs are located off the gable end of the barn to avoid runoff. Runs are located on the south side of the barn for winter drying and warmth and to be visible from the house.

HAY SHED: 16'x48', which stores six hundred standard-sized 70 lb. bales. Hay shed is located with easy access for loading and unloading and separate from other buildings for fire protection.

ROUND PEN: Post and board, sand, 60' diameter. Visible from house for trainer safety.

ARENA: Post and board, 70'x200'. Electric fence can be added to pasture side of arena to prevent chewing.

SOUTH PASTURE: 110'x500'.

NORTH PASTURE: 140'x460'.

PERIMETER RUNS: North run is 34'x680'; south run is 16'x680'; west run is 20'x300'; 16' gates at ends of run can be opened to provide access for mower or horses. Long runs can be used as part of the rotational grazing and to separate horses, providing insulation between neighbors' animals when they are present.

LANDSCAPING: Hedge along entire north border provides windbreak and visual privacy. Electric fence in north run protects hedge from horses. Evergreen trees on west and north borders provide shelter for horses.

GATES: Two driveway gates ensure that loose horses cannot get onto roadway.

SAMPLE LAYOUT: 3.5 acres, suburbs

EXISTING CONDITIONS

SITE: 3.5 acres, irregular shape, suburb of midsized city.

HOUSE: 30'x70' with attached two-car garage.

SHOP: 60'x80' with 20'x75' addition.

WELL HOUSE: 5'x8' with heat tape on pipes.

SHED: 5' opening and sharp metal roof edge; unsafe for horse housing.

FENCING: Marginal. Barbed wire on cattle ranch border.

HORSE FACILITIES: None.

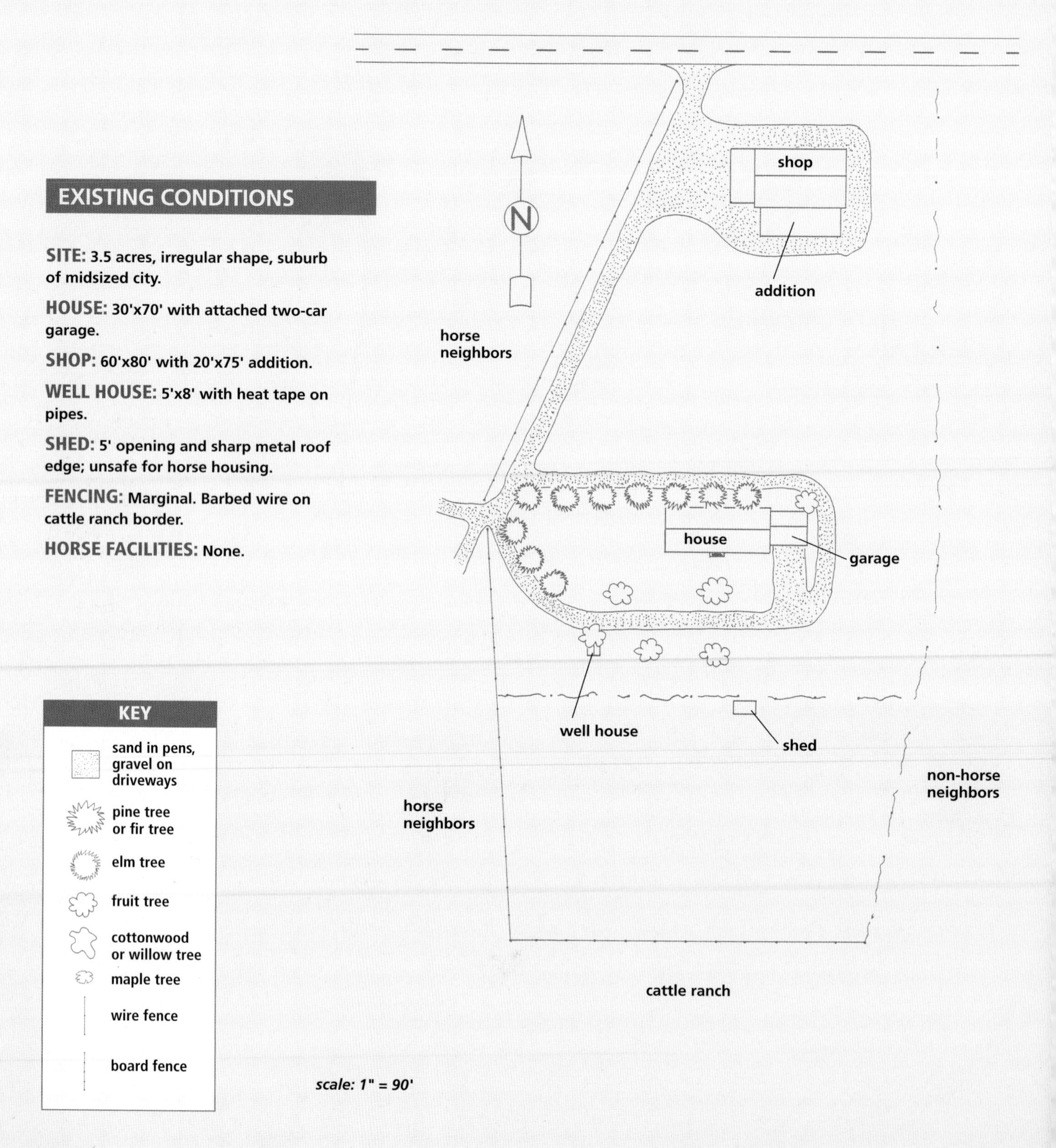

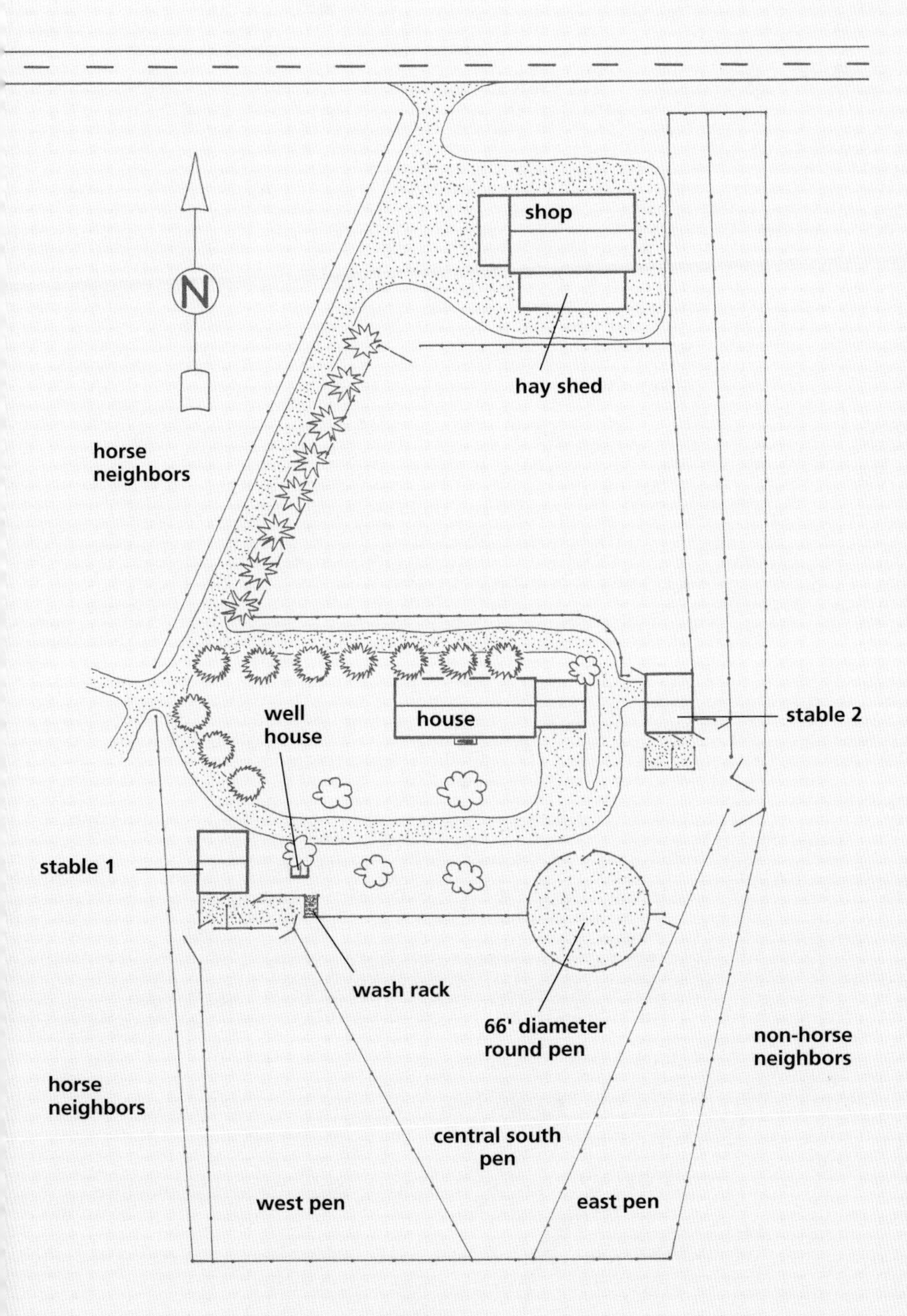

DESIGN SOLUTION

SHOP: Floor of 20'x75' shop addition lined with pallets and used to store one thousand bales of hay.

WELL HOUSE: Installed heater with thermostat to prevent freezing.

SHED: Removed.

STABLES: Stable 1: 24'x28' (see page 88 for details); two 12'x16' stalls open to the south with pea gravel runs; 12'x24' tack and feed room. Stable 2: Same as stable 1 except 12'x24' tool and feed room instead of tack.

WASH RACK: Built on existing concrete pad, which was located in close proximity to water source (well house).

PENS: 66' diameter round pen, post and board, sand footing. The central south pen separates the west and east pens.

RUNS: The long east runs can be joined to form one run the length of the property. A run insulates the north pasture from the east run to separate horses if necessary. The small run on the southwest border acts as insulation from neighbors' horses for health and safety reasons.

FENCING: Cattle fences (barbed wire) were lined on inside with V-mesh wire using existing fence posts and were fortified on top with electric wire. All new fences are smooth wire or board.

LANDSCAPING: Pines along driveway for privacy, wind protection, and shade.

SAMPLE LAYOUT: 10 acres, farm community

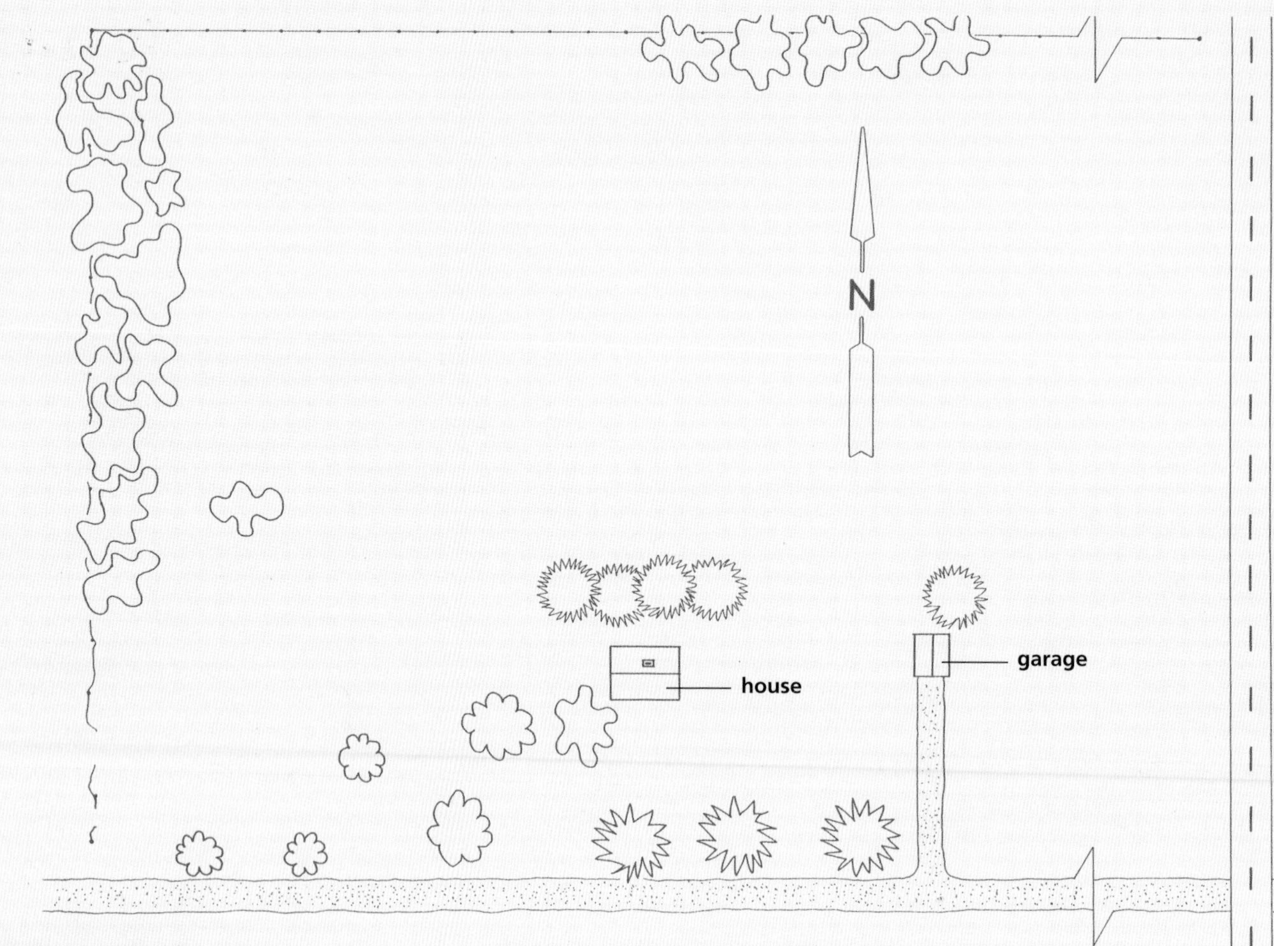

scale: 1" = 100'

KEY

- sand in pens, gravel on driveways
- pine tree or fir tree
- elm tree
- fruit tree
- cottonwood or willow tree
- maple tree
- wire fence
- board fence

EXISTING CONDITIONS

SITE: 10 acres, farm community.

HOUSE: 30'x40'.

GARAGE: 20'x20'.

LAYOUT: Two adjacent 5-acre pieces with driveway to garage approximately at center line.

ROADS: County gravel road frontage along south border. County paved road frontage along east border.

FENCING: One good fence along north border.

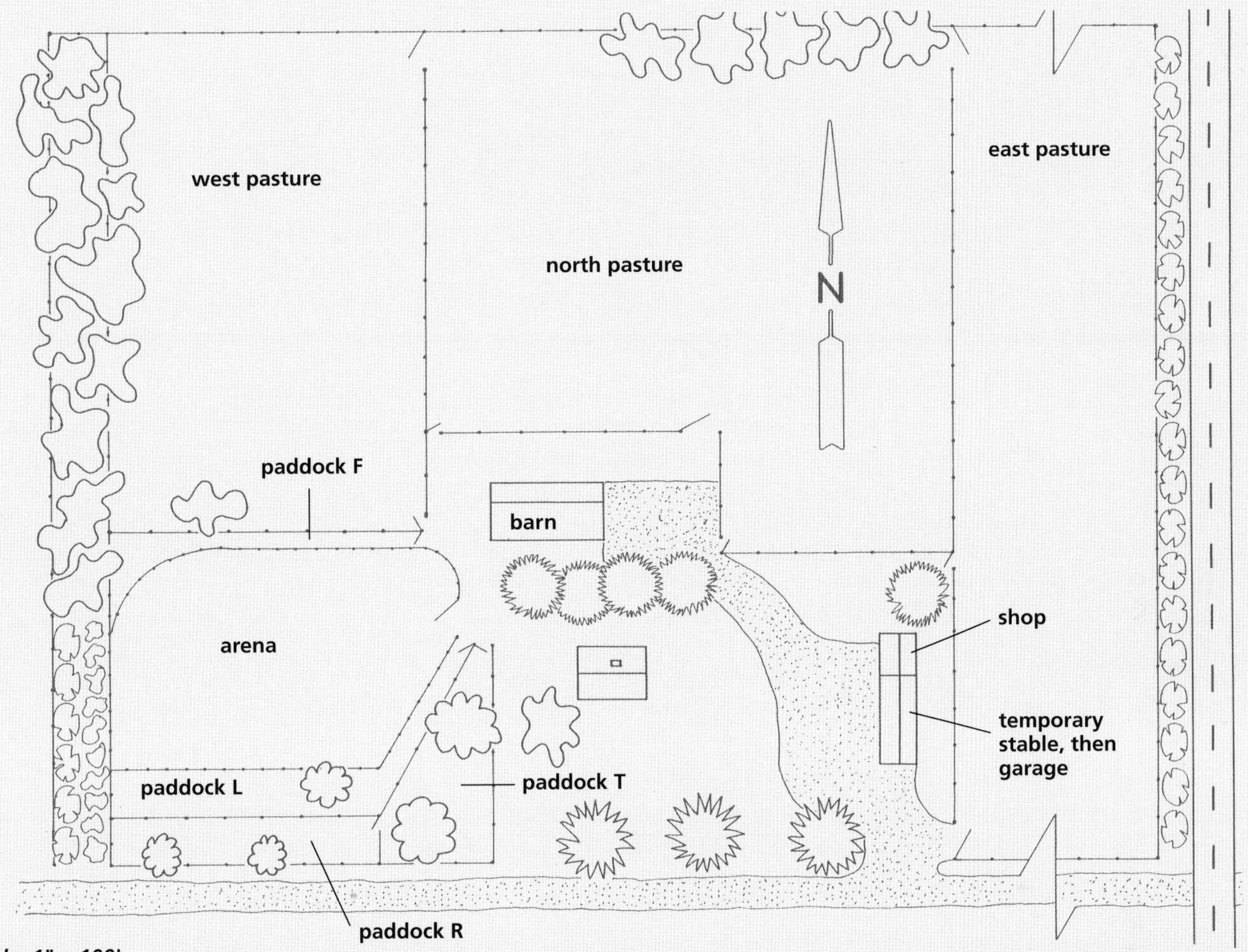

scale: 1" = 100'

DESIGN SOLUTION

GARAGE: Expanded to 20'x70' and used temporarily as stable and shop, then later as shop and garage.

ARENA: 140'x110'.

BARN: 30'x40'. See pages 86 and 87 for details.

WEST PASTURE: 170'x260'.

NORTH PASTURE: 275'x275'.

EAST PASTURE: 430'x525' (5 acres). Used as hay field for first cutting, then grazed the balance of the year.

PADDOCKS: Paddock-T (triangle): 100'x60'x75'.

Paddock-R (rectangle): 25'x140'.

Paddock-L (L-shaped): 25'x225'.

Paddock-F (flask): 10'x170'.

LANDSCAPING: West wildlife strip: 30'x430'. Maples planted along entire east border were all killed by county spraying. Post signs that say NO SPRAYING!

SAMPLE LAYOUT: 20 acres, ranch country

EXISTING CONDITIONS

SITE: 20 acres, ranch country.

HOUSE: 20'x50' cabin.

BARN: 24'x48' steel-sided pole barn (shell).

FENCING: Three-strand smooth wire and metal T-posts around entire perimeter.

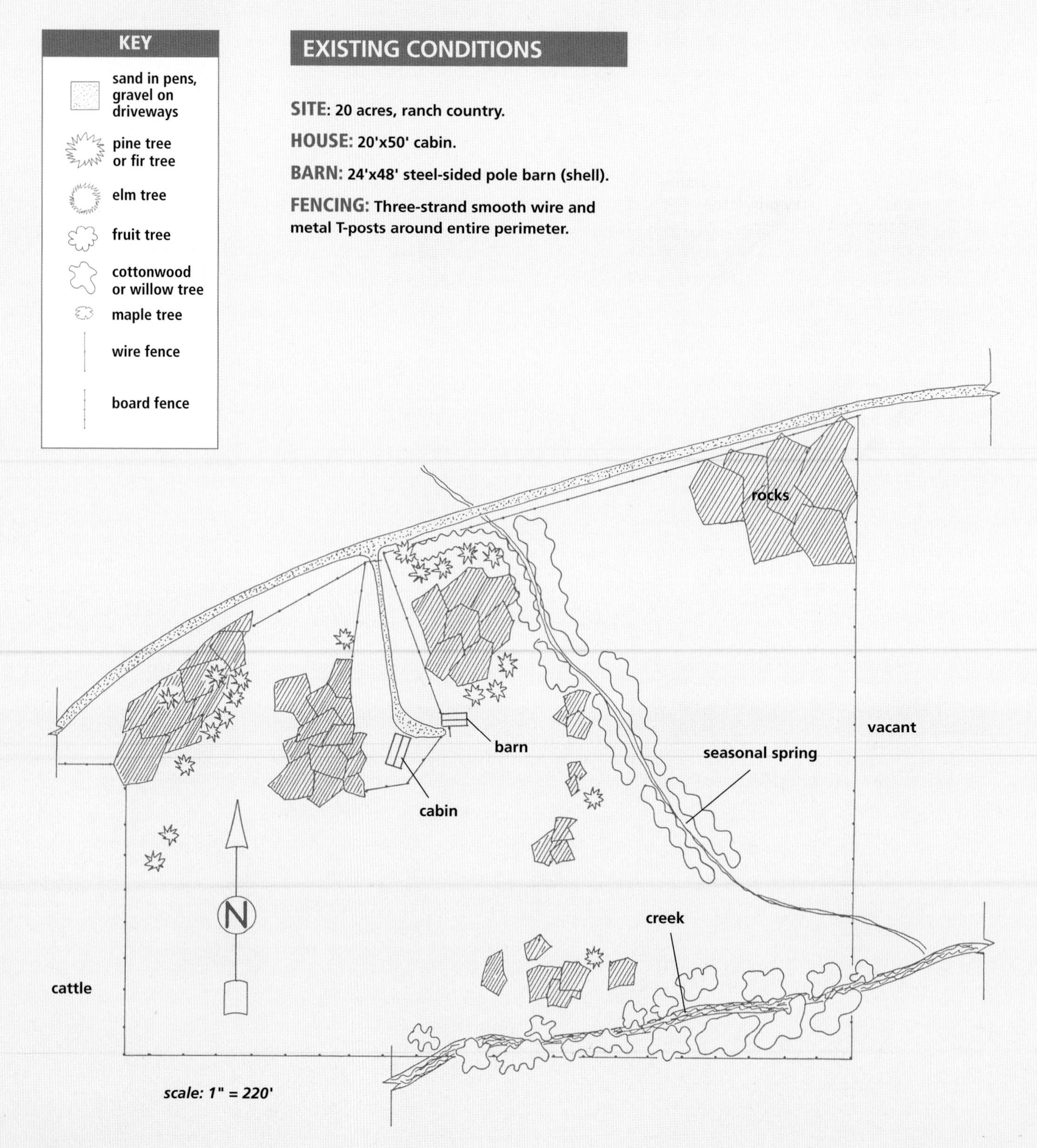

DESIGN SOLUTION

HOUSE: 35'x60' log home with approximately 5 acres of non-pasture area. House is located to obtain maximum wind protection from rocks.

GARAGE: 24'x24'.

BARN: See page 89 for detail. Added tack room (see page 115 for detail) and stalls inside; added 16'x48' hay storage area.

SHOP: 48'x100'; also accommodates tractor storage.

FENCING: Cross fences of various types: metal T-posts with smooth wire and/or electric, buck fence where metal posts cannot be driven due to rocky ground. V-mesh wire on the inside of existing smooth wire fence on the east and southeast fences to keep out neighbors' cattle. Also some post and board, and metal panels used for training and holding pens.

ARENA: 160'x100' square post and board with native soil (decomposed granite) footing. Location was dictated by the presence of rocks and the lay of the land. Excavation into hillside was required, leaving 16' banks on the east and west sides. Pasture fences skirt these banks to prevent erosion from horse traffic.

PENS: Round pen, 65' diameter round post and board with sand footing. Four metal panel pens 16'x32'.

PASTURES: Used primarily for exercise, as fragile mountain grasses afford only limited grazing. South: approximately 6 acres; provides water from creek. West: approximately 3 acres. East: approximately 6 acres; provides seasonal water from spring.

LANDSCAPING: Wildlife area: entire central area (between east and west pastures) is off limits to free horses; thus it is a protected habitat for wildlife.

PRIVACY BORDER: Trees planted along the north fence between the spring and rocks provide visual screen from road.

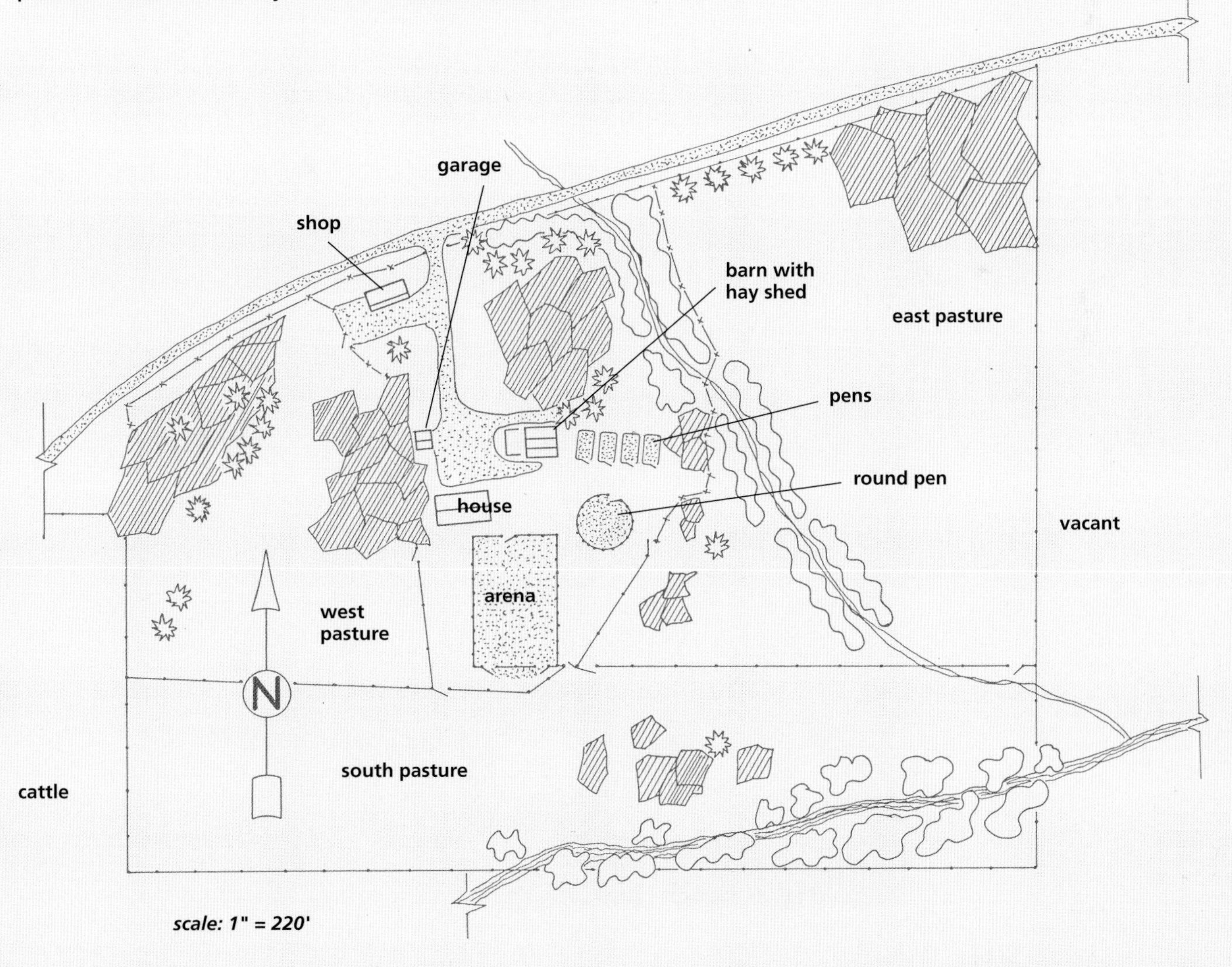

7 Barn

A barn should provide a safe, comfortable, and healthy home for your horses. Keep this in mind as you design it. Decide what features you need in your barn and then select a plan with those components or choose a builder who can create a custom barn for you. A common mistake is to accept a package deal that may not have everything you want, yet includes features that you may never use.

Barn Site

The site for your barn should be properly prepared. The barn floor should be 8 to 12 inches above ground level and located on well-drained soil. The addition of 6 inches of crushed rock covered by tamped clay is a traditional barn favorite if the existing subsoil is well drained. Poorly drained soils should be excavated to between 3 and 10 feet, and several feet of large rock should be laid at the base of the excavation. Crushed rock of decreasing sizes should follow in layers, leaving about 1 foot for the barn's topsoil. This can be tamped clay or a mixture of three parts clay to one part sand.

If the soil is too soft, loose, or weak, and its bearing capacity is inadequate for the footings (deepest support structures), the design engineer of the barn will have to make adjustments in the location or depth of the footings, or in the size of the concrete forms for the footings. The barn should have a strong foundation made of poured concrete, concrete block, or pressure-treated wood.

There should be plenty of windows or doors to let in the sun and air but keep out the cold wind, rain, and snow. A temperature range of 30 to 80°F is best for horses, with 55°F being the ideal. A humidity of 35 to 60 percent is good, with 60 percent optimum; however, it is better to be a little too dry than

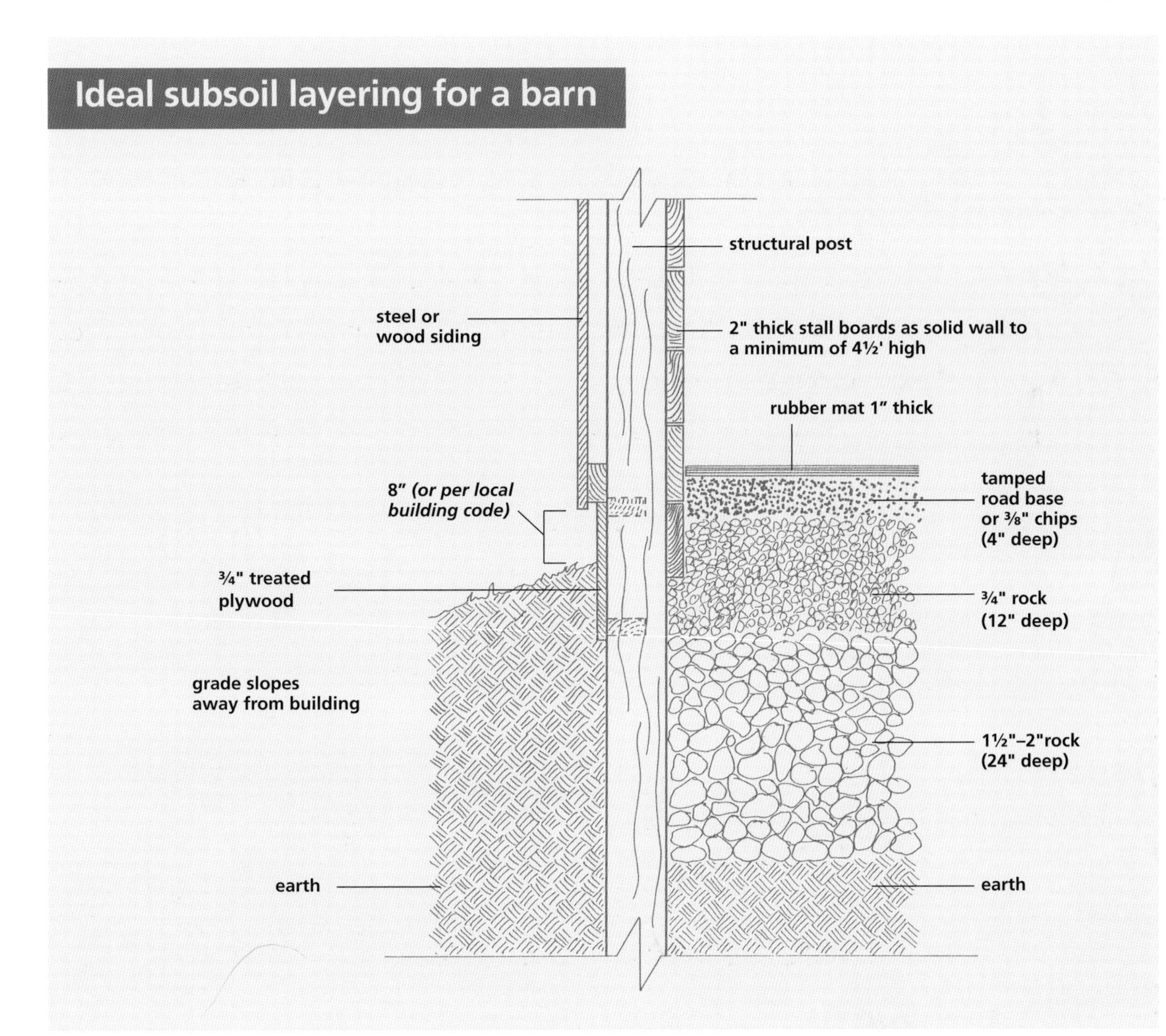

Plan for enough doors so you have access to your barn during extreme weather such as high winds or a blizzard.

damp. Horses need adequate ventilation but cannot take cold, damp drafts.

Because horses roll, kick, and sometimes buck while in their stalls, the structure must be very strong. In addition, all hardware, bolts, doors, handles, latches, locks, and hinges must be heavy duty to withstand horse use. Stalls, alleyways, and doorways should be safe, with no protruding parts or narrow openings. Heavy traffic areas should be well sloped and well drained and have a durable, nonslip surface that is appropriate for the use and the locale.

Think of your fire plan as you design your facility. Design some lockable areas, such as a tack room and office, for security and insurance purposes.

The barn should be located where it can be provided with electricity and water and situated so that there is room for a future addition if desired. There should be convenient access from feed and bedding storage buildings to the barn and from the barn to exercise and training areas. Many traditional designs and techniques have stood the test of time, but new materials and innovations are worth considering.

Barn Types

You must make many decisions when planning your barn. In warm climates, an inside aisle isn't essential, so many southern barns are simply single rows of covered stalls that open to outside pens or runs. In hot climates, you may forgo a building and opt for a large roof for sun protection.

Cold climates require more shelter for the stalls and inside access to the stalls. A very simple and popular style of cold-weather barn consists of two rows of stalls that face each other, separated by a center aisle. Often individual runs are attached to each stall to provide turnout areas.

Pole, frame, or masonry construction is commonly used in horse barns. Pole barns are quick, economical buildings to construct. They usually consist of 6- to 8-inch-diameter pressure-treated posts set 3 to 6 feet below the ground with the bases fixed in concrete. The poles are set at 8- to 16-foot intervals and have trusses attached to support the roof. Because the need for vertical supports in the center of such a building is eliminated, the result is a clear inside span that makes for flexible barn planning, the possibility of indoor riding spaces, and ease of expansion.

Frame or masonry barns require footings and foundation walls that extend out of the ground and support the barn walls. Where the outer walls of the building will be, a trench is dug to below the frost

A shed row–style barn, appropriate for warm climates, has no center aisle; the stalls open to the outdoors.

Barnmaster, Inc.

Covered pens, called *mare motels* on southern breeding farms, are generously sized and provide shelter from the sun.

Stalls with individual runs may be used as stalls only, runs only, or as in-and-outs, as shown here.

Modular barns can be erected quickly because they are assembled from previously fabricated components.

Barnmaster, Inc.

Profile of barn

Profile of barn on page 87. Note loft for hay storage and hay chute to lower level. Translucent panels in south-facing short wall are formed at the junction of the two roofs.

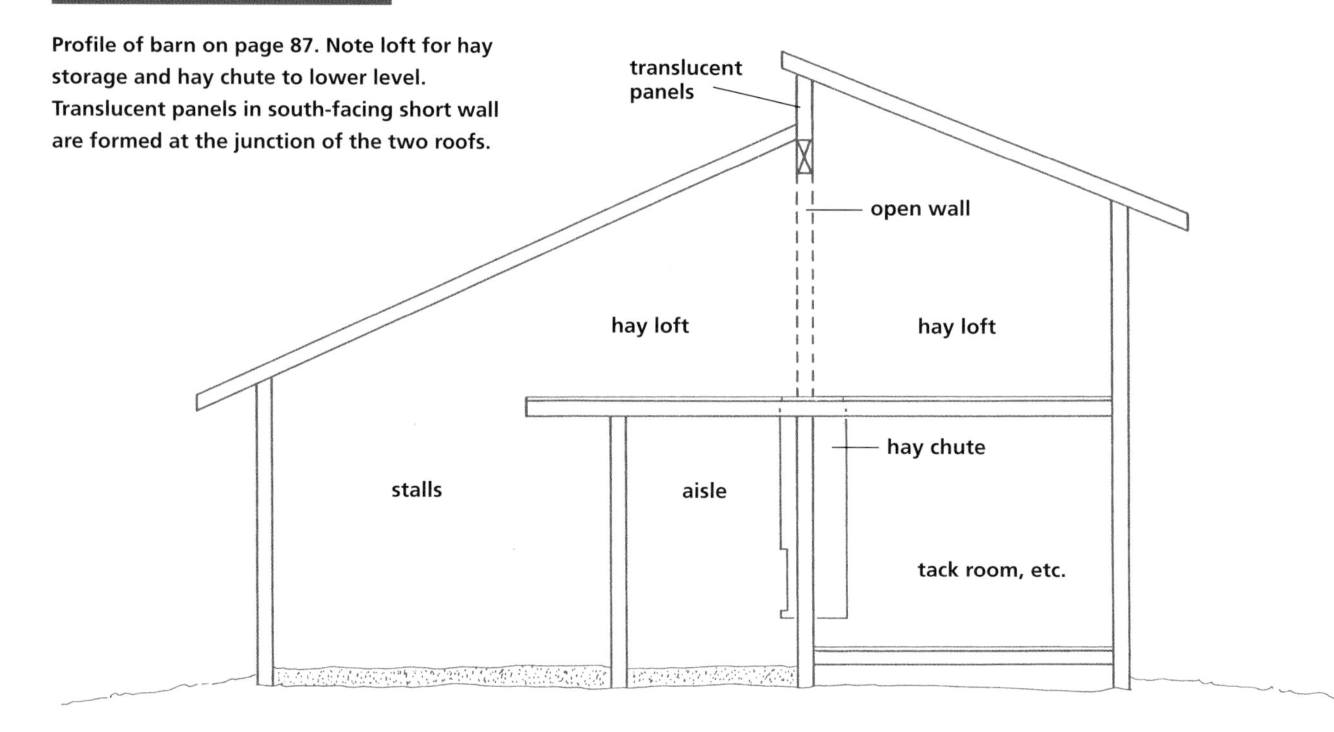

line (the maximum depth the ground freezes in the winter) or according to the applicable building code. Concrete footings are formed and poured in the bottom of the trench to transfer the load of the structure to the soil. The foundation walls of concrete block or poured concrete sit on the footing and extend about 16 inches above ground level.

Closed barns can be uninsulated, insulated, or insulated and heated. Heated barns are expensive and an unnatural environment for horses, tending to result in more respiratory illnesses, and therefore are not recommended. Insulation prevents condensation by keeping the temperature of the interior walls and the ceiling surface the same as the temperature of the air inside the structure. Insulation is placed in the space between the inner and outer walls. It can include blanket, rigid, sprayed-on, and foamed-in-place products. Blanket type is usually foil-backed or kraft-paper-backed and comes in rolls or in short lengths called *batts.* Rigid insulation is usually a sheet of pressed fibers or molded foam, often with a vinyl-coated side that can serve as the inside surface of the building. Spray-on cellulose fibers, though inexpensive, can absorb moisture in a humid climate and cause condensation and corrosion problems. Spray-on plastic foam can be useful for both roofs and walls but, according to most building codes, must be covered.

A vapor barrier prevents or minimizes the flow of water vapor into the walls. The vapor barrier should be applied on the warm side of the insulation, usually between the insulation and the inside wall covering. Special paints or waterproof membranes such as treated paper, plastic, and aluminum foil are commonly used for vapor barriers.

The roof type and whether you plan to have a loft in the barn usually decide the overall shape of your barn. Storing hay or bedding in a loft over the stalls can be convenient and does provide some insulation, but is such a fire hazard that it is strongly recommended to locate the bulk of your hay and bedding storage in buildings separate from the stable.

Barn plan from 10-acre layout

Barn plan from sample layout on page 79. Vents are 5' high from floor and are covered with heavy mesh panels.

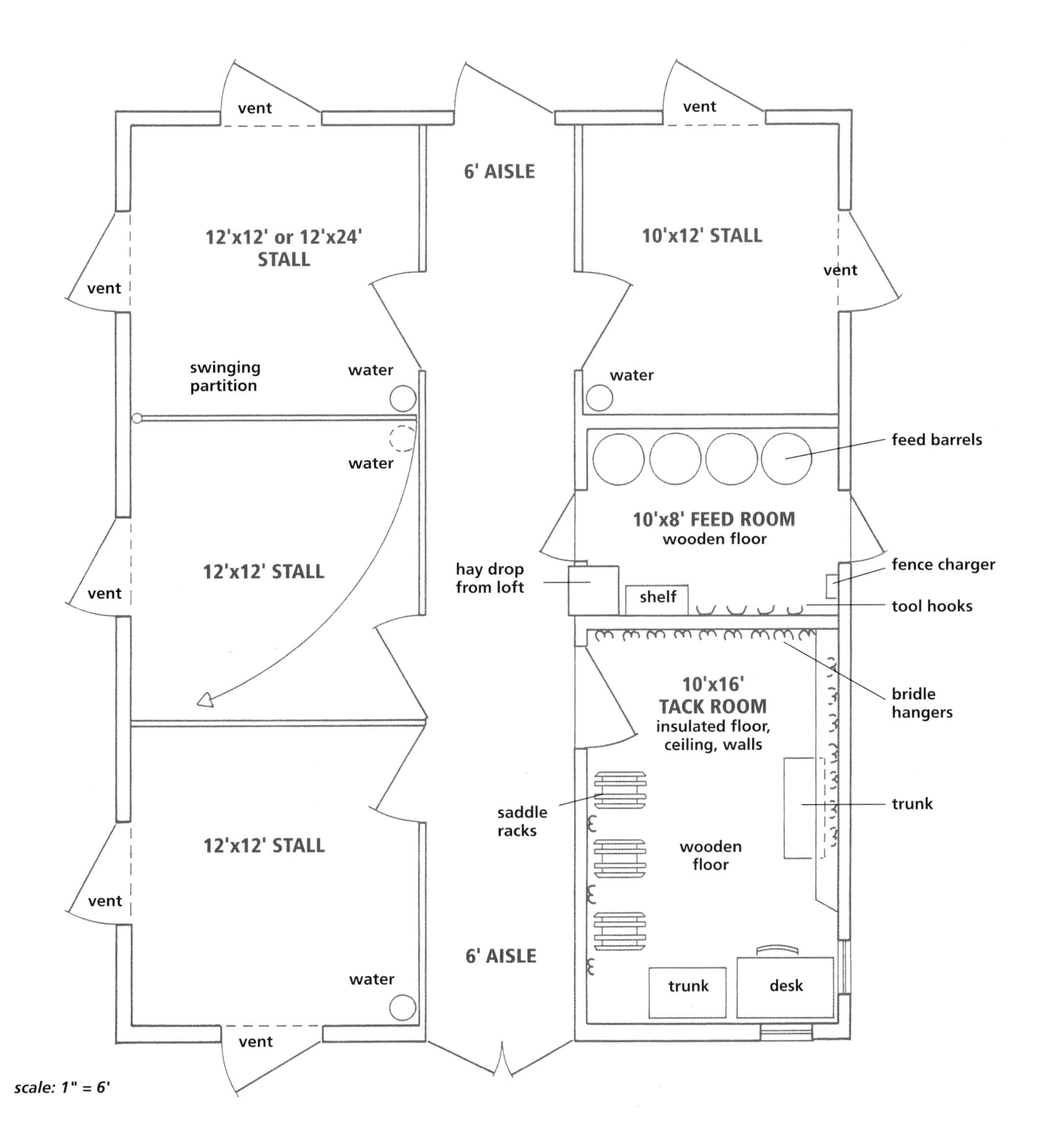

scale: 1" = 6'

Stable plan from 3.5-acre layout

(see page 77)

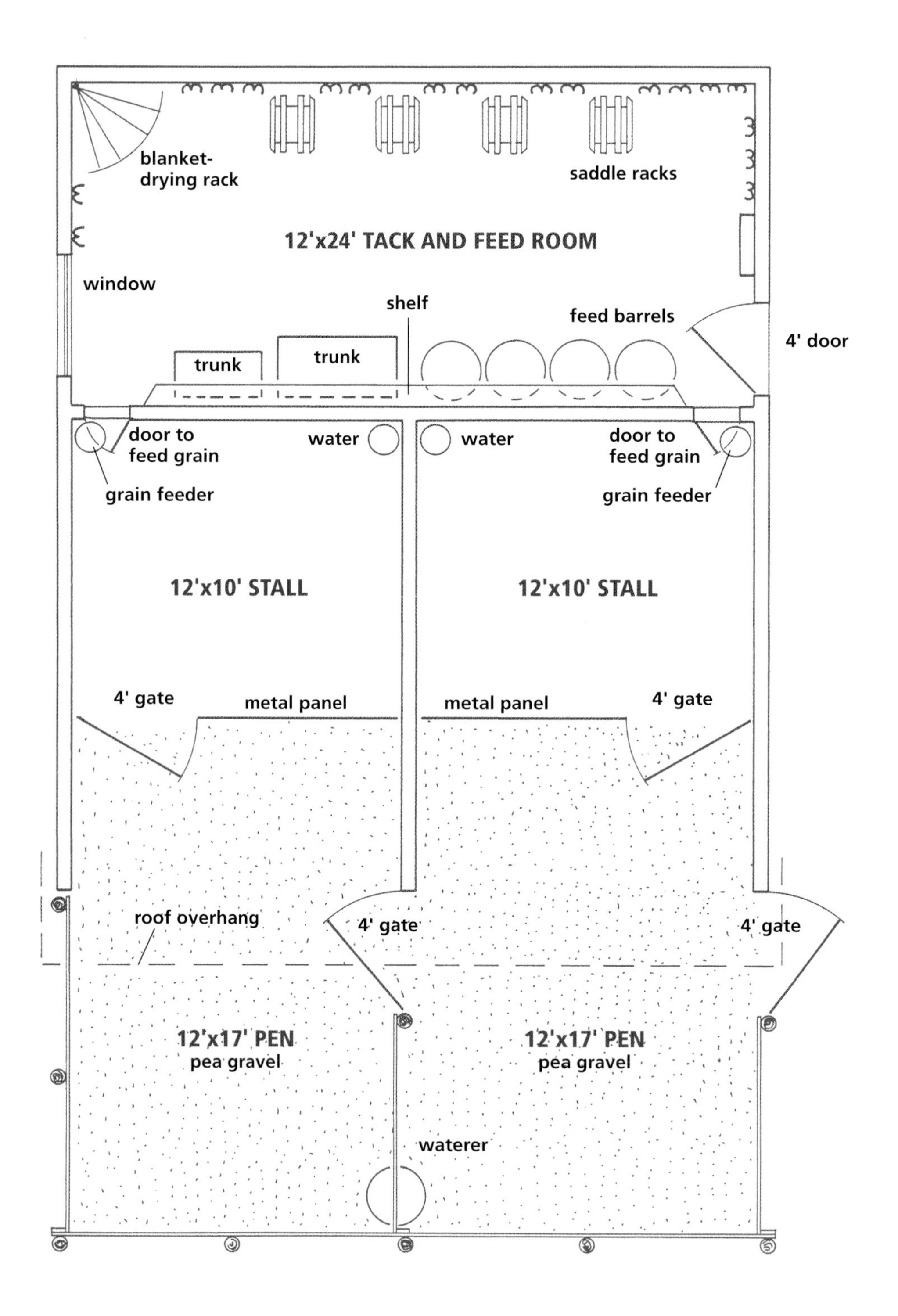

Barn plan from 20-acre layout

(see pages 81 and 115 top)

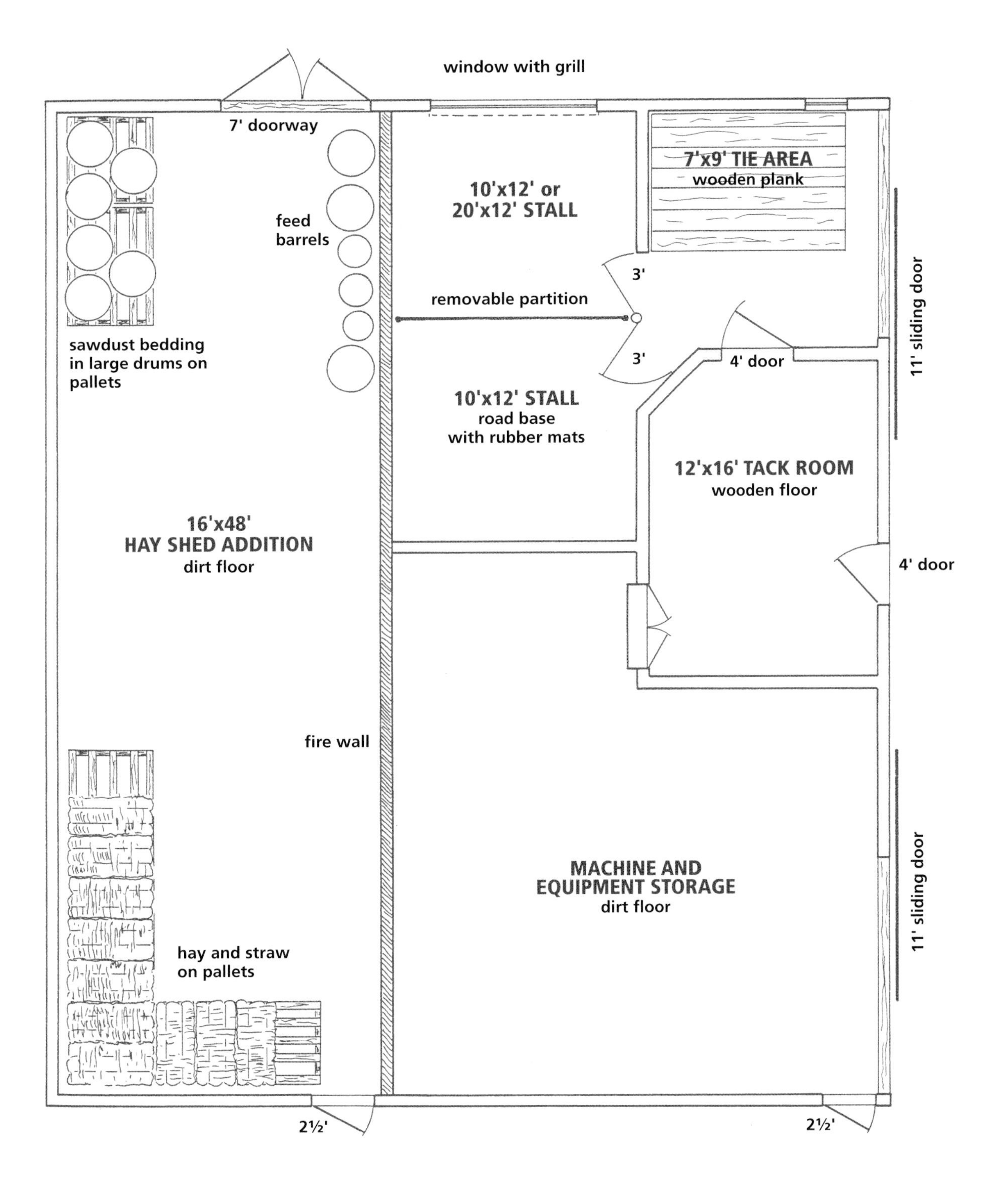

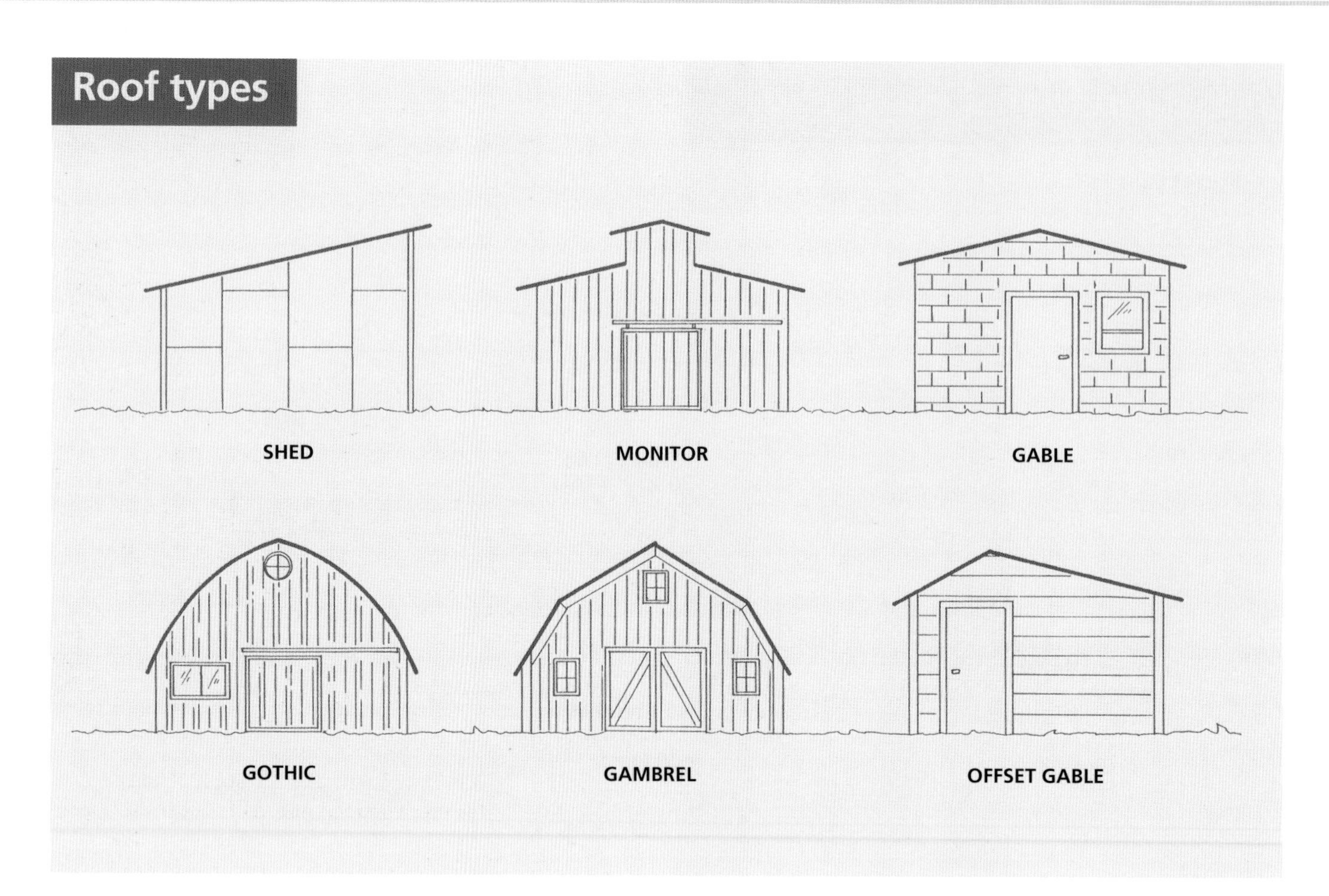

Roof Types

The gable roof is very popular and allows great flexibility in layout. The shed roof is often used for three-sided shelters or small stables or as an addition to an existing building that has a gable roof. The monitor is essentially two shed roofs with a gable in the middle. This is good for long rows of stalls. The area under the upper gable roof can be windows, vents, or clear panels.

Roofing Materials

If you live in wildfire country, choose a roofing with a Class C or better rating. Metal roofs are quick to install, relatively inexpensive, very low maintenance, and fire resistant. However, they can be very noisy, may leak around the fasteners, are prime candidates for condensation problems, and can make the barn hot during sunny weather unless insulated and vented. Zinc-coated (galvanized) steel panels are durable and economical but not particularly appealing. Steel and aluminum sheets with baked-on colors are more attractive, with steel being the stronger of the two. Translucent fiberglass panels can easily be incorporated into metal roofs for light. Snow that accumulates on metal roofs often slides off in huge sheets and without warning. Plan doorways and runs with this in mind.

Redwood or cedar shakes make very attractive roofs that can last 50 years or more. Shake roofs are warm even without insulation; they are cool in summer and condensation is generally not a problem if the barn is not heated. But they are expensive, and the labor cost to lay wood shingles is very high. In addition, they are considered a fire hazard and may affect your insurance rates. Shakes should be used only on a roof with at least a 1/4 pitch so that water drains off and the shingles dry out quickly.

Concrete and clay roof tiles and quarried slate tiles come in a variety of colors and designs that will last almost indefinitely. They are more expensive and

Barnmaster, Inc.

Top: This raised-center-aisle (RCA) barn features windows and vents under the upper eaves. _Left:_ Roof overhangs can provide added protection for turnout pens. During this record snowfall, however, the snow that slid off the roof blocked Zipper's access to his pen until we fired up the tractor.

more difficult to install than most other roofing materials. Masonry roofing products are heavy and require more support than other types of roofing; however, they are stable and fireproof.

Fiber cement roofing products are made to look like shingles, slate, and shakes. Products manufactured before 1990 contained asbestos fibers, which make them very durable. Since then, cellulose fibers have replaced asbestos.

Asphalt roofing, both shingles and rolled roofing, will last 20 years if properly installed. It is relatively inexpensive, comes in a range of colors, is one of the least complicated types of roofing to install, and is quieter than steel roofing. It is more susceptible to damage by wind, cold, ice buildup, and extreme heat than is steel roofing.

Felt roofing paper is useful as an underlayer but not as an exterior roofing; it simply will not last.

My barn

PERSPECTIVE

The siding and roof are low-maintenance ribbed steel panels.

A pressure-treated plywood skirting extends 1' above the ground and 1' below the ground around the entire perimeter. This skirting prevents burrowing rodents from entering under the walls and protects the siding from contacting the earth.

The roof has a 4/12 pitch, with the lower edge at a height of 9' and the peak at 17'.The barn can provide shelter for eight horses, four in stalls and four in pens.

The four 10'x12' stalls with rubber mat flooring are used for foaling, convalescence, or severe weather.

The partition between each pair of 10'x12' stalls hinges open and is secured against the wall to make two foaling or recovery stalls, one 10'x24' and one 12'x20'.

The generous roof overhang along the east side and on the south end provides partial shelter for four outside pens, each approximately 400–500 sq. ft.

The concrete floor in the aisle, hay area, tool room, and wash rack has a rough-textured finish for traction. The concrete feed room floor is finished smooth for ease of cleaning.

Translucent fiberglass panels on the ends of the barn and upper 3' of the walls (except for the tack room walls) let in much light and eliminate the need for electric lights for normal chores during the day, including in the enclosed feed room.

The well-lighted 8'x12' wash stall provides a convenient place to bathe horses, wash tools and buckets, and hose off muddy blankets. With hot and cold water and a sink, it also serves as an excellent place for a veterinarian to work. Guardrails along either side of the wash stall protect the water faucets and tools and buckets placed against the walls. The textured concrete floor slopes to a centrally located drain.

The 8'x10' open tool room stores carts, cleaning tools, muck buckets, supplies for fence repair, and other items that might otherwise be in the barn aisle.

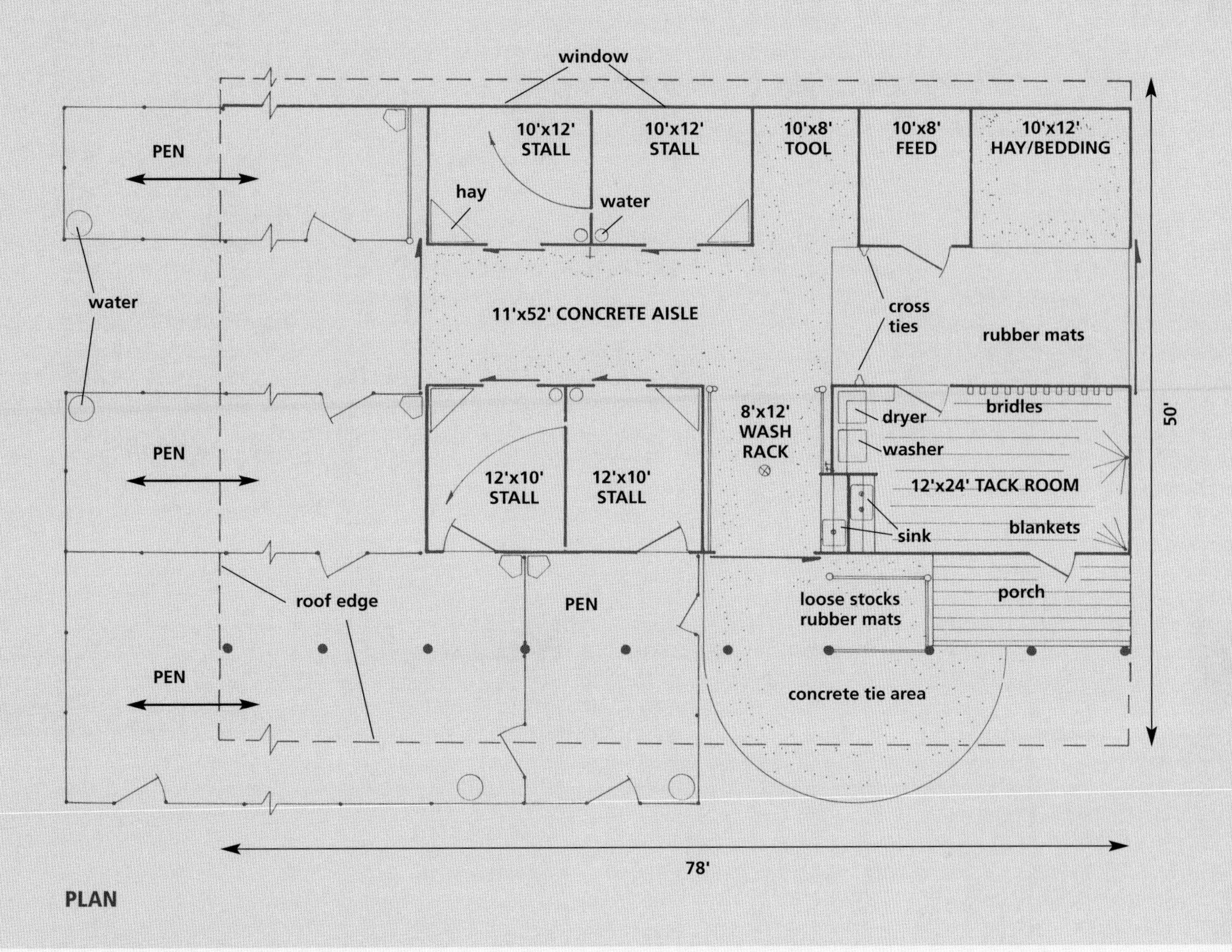

PLAN

An 8'x10' feed room keeps grain dry and safe from rodents and has a horse-proof latch on the door to prevent a loose horse from foundering on grain.

The 12'x10' hay area is used to store a week's worth of hay and stall bedding. It is located next to a large sliding door for convenient unloading.

The 11' aisle provides ample room to maneuver a horse anywhere in the barn and for safely working on a horse in cross ties. It also is wide enough to back in a pickup to deliver grain or to use a small tractor and wagon for cleaning stalls.

11' wide sliding doors on either end of the aisle, as well as the 8' wash rack door, can be opened for cross-ventilation.

The 12'x24' insulated tack room has a sink with hot and cold water, a washer and dryer, storage cabinets for supplies, and built-in bridle and blanket racks. There is plenty of room for saddles, tack, and blankets for at least eight horses.

Cross ties in the aisle are placed so the horse is conveniently located with his near side next to the tack room door. The grooming area is well lighted, has plenty of electrical outlets, and has recessed shelves in the wall to hold grooming supplies and totes. The concrete floor in the grooming area is covered with rubber mats.

Loft areas above the tack room and feed room can be used to store boxes of seldom used or seasonal tack, extra feeders, and other items.

Two tie areas are just outside the wash rack door. A loose stocks area is completely protected from the sun and wind, and a three-rail tie area lets a horse stand in the morning sun. These are good places to let a horse dry after a bath or cool down after a workout.

The wooden porch off the tack room provides a relaxing, shaded area for cleaning tack or just enjoying the moment.

Barn plan with attached indoor arena

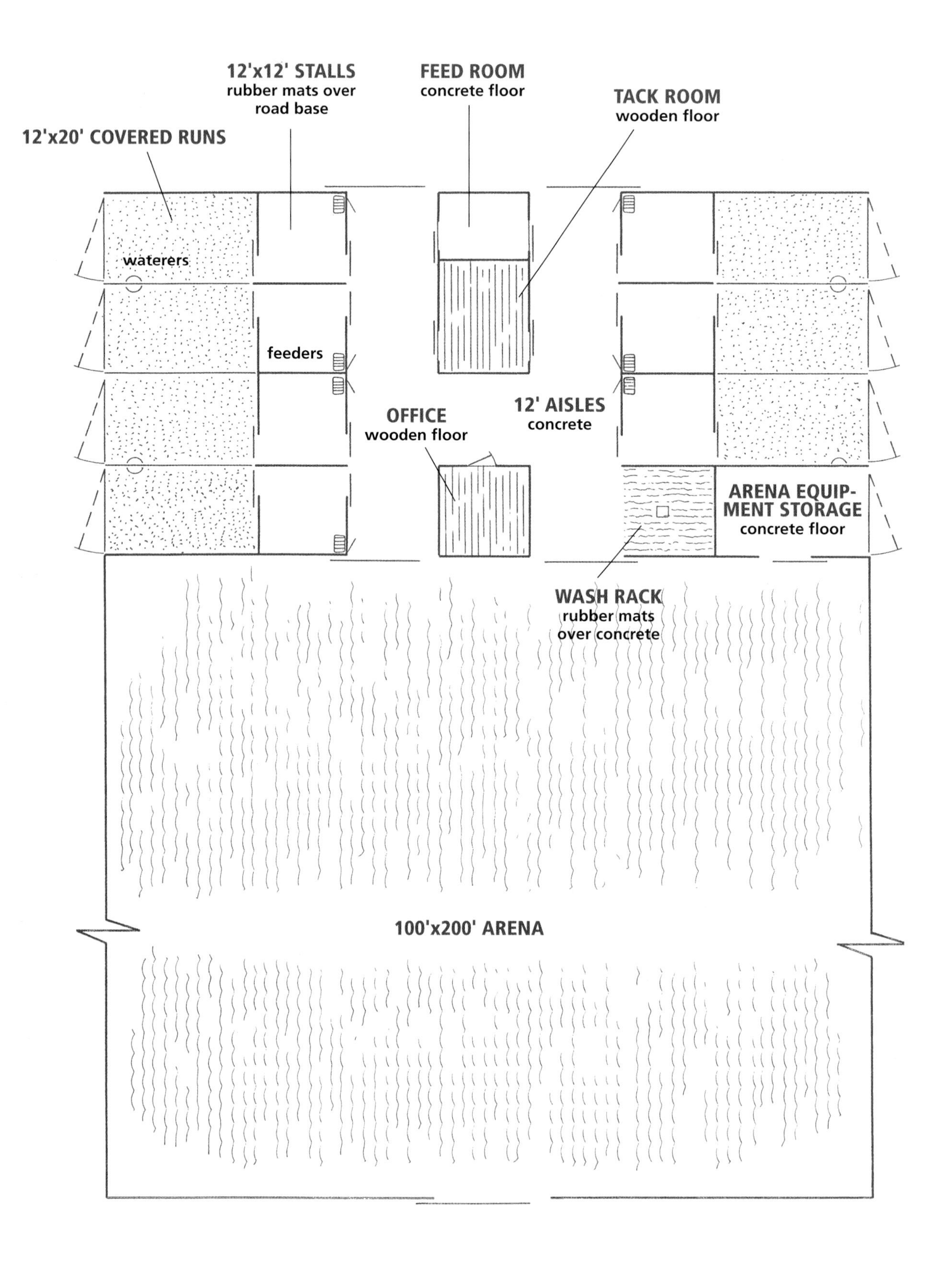

Barn Walls

When choosing the materials for your barn walls and roof, consider cost, durability, maintenance, fire resistance, and aesthetics. Metal (steel or aluminum) buildings are quick to put up, are less expensive than wood, and require minimum maintenance. They can be noisy, however, during windy or rainy weather; if not insulated, they can be cold in winter and hot in the summer; and although they are neat in appearance, they may not be thought to be as aesthetically appealing as other choices. In a metal building, the stalls must be lined with planks or other sturdy material at least 4 feet up from the ground to prevent damage to the metal siding by the horse.

Wood buildings are traditional and attractive and provide good insulation. However, they are expensive, can require more labor for construction, and need frequent maintenance. Horses chew wood, so all surfaces that horses can contact must be treated with an antichew product or covered with metal. Wood buildings are a fire hazard and should be built of fire-retardant materials and have a fireproof lining where possible. Wood or fiberboard shiplap siding or vertical boards give a traditional look to the exterior.

Though plywood siding may not be as attractive as boards, it is very strong. It is a manufactured wood product with a solid or veneer core covered by thin layers (plies) of wood. Each ply is placed with its grain at right angles to the next ply, and all of the layers are laminated together with glue. Plywood has a much higher strength-to-weight ratio than lumber has. Plywood won't warp, split, or shrink the way boards will.

Masonry-type buildings include brick, concrete block, poured cement, and stone. These buildings are cool in warm climates but can be cold and damp in cold climates. Generally, there is a high construction cost in both labor and materials. Because such structures are virtually fireproof, they often qualify for lower insurance rates. Masonry-type buildings must usually be mechanically ventilated in order to produce a satisfactory environment for horses.

DT Industries, Inc.

Top: **Locating runs off the end of a barn eliminates problems associated with rain runoff and snow dump.** ***Middle:*** **This wood barn has a mighty loft under its gambrel roof.** ***Bottom:*** **Stucco barns are popular in hot climates.**

Barn and Stall Kits

You can buy packaged kits for an entire barn or for a specific number of stalls. But before you consider buying a kit, check with your local building department to see if the kit you are considering meets the standards and codes required. You might be required to obtain blueprints or engineering calculations from the kit manufacturer for the structure before it can be approved. If you erect a kit that does not meet local building codes, it becomes your responsibility, not the manufacturer's, to bring it up to code. Also, check with your insurance company to be sure the building can be insured.

Generally available in wood and metal, kits are priced according to the size, style, quantity, and quality of materials, and the amount of prefabrication completed by the manufacturer. Barn kits are available for post-and-beam, free-span, and modular construction.

Post-and-beam barn kits are wood. Vertical posts are anchored to concrete piers or footings. Horizontal beams connect the posts and provide support for the roof. Kits contain marked, precut lumber plus some or all of the following: siding, flooring for the loft, roofing, doors, windows, fasteners, and hardware. To assemble a wooden post-and-beam barn, you need to be able to handle a 50-pound sheet of plywood and have some means of raising heavy beams.

Free-span barn kits can be made of wood or steel framing. They have no interior weight-bearing posts, so they provide the most flexible layout plan. However, after the construction of the trusses, it requires three or four people or two people and a crane to hoist the trusses into place. Usually the exterior wall panels are laminated wood-core encased in steel that are assembled with bolts.

Modular barn kits are steel-framed panels that are erected in modules. You can link together as many of these modules as you want. The panels are often a steel laminate over a plywood core. These are the fastest kits to set up and require only two people.

Stall kits are available in steel and wood and come in several designs, from the traditional box stall to something that looks like a panel pen with solid lower walls. Stall kits can range in difficulty, from prefabricated components that go up quickly (such as panel stalls) to very labor-intensive kits. Some kits require you to handle up to 100 pounds and work with saws and power tools, so they might not be as quick and easy as you would expect.

Kit essentials

If you are considering purchasing a barn or stall kit, get the following information in writing:

- Where you can see a sample building or stall in your area.
- References from previous customers.
- Finished dimensions, inside and out.
- Snow-load rating (maximum weight of snow per square foot on the roof), live load (weight of human and other temporary loads per square foot on the roof), and wind load (speed of wind against the building).
- What is included.
- Description of materials including dimensions, gauge, brand names, finish, insulation.
- What else you will need to finish the project.
- What tools and equipment are required for assembly.
- What needs to be done to prepare the building site (e.g., grading, foundations or pad).
- What instructions are provided and whether there is someone available to answer questions.
- Cost including shipping.
- Delivery date.
- Payment options.
- Warranty.

Exterior Features

To handle water from the roof during a rain, you may want to think about including gutters, downspouts, and concrete splash pads.

To keep entryways from becoming muddy, consider attaching overhangs to the roof to shelter the doorways. Such a covered area can provide space for a sheltered tie area, loose stocks, and a porch. If you plan to attach exercise runs to the sides of the barn, plan the roof so that snow accumulation does not block doorways.

To prevent fire by lightning, consider installing a lightning conductor to the most prominent roof. (See chapter 16, Disasters, on fire prevention, for more details.)

Left: **This building overhang provides cool, airy, loose stocks to tie Zipper and a shady place for Richard to work on tack.** ***Above:*** **Local building codes require that structures be built to withstand the area's weather extremes; snow-load ratings for roofs are one example.**

Flooring choices

TYPE	INITIAL COST	LONGEVITY	CUSHION	DRAINAGE	ODOR
DIRT	Low	1–2 yrs. in stall	Good	Forms mud	Yes
CLAY	Low	1–2 yrs. in stall	Good	Slow drying	Yes
SAND	Low	Not recommended	Very good	Good	Mild
GRAVEL	Low	5 yrs. to indefinite (add periodically)	Shifty, can be rough	Good	Mild
ROAD BASE (gravel/limestone/dirt mix)	Low	2 yrs. to indefinite (add periodically)	OK to none	Poor to good (varies due to product and tamping)	Mild
CONCRETE	Medium	Indefinite	None	Needs drain or slope	No
BRICKS	High	10+ yrs.	None	Needs drain or slope	No
RUBBER MATS	High	10+ yrs.	Good	Needs base or slope	No (base may have odor)
RUBBER BRICKS & TILES	High	10+ yrs.	Good	Needs base or slope	No (base may have odor)
WOOD	High	3–30 yrs.	OK	Needs base and spaces	Yes
ASPHALT	Medium	5+ yrs.	Very little	Needs slope or drain	Mild
INDOOR/OUTDOOR **CARPET**	High	5+ yrs.	OK	Needs slope or drain	Mild

Flooring

You will probably end up with three or more types of flooring in your barn. Stalls, aisles, the tack room, the feed room, the wash rack, and the office all have different flooring requirements. Things that you need to consider for any floor are cushion, absorbency or drainage, cleaning convenience, traction, odor retention, moisture retention, appearance, and, of course, that pesky item, cost.

Some floor areas will need to be sloped to encourage surface drainage, but not more than 1 inch in 5 feet. Other areas should provide level spots for work, such as the farrier's slab. Too much slope or irregular footing here makes accurate work difficult and is uncomfortable for both the farrier and the horse. Most floors must provide adequate traction, but especially those that get wet, such as stalls, wash racks, and grooming areas. When choosing floorings for various areas in the barn, remember that they must be raked and/or swept regularly and occasionally washed and disinfected.

CLEANING EASE	SANITATION/ MAINTENANCE	TRACTION	SUITABILITY	OTHER COMMENTS
Difficult	Difficult	Good	Stalls	Frequent maintenance and replacement
Difficult	Difficult	Good	Stalls	Frequent maintenance and replacement
Difficult	Difficult	Fair to good	Laminitis treatment — stall only	Problems with sand colic; sand cracks, unstable, dusty
OK	Difficult	Poor to OK	Under mats in stalls; pens, driveways	Unstable alone
Difficult	Difficult	Good	Under mats in stalls and aisles	Can be dusty, muddy, hard when tamped
Easy	Easy	Good if texturized	Aisles, wash rack, feed room, shoeing area	Use appropriate texture for location
OK	Easy	OK, but slick when wet	Aisles, tack room, office	Attractive; needs stable base
Easy	Easy	Good when dry; good when wet if texturized	Stalls, wash rack, shoeing area	Best choice; good over concrete except in stalls; quiet; save bedding costs
Easy to OK	Easy	Good, depending on texture	Aisles, tack room, office	Attractive, quiet
OK to difficult	Difficult	Slippery when wet	Tack room, office	Should be kept dry
OK	Possible	Slippery when wet	Aisles; driveways	Sticky when hot; slippery when wet or cold; cracks with freeze/thaw cycles
OK	Possible	Good	Tack room, office	Best for low-traffic areas

As you look over the information in the flooring chart and begin talking with dealers, realize that some floorings are made to be used in conjunction with others, so you may end up choosing two floorings for one space — for example, rubber mats over tamped road base for stalls or indoor/outdoor carpet over wood for the tack room.

Our barn has a concrete aisle and a rubber-matted grooming area where Zipper stands tacked up and ready to go.

Lighting

Incorporate a combination of natural and artificial light in your stable plans. Natural light should include some direct sunlight, a source of vitamin D. Sunlight is also one of the best sanitizing agents and can help keep a barn from developing a rank odor. Sunlight also adds warmth, however, and although it may be greatly appreciated in the winter, extra heat can become a big problem in the summer. Natural light can enter the barn through windows, large sliding doors, and panels and vents in the roof.

The three main types of electric lights used in barns are incandescent, fluorescent, and high-density discharge (HID).

The common household lightbulb produces incandescent light by heating tiny wire filaments made of tungsten within a glass vacuum bulb. Such bulbs convert only 6 percent of electricity to light, and the rest goes to heat. They burn out much more quickly than other lamps, but they are cheap and easy to replace. They are satisfactory for tack rooms, feed rooms, and other similar spaces.

Quartz halogen lamps produce incandescent light by heating tungsten filaments in a clear quartz tube filled with halogen gas. Quartz lamps last up to four times longer than common lightbulbs, cost more, burn brighter, and are 10 to 20 percent more efficient. They get very hot and must be used in an enclosed fixture. The light from incandescent lamps has little effect on the true color of objects. Quartz floodlights are a good choice for lighting barn aisles, stalls, and pens.

Fluorescent lights are the coolest because instead of heating a filament, they use an electric arc to heat a small amount of mercury, which gives off UV radiation and causes phosphors on the inner surface of the lamp to glow. A 40-watt fluorescent bulb puts out twice the light of a 95-watt incandescent tungsten bulb and uses less than half the electricity. Fluorescent bulbs typically cost more but last ten times longer. The color of fluorescent light varies sig-

Ambient light illuminates this barn, so electricity is not needed during the day.

nificantly depending on the type of phosphors used in the lamp. They are best used where they won't be turned off and on frequently and in a work area where you might need concentrated lighting.

Fluorescent fixtures contain a part called a ballast, which regulates current and determines how well a fluorescent fixture performs. Preheat ballast fixtures are the most readily available, are the slowest to start, and work best above 50°F. Instant-start (cold-start) fixtures use a higher starting voltage to provide full light output instantly but sometimes have a loud hum. Rapid-start fixtures come on quickly and make less hum than instant-start fixtures. Cold-temperature, or zero ballast, fixtures work well to 0°F but usually need to be special ordered.

Compact fluorescent lamps (CFLs) that screw into a standard light socket are relatively expensive but can provide energy savings of 60 to 75 percent over a standard incandescent bulb and last ten times longer. They are often too large to fit into some fixtures, may not start or light well in temperatures under 50°F, can affect the accuracy of colors, and might produce an annoying flicker or radio static.

HID lamps, originally developed for street lamps, have the longest life of all, are the most efficient to operate, and are the most expensive to buy. They need at least 16 feet to diffuse evenly, so are best for an arena, a yard light, or a barn — an area where lights are higher and used for longer periods, since the fixtures need several minutes to warm up. Metal halide lamps have the most natural light, mercury lamps produce a blue-green light, and sodium lights produce a golden light.

All lights, switches, and receptacles must be out of the reach of your horse's curious lips. Bulbs in stalls should be fitted with wire baskets or protected by a heavy glass "jelly jar." Depending on the receptacle and ceiling, a 100-watt incandescent bulb should illuminate a 12-foot by 12-foot box stall sufficiently for you to clean it. If you locate the lights above the stall dividers and toward one corner, you may have fewer dark spots in the stall. A 150-watt halogen lamp is a good choice, providing it is well out of a rearing horse's reach.

All electrical wires should be threaded through metal or heavy plastic conduit pipe to prevent rodents from chewing the wires, one of the leading causes of stable fires. You should have a separate electrical service for the stable to prevent domestic overload. Check your local building code for the required number of receptacles for your structure, but plan for more electrical outlets than you think you will need. Consider that you may use any or all of the following: water pump, extra flood lamps, clippers, vacuum cleaner, heaters, battery chargers, electric waterers, hot plate, coffeepot, heat lamps, radio, refrigerator, and so on. Try to avoid using extension cords, and never overload plugs. Plan ahead and install three-way switches wherever practical. These allow two switches to control the same light and will help ensure that lights are turned off when they should be.

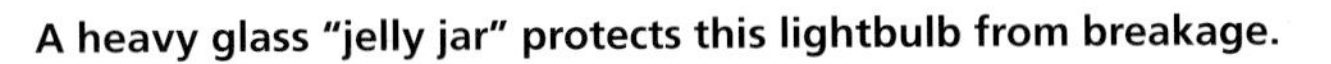

A heavy glass "jelly jar" protects this lightbulb from breakage.

Ventilation

A 1000-pound horse releases 2 gallons of moisture into the air each day through respiration. A four-horse barn must thus deal with more than 8 gallons of water vapor per day, not counting the additional moisture created by the evaporation of urine and manure.

Warm air can hold a certain amount of water vapor, but as the temperature drops, it loses as much as half of its water-carrying capacity. Moist air rising from a horse's stall condenses on the underside of uninsulated roofs, causing dripping, dampness, and sometimes ice formation. Damp air contributes to respiratory ailments, stiffness, and bacterial and fungal growth.

Ventilation, the movement of air through the barn, can be used to regulate the temperature and humidity. The goal is to make the air in the stalls and the aisles fresh. Check your design to ensure that ammonia fumes from manure and bedding decomposition can move up and out of the stalls. Also plan your structure to prevent condensation, which occurs when the difference between indoor and outdoor temperature is too great. Condensation does not occur as readily in moving air as it does in stagnant air.

Ventilation can be provided by natural and/or mechanical means. One of the easiest methods is to open windows, barn doors, and the tops of stall doors. Another is to install a ceiling fan over every stall to move the stagnant air trapped between the stall walls. Of course, you must beware of drafts, as horses are very susceptible to chill. Stall windows should be outfitted with a metal grill with spaces no more than 3 inches apart to protect the glass from horses. Metal window frames are better than wood, as they are fireproof and less susceptible to chewing damage. If a window is set high in the stall, it can have hinges on the bottom and open inward.

Doors should be a minimum of 8 feet high and 4 feet wide. If a door is to be left open, it should have a latch or hook to keep it there. Sliding doors should be outfitted with proper tracks and with bumpers to prevent them from rolling off the tracks.

Various vents can be placed in the barn to allow foul air to escape. All sorts of variations of louver boards can be implemented in the roof ridge or under the eaves in the soffitt. A spinning cupola ventilator can be added on the roof as a supplement to ridge vents. Other nonmechanical means of ventilation include adjustable vents 6 feet from the stall floor (to let in cool air) and vents at the top of the back wall where it joins the roof (to let out warm, moist air).

Mechanical ventilation forces air in or out of the barn, usually with electrical fans or blowers. When figuring either for a pressure or an exhaust system, figure 1 square foot of vent or inlet space for every 750 cubic feet per minute fan capacity. More than this will create a draft, and less simply will not ventilate adequately.

Ventilation

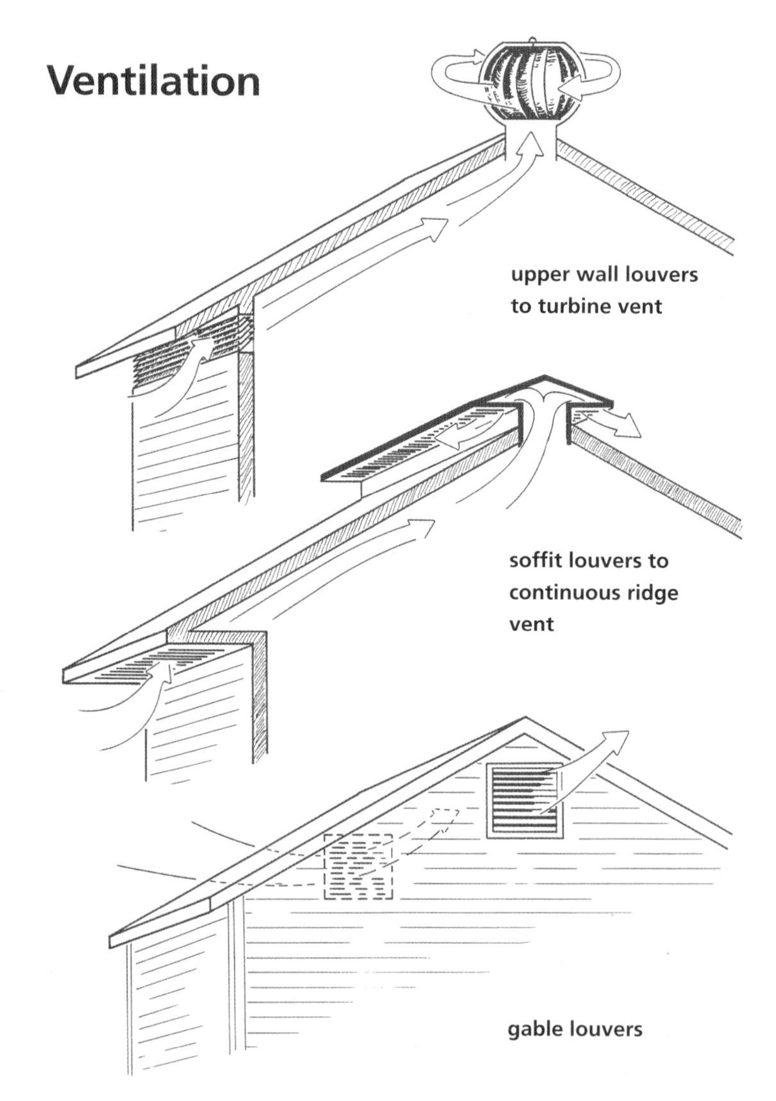

Above: Sliding doors can be opened wide or a crack to regulate ventilation. *Right:* A cupola set on top of a barn helps evacuate foul air.

French drain

A trench is located so that water dropping from the eaves falls 3" to 4" inside the trench.

Geotextile fabric (landscape cloth) lines the sides of the trench and overlaps over the top of the gravel that fills the trench. This keeps dirt and silt from clogging the gravel.

Portions of the drain that will have horse contact should have an extra membrane of rigid draining material to prevent horses from pawing through the geotextile layer.

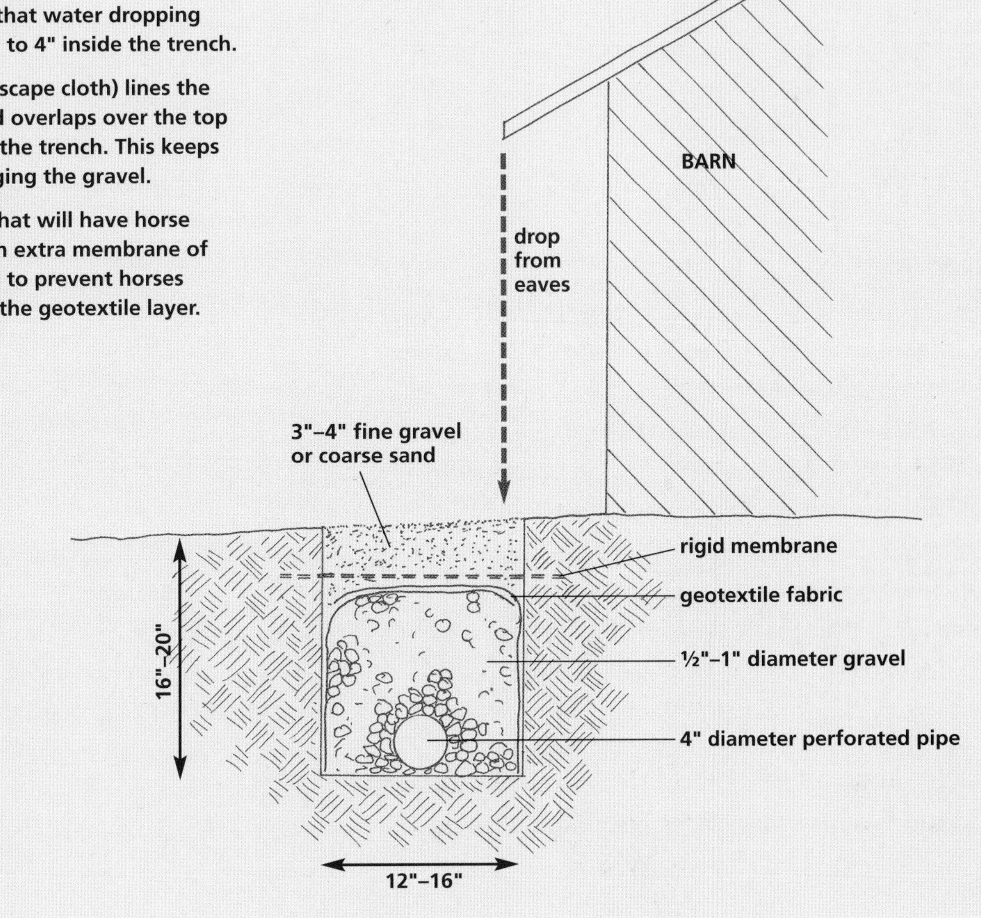

Building on a budget

When planning to build a horse barn, consider immediate, short-term, long-term, maintenance, and associated costs and savings.

- *Immediate costs* (or savings) are the initial costs of materials and labor to construct the barn.
- *Short-term costs* (or savings) include replacements that will need to be made in the first year or two of service, such as flooring that was a poor choice or wooden rails that weren't protected from chewing.
- *Long-term costs* (or savings) are replacements made after the barn is several years old, such as replacing the shingles on the roof and gutting the stalls to replace dangerously deteriorated or damaged stall walls.
- *Maintenance costs* (or savings) are regular-upkeep items, such as painting and weatherproofing, as well as the amount of bedding that is required daily.
- *Associated costs* (or savings) include the amount of feed wasted or optimally used due to feeder or stall design. A significant associated cost can be veterinary bills due to management-related mishaps, such as colic (automatic waterer malfunction or horse escaping from stall and getting into grain room) or injuries (unlined stalls, dangerous projections in aisles).

ESSENTIALS

Construction is serious business and you want to build a structure that will serve you and your horses well for years to come. Don't try to cut costs for these essentials.

SITE PREPARATION. The site must be level where the building sits and well drained with grading so water flows away from the building. Know the percolation rate of your soil.

CONCRETE. Don't mix batches of concrete for large areas in a small, home-sized cement mixer. It takes too much labor, time, and electricity, and the result will be an inferior pad that looks patchy. Order cement by the yard and have it delivered by truck. One way you can save costs is by having your own crew of neighbors and friends help. (See under Strategies to Reduce Costs, Concrete [next page].)

LUMBER. Use 2x6s, *not* 2x4s, for stall framing and any other areas of the barn that horses will contact. Use rough-sawn (RS) boards, which are full-dimension boards. (A 2-inch RS plank is 2 inches thick; a 2-inch planed board is 1½ inches thick.)

STALL LINING. If the barn is metal, line stalls at least 4 feet up from ground level with 2-inch boards or a double thickness of ¾-inch plywood or similar material. Nasty wounds result if a hoof punches through a metal wall or if a pawing or rolling horse gets a hoof caught under the sharp bottom edge of a steel wall.

HARDWARE. Bolts, hinges, handles, latches and locks should be heavy duty. Many home latches and hinges are too light for use in a horse barn. If your horse breaks through them, he could get into the grain room and founder or escape to the highway and get killed. If you can't find heavy-duty hardware in a store or on the Internet, find a blacksmith or welding shop and have someone make what you need.

ANTICHEW STRIPS. Don't use drywall corners, which will last about 10 minutes. Instead, get 14-gauge or heavier angle iron from a metal fabricator or a sheet metal or welding shop, or buy ready-made antichew strips online.

FIRE SAFETY EQUIPMENT. Buy it, maintain it, and know how to use it.

LOW-PRICE PACKAGE DEAL. Don't take a package deal for a "lower price" if the layout is not what you want or need. Don't accept a building plan that features narrow aisles or doors or light-duty materials or hardware.

PLUMBING. Unless you have plumbing experience and access to wholesale plumbing supplies and tools, hire a professional to install water pipes, drains, sinks, washers, and wash racks. With plumbing, it is sometimes difficult for an amateur to do things right. There is much to consider: the slope of the drains, plumbing codes, what materials you can and can't use, depth of pipes, various means of planning for freezing weather. If you do it wrong, it is a big, expensive problem to correct.

ELECTRICAL. Hire an electrician. Don't skimp on the number of outlets, switches, or light fixtures inside or outside the barn.

STRATEGIES TO REDUCE COSTS

If you make decisions based on your needs rather than wants, work to stay within your budget, and are willing to do some work, it is possible to reduce construction costs.

MANAGEMENT AND PLANNING. Be your own contractor. Organize and coordinate the project yourself. Find out the required format for plans that need to be submitted to the planning commission's building inspector. Draw the plans yourself, hire someone to draw them to your specifications, or purchase preapproved plans from a barn builder. Find and purchase materials yourself, hire (and fire) workers, provide insurance, check quality of work, and keep the project on track.

STYLE OF BARN. Choose a barn design that is appropriate for your needs and budget. A run-in shelter is the least expensive option. A shed row barn is less expensive than a fully enclosed barn with central aisle.

OPTIONS. Added expenses include wash racks, hot and cold water, heated (or air-conditioned) insulated tack rooms, and restrooms. If you don't need them right away, plan to add them later.

LABOR. Ask if you can help prepare for or work with any crew that you hire. Most insurance regulations do not allow it, but you might be able to do prep work or clean up that could save the crew time at the beginning and end of each day, which in turn might save you some money.

CONCRETE. Find three to six helpers (two of whom are strong and experienced) and run your own concrete crew. All helpers should have good hand protection and tall rubber boots, and wear old clothes. As the concrete truck unloads, initially there will be heavy work moving and floating the concrete. After it is roughly leveled, there will be a lot of edging and finishing to do.

TRENCHING. Pipes and wires to the barn and under the barn must be buried a specific depth (as indicated by your county code). Rent a trencher and prepare the small ditches for your plumber and/or electrician ahead of time.

ELECTRICAL. Never tackle something as important as wiring unless you are experienced, because faulty wiring is a fire hazard. Check a book on wiring out of the library and study the county building code; if it seems doable to you, you might want to do some of the work yourself. If it sounds too complicated, hire an electrician to do the whole job.

You will need to hire a licensed electrician to make the connection from the power company's source to the barn site. If you want to do some of the work yourself, ask the electrician if he will be willing to act as a consultant to advise you regarding the materials you need for the interior and exterior wiring, including plugs, switches, and light fixtures, and to supervise your work.

In a barn, all exposed wires must be run through metal conduit pipe. Cutting the pipe and feeding the wire through can be time-consuming and costly in terms of labor. Your barn may have as many as one hundred switches, plugs, and lights, each of which must be connected. Working alongside an electrician or doing the work yourself and having the electrician check it can result in a significant savings. Electrical work must be inspected before the walls are sealed and the power is turned on, so be sure this is done before you call the county electrical inspector.

SHEATHING. Save labor costs by installing the sub-roof (plywood or insulation board) and sub-wall (insulation board) yourself. If you don't mind climbing around on a roof or handling 4-foot by 8-foot sheets of material, you can put the "skin" on your barn. Nailing these large areas can take a considerable amount of time.

MATERIALS. Pole barns are generally less expensive than post-and-beam, block, or frame construction. Trusses are less expensive than rafters and beams. Metal roofing is less expensive than shake roofs. The wall framing for non-horse areas (such as tack rooms, feed rooms, etc.) can be made of 2x4s and covered with inexpensive siding or paneling or not finished inside. To cut down on the number of windows for light, you can install fiberglass panels at the top of the walls (clerestory) where they join the roof and at the triangular portions of the gable ends.

FINISH WORK. You can do most of the finish work yourself, including landscaping, lining the stalls, laying stall flooring or mats, painting and finishing the tack room, painting and treating wood areas, and installing metal antichew strips.

CONSIDER ADD-ONS. To help the immediate budget, with carefully designed plans you might be able to build the barn in stages. Start with two stalls, a tack room and feed room, and plan to add two or more stalls later, or perhaps a wash rack or breeding area.

A barn with living quarters can be an economical dual-purpose building.

8 Interior

A horse barn can be simple or elaborate. The most expensive barn is not necessarily the best environment for horses, nor does it ensure efficiency and convenience of routine management. Decide what the main purpose of the barn will be, where you will be spending the majority of your time, what work areas need to be roomy and well equipped, and how many stalls you will realistically need. Spend time with a pencil and some graph paper sketching out your ideas.

Whether you are designing a new barn, remodeling an old one, or making a few changes in your horse's present stall, keep your horse's comfort and safety foremost in your mind. Considerations include durability, sanitation, and convenience of cleaning and feeding routines.

Stalls

Stalls are contrary to a horse's natural and preferred habitat, but a horse can learn to enjoy the comforts of a well-designed and well-managed stall. Make your horse's stall a safe and pleasant home. Doing so will help ensure that when you arrive to take him for a ride, your horse will be healthy, well rested, and in a good state of mind.

Horses are usually kept either in box stalls, where they can move about freely, or in tie stalls, where they remain tied. Tie stalls are only a little larger than the traveling space in a conventional horse trailer. Tie stalls are fine when separating horses at feeding time, but they are not the best choice for long-term horse housing. A horse cannot move around much in a 5-foot by 10-foot enclosure. Some horses are hesitant to lie down at all when their heads are tied to the manger. In order for a horse to lie down comfortably, the lead rope must be long, which then presents a potential hazard: the horse may get tangled in the rope. Box stalls, which are roomier and more comfortable, are popular for horses that are kept inside for training, showing, rehabilitation, or foaling.

BOX STALL SIZE

A 12-foot by 12-foot box stall is appropriate for most horses. The 144-square-foot area that such a stall provides seems to be optimal when considering the horse's comfort and the stall's maintenance. This size offers enough room for a horse to confine his defecation and urination to a certain portion of the floor and still have plenty of clean space for eating and resting. A smaller stall can result in a horse inadvertently churning manure into the bedding and his feed every time he moves. This means more labor for stall cleaning, greater waste of feed and bedding, and a greater potential for parasite infestation. Stalls much larger than 12 feet by 12 feet may allow more space

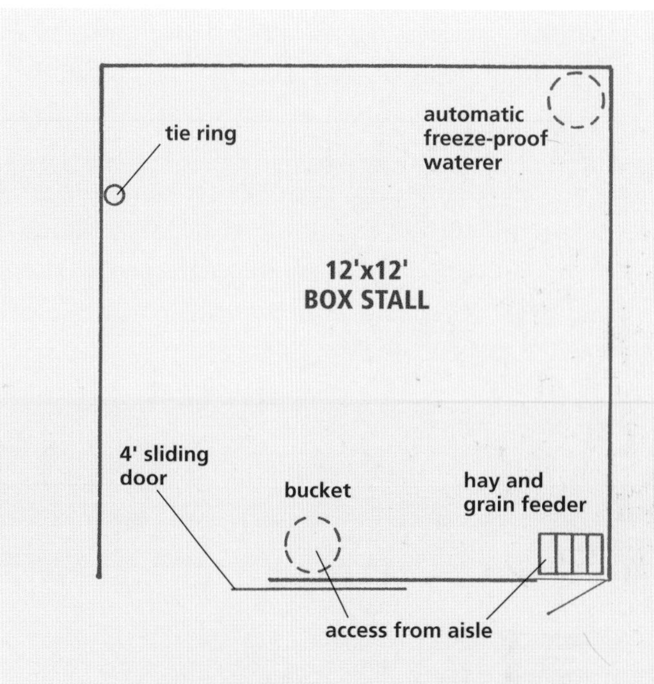

BOX STALL WITH TWO WATERING OPTIONS

Aria's stall has plenty of air and light. The window on the back wall can be opened for cross ventilation. Her halter and turnout sheet hang conveniently on the front of the sliding stall door.

between defecation, feed, and rest areas, but the cost of building space and bedding may be prohibitive.

An important aspect to consider when determining stall size is the behavior of each individual horse. It may be to your advantage when planning a barn to have a few 10-foot by 10-foot stalls for small horses or ponies or for the horse that defecates anywhere and everywhere and minces manure into the bedding regardless of the size of the living quarters. In addition, you may wish to consider a few oversized stalls for the very large horse, for the horse that rolls frequently, or for the rehabilitating horse.

The foaling or nursing mare requires a double stall. You can plan your barn so that two of the regular-sized stalls have a removable or hinged partition between them. This will allow you to use the space either for two single horses or for one mare and foal. You may want to locate the foaling stall in a position in the barn that allows observation, such as from a tack room. You might consider an isolation stall, separated from the others, such as for stallions or for quarantining incoming or sick horses.

Space requirements

ANIMAL	ANIMAL SIZE	BOX STALL SIZE	TIE STALL SIZE*
Mature animal (mare or gelding)	Small Medium Large	10'x10' 12'x12' 12'x14'	 5'x9' 5'x12'
Broodmare		12'x12'	
Broodmare with foal		12'x24'	
Foal to 2-yr.-old	Average Large	10'x10' 12'x12'	4½'x9' 5'x9'
Stallion		14'x14'	
Pony	Average	9'x9'	3'x6'

*Including manger.

STALL WALLS

Stall walls are often made of wood, metal, or cement block. They must be specially designed to withstand kicking, rubbing, wood chewing, and the rotting and corrosive effects of urine and manure. Be sure the interior, especially, is smooth, free of any projections, and durable and has no exposed wood edges. Even the end of a bolt protruding to the inside of the stall can be a potential hazard, because, as you probably know, if there is a way to get hurt, a horse will ultimately find it.

Sassy, in the weeks before foaling, enjoys a double stall: half deeply bedded with straw and half swept clean for feeding. The hinged solid stall divider is open and fastened on the right.

Far left: Stall walls lined with 2" thick lumber are safe and durable. *Left:* Any wooden edges that a horse might chew should be covered with metal strips (at least 14 gauge).

The lower 4 to 5 feet of a stall wall should be solid. If lumber is used, it should be a full 2 inches thick to withstand kicking, and this wall lining should have no spaces for legs to get caught in when the horse rolls. Unless lumber is ordered rough sawn (RS), it is planed so its actual dimensions are less than the dimensions by which it is ordered. An RS 2 × 8 is a full 2 inches thick, whereas a dressed 2 × 8 is only 1½ inches thick.

Most lumber for construction comes from softwoods, pine and fir being the most common. Fir is the stronger species but is harder to drive nails into, splits more easily, and will splinter, posing a possible hazard to horses. Two-inch pine boards would be fine for the stalls. Choose boards that are straight-grained, free of large knots (where breaking usually occurs), and not warped. White oak, a hardwood, is more expensive than pine but is a very strong wood. Plywood, which comes in 4-foot by 8-foot sheets, can also be used for the lower portion of stall walls if sufficient backup framing is provided and the total plywood thickness is at least 1½ inches.

The top portion of the front stall wall is generally made of mesh, pipe, or bars to allow the horse to see out and to ensure proper ventilation. Spaces larger than 2 inches between bars or in the mesh can be dangerous. A nibbling horse can get his teeth or jaws caught and inflict serious damage to himself. Because of confined horses' tendency to mouth and play with stall fixtures, all exposed wood edges should be covered with a chemical chewing deterrent and/or sturdy metal strips.

Because mesh or grill partitions allow horses to see one another, they sometimes play and fight. To discourage fighting between stalls, the dividing partition can be solid and a minimum of 8 feet high. The ceiling of a stall should be at least 11 feet high.

These 8-foot-high stall dividers are solid to 4 feet to prevent injury and damage from kicking. The bars on top enhance ventilation and illumination.

STALL DOORS

Stall doors should be at least 4 feet wide and hinged or sliding. Hinged doors that open outward can block the alley, making it awkward to take a horse in or out unless the aisle is absolutely clear. Hinged doors that open inward crowd both the horse and the handler and can cause either to get wedged if the door should happen to catch on horse or handler on the way out. Dutch doors, those composed of two half-doors, are traditionally hinged to open outward and are most common on the outside wall of the barn. They should be able to be opened 180 degrees so they can be securely fastened flush against the wall. The top can be fitted with a wire-mesh panel if the horse lunges at passing horses or handlers. Otherwise, it can be fastened in its open position so that the horse can put his head over the lower door and see other horses and perhaps get some fresh air and sunlight. Sometimes a horse will develop a habit of leaning on the lower door, which can be damaging to the hinges and latch.

Our sliding stall doors have blankets bars with halter hooks. A horseshoe stall latch such as the one shown should be used only where it is out of reach of the horse or he could injure his jaw on it.

Sliding doors are convenient and space efficient because when open they fit closely along the front of the stall wall. Therefore, they are suitable for the aisle side of a stall. It is important to be sure the sliding door is secured at both ends of the bottom when closed to keep the bottom of the door from being pushed outward when a horse rolls against it. Sliding doors are also available with a Dutch door–type feature. A drop panel in the top of the sliding door will allow the horse to put his head out. Whatever type of door is chosen, it should be a minimum of 4 feet wide and at least 8 feet tall. Stall doors that allow a horse to put his head out should have two latches, one at the bottom of the door, out of the horse's reach.

THE STALL FLOOR

Because a horse can produce up to 50 pounds of manure and 10 gallons of urine daily, a good deal of thought must be given to the stall flooring and bedding. As you make your decision, compare the initial cost of installation versus the durability and longevity of each product. Weigh that along with the margin of safety and comfort for the horse provided by each type of flooring. And finally, be sure to consider what type of bedding you plan to use, as some flooring-bedding combinations work well and some can be undesirable. (See chapter 15, Sanitation.)

The Subbase and Base

If the barn site has been properly excavated and prepared with well-draining materials, the stalls will function best. All you'd need to do is add at least 2 inches of base material, tamp it well, and apply your flooring.

If a barn site has not been properly prepared or if the barn has been in use for some time and the soil under the stalls is no longer draining well, excavate the soil in the stalls at least 6 to 8 inches, taking note of the type and smell of the material you uncover. Depending on what you find, you may be able to use the native material as is, disinfect and dry the material before reusing it, mix the material with some gravel, or replace it with about 4 to 6 inches of new subbase material. Something like 1½ to 2 inches of gravel works as a subbase. Top it with 2 inches of a

Stone dust

Stone dust is known by various names according to the locale:

Screenings	Decomposed granite
Crushed stone	Crusher dust
Crushed limestone	1/4" material
Blue stone	

base material. Dampen the base material, then tamp with a hand tamper or a mechanical compactor from a tool rental store.

Stone dust (see other names in the box) works well for the base. Don't use topsoil, sand, or gravel for the base. Topsoil tends not to drain well and encourages mold growth. Sand and gravel are too shifty. Some commercially prepared products that are used for sports fields (such as baseball diamonds) may be suitable for base material.

Flooring

Stall flooring must be comfortable and safe for your horse, easy to clean and disinfect if necessary, and dust-free, and work well with the type of bedding you plan to use. Avoid flooring that requires constant maintenance, mixes in with your horse's feed, and absorbs urine or allows urine pooling.

Tamped clay. Tamped clay was a longtime flooring favorite for stalls because it provides cushion and good traction and is warm and quiet. However, clay does not percolate well, and stall floors must slope to allow drainage. In addition, urine pools soon become potholes of enormous proportions, requiring that the clay floor be leveled routinely and overhauled annually. Two to 4 inches of the original 6 to 12 inches are removed each year and replaced with fresh clay and retamped. Pure clay may be difficult to buy in some areas, so if you decide to go with clay, you can extend the life of the stall by allowing it to rest periodically until dry. If you have at least one extra stall in your barn, rotating horses will allow one stall to be empty at all times.

Mixtures of clay and crushed rock. Mixtures of clay and sand or crushed rock may be more readily available than pure clay and will have improved drainage while retaining most of the clay floor's desirable features, except that the mixtures tend to be shifty, which results in bedding and feed being mixed in with the flooring. Road base in your locale may be such a mixture — a blend of crushed limestone and clay. These blends may result in better sanitation and comfort than clay, but because they are soft, they invite pawing. Generally, these materials are better used as a base for other types of flooring, such as rubber mats.

Concrete. Concrete makes a permanent, low-maintenance floor that is fairly easy to sanitize. However, it requires very deep bedding (more than 12 inches) because it is hard, cold, and abrasive. Concrete floors must be designed with proper slope for drainage and should be finished rough or scored to ensure good traction.

Asphalt. Asphalt has the same drawbacks as concrete and is not as durable.

Wood. Wood, an old-time favorite for tie stalls, is not appropriate for box stalls. Although it is warmer than concrete and fairly durable if appropriate wood is used, it can be slippery, is difficult to sanitize and deodorize, and can be noisy under a nervous horse. Because of its hard surface, it requires deep bedding to provide comfort and prevent sores.

Rubber stall mats. Rubber stall mats can be thought of as part flooring, part bedding. The mats are usually 5/8 to 3/4 inch thick and 4 by 6 feet or 5 by 7 feet and made of a combination of rubber, clay, nylon, and rayon. Some mats are made of recycled rubber tires. If the rubber particles are revulcanized, it means the material has been remolded with heat.

Stall mats act as an intermediary between the soil and the bedding. In this way they prevent horses from ingesting dirt or sand with feed they eat off the stall floor. Mats have superior cushioning for comfort, can be easily sanitized, make stall cleaning easy, decrease dust so horses stay healthier and cleaner, and decrease the amount of bedding required by up to half. They also prevent a pawing horse from digging holes in the

Flooring layouts

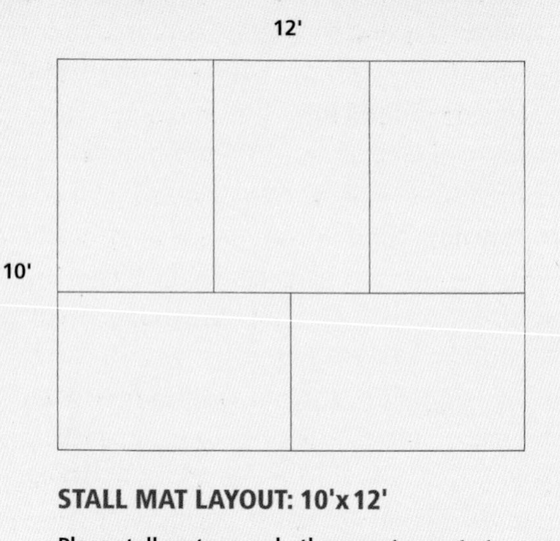

STALL MAT LAYOUT: 10'x12'

Place stall mats so only three mats meet at any one junction

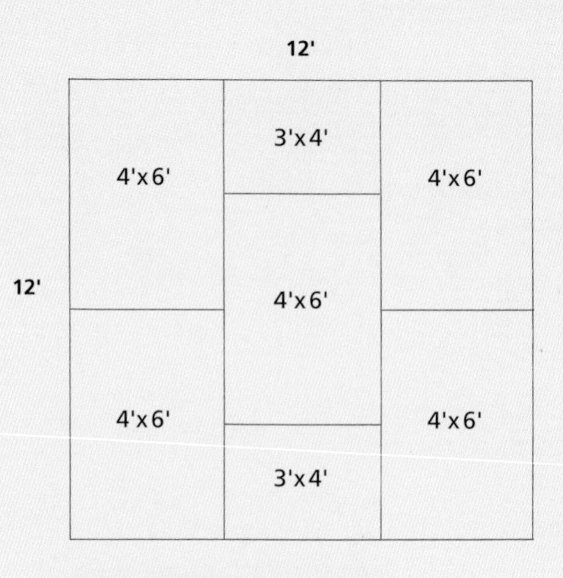

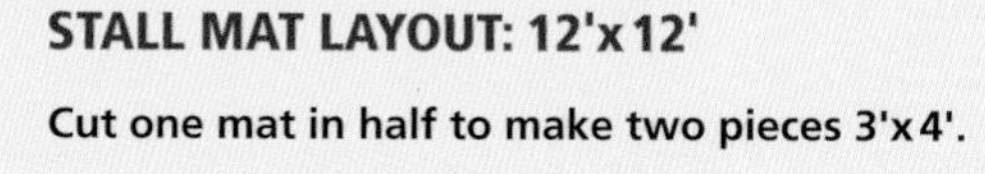

STALL MAT LAYOUT: 12'x12'

Cut one mat in half to make two pieces 3'x4'.

stall. Mats with textured top sides provide added traction. The seams between mats can become a source of trouble if feed, bedding, or manure works between and under edges and results in bulging and curling. To prevent this, cut and fit mats tightly and/or use interlocking mats with puzzle-piece edges, which allow you to fasten the mats together, making one solid floor.

It is important to note that stall mats work best if they are topped with about 8 inches of bedding and the damp and soiled bedding is removed regularly to prevent drainage between the seams or along the edges down to the base, where the urine could pool and cause an odor problem. Stall mats are initially expensive but an investment well worth their price.

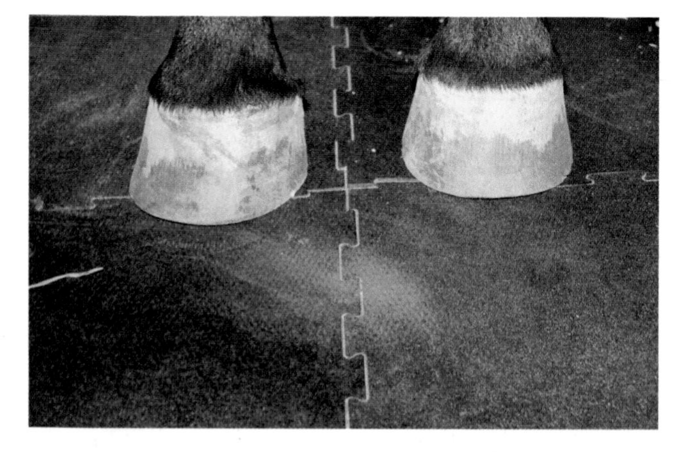

Interlocking rubber stall mats reduce problems with bulging and curling. As with any mat, the base must be tamped and leveled for best results.

Old conveyor belts don't make suitable stall mats because they are too hard and slippery. They would be better put to use for a stall lining or pen windscreen.

Drainage flooring. Grid-style flooring has holes of various sizes in it to allow urine to drain through into the base and subbase. Therefore, consider drainage-style flooring only if your barn is on a very well-drained site that is engineered for such use, or use it in stalls that have drains. Grid flooring is available in large sheets that are rolled up like linoleum, as 1-by-3-foot sections or 1-foot by 1-foot tiles.

Stall mattresses. Stall mattresses have been used in the dairy industry for years but have only recently been modified for use in horse stalls. Usually, a stall mattress system consists of two parts: a mattress and a cover. The mattress is a thick rubber cushion filled with recycled rubber. The cover is made of latex or urethane-treated rubber and fastened to the side of the stall with plastic strips. Plenty of absorbent bedding must be used to soak up urine. If a cover needs to be sterilized, it can be removed, washed, sanitized,

Relative cost of floorings

Note: X is used for relative comparison; stall is 12'x12'.

Clay = $X	Grids = $5X
Wood = $3X	Mattress = $5X
Rubber mats = $4X	

and dried within a day. Although the covers are designed to withstand shod hooves, they can tear, which then lets urine between the cover and mattress.

When purchasing stall flooring, note the type of warranty that comes with it. A limited warranty might not cover damage from pawing or hooves. A pro rata or prorated warranty decreases over time. A warranty that covers wear and tear is better than a workmanship warranty.

If flooring needs to be disinfected, ask your veterinarian what you should use and check with the flooring manufacturer to verify that it is safe to use on the flooring. Nolvasan is usually safe.

STALL DETAILS

Ventilation should have been considered in the overall plan for the barn, but certain stall features will help ensure that horses get adequate airflow without drafts. If part of the stall front is fitted with bars or mesh and a 2-foot by 2-foot window is located on the back wall of the stall, the stall will be set up to take advantage of additional light and warmth from the sun or cool breezes, depending on the season and time of day. All windows should be covered with heavy wire mesh or close-fitting bars on the stall side for safety. Translucent panels can be used in the roof to increase natural light and take advantage of heat from the sun. Unless the panels are fitted with shades, however, this may not be the best choice for the summer season in a very sunny or hot climate.

The feeding and watering features of a stall should be convenient for the manager and should allow the horse to eat and drink easily and safely. Feeders should be located for easy filling and have a capacity of up to 20 pounds of hay. It is best for your horse's health if he can eat from a feeder at ground level. Feeders higher than the withers often cause dust or leaf particles to fall into the horse's nostrils or eyes and can cause respiratory problems and clogged tear ducts. Hay nets can be a good temporary way to feed hay, but care must be taken to tie the hay net so that when it is empty it does not hang dangerously low where a horse can get a leg caught. As an added safeguard, hay nets can be fitted with a breakaway fastener so that if the horse does become entangled, he will not hang himself. Feeding hay at ground level on stall mats is an excellent natural choice, as long as the mats are swept periodically to remove tracked-in dirt and sand.

Grain fed in shallow tubs on the ground is a natural way for horses to eat, but some horses tip the tubs over and spill the grain. A good solution is to feed grain in a shallow, corner-mounted grain tub that is located at a level about two-thirds the height of the horse or about 38 to 42 inches off the ground. Feeding grain in buckets is OK, but horses tend to gulp when fed grain in a deep bucket, and buckets tend to be a toy that horses chew on, rub on, bang against, and tip over.

Since a horse drinks from 8 to 12 gallons of water per day, it is essential to find an efficient watering system. A water bucket system probably will not work well, unless you use a 3- to 5-gallon bucket and fill it two or three times per day. The water supply will stay cleaner if it is located away from the feeding area. If you have turnout pens attached to each stall, you may want to locate the water out there in barrels, tubs, or troughs.

Automatic waterers are much more convenient than buckets, but they have drawbacks as well. They are expensive and should be installed while a barn is being built, not afterward. You never really know if a horse is drinking and, if so, exactly how much. However, some new models can measure consumption. Also, with some models, horses learn how to keep the water running and flood their stalls. In cold climates you'll need to outfit automatic waterers with heating accessories to keep them from freezing. These waterers are safest if they are round and set in a corner. If they are square, the corners should be covered with a protective edging, or, better yet, the entire waterer can be installed flush with a wall. Check automatic waterers daily to be sure that they are clean and functioning properly.

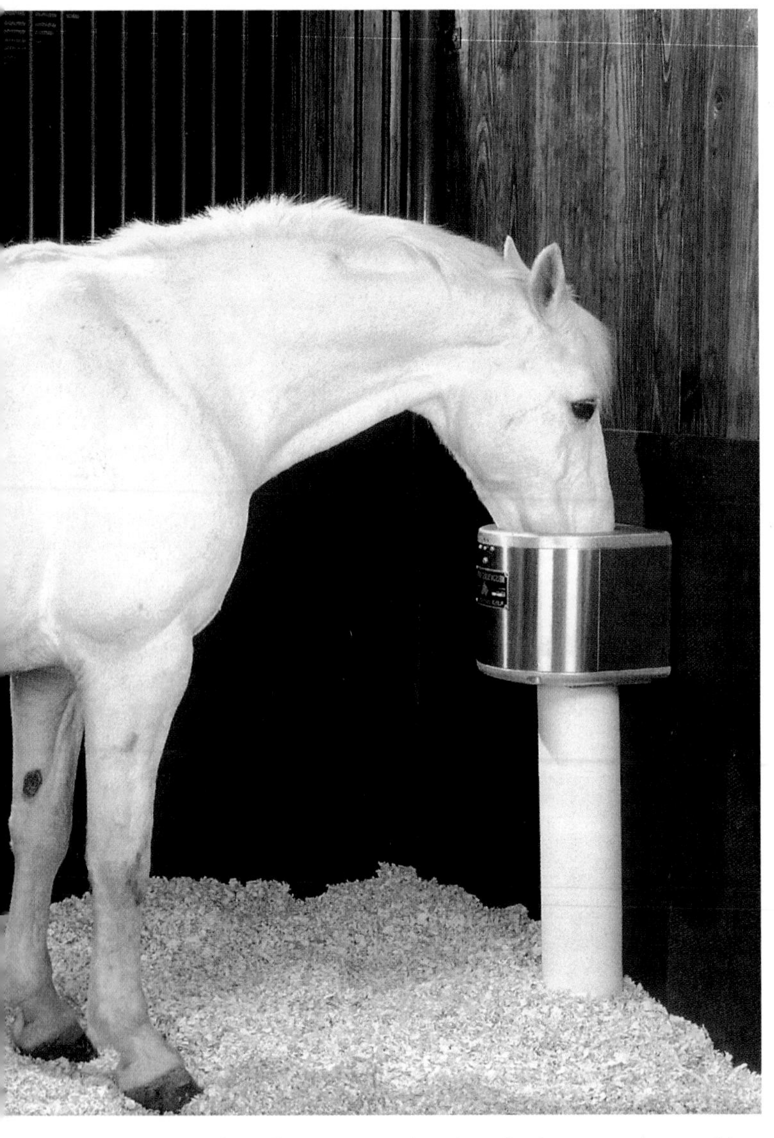

Automatic stall waterers, when functioning properly, provide a horse with a constant supply of fresh water, which reduces daily labor. However, potential problems include frozen pipes, flooded stalls, and no water.

Tack Room

Because a tack room is often the hub of human activity in a barn, it should be roomy and well designed. If you are building a new barn or remodeling an old one, first consider the main purposes your tack room will serve.

Depending on the scope of your horse operation and the availability of other buildings and rooms, a tack room can end up being like a feed room or lounge. Limiting the purposes for which it is intended will help keep a tack room neat and functional. Do you need an area to organize your everyday working gear? To arrange your grooming and medical supplies? To safeguard your records? Do you need a place to store winter blankets? Show saddles and bridles? Do you require an area for laundry, saddle cleaning, and repair? Does your tack room need to double as an office or trophy showcase?

A tack room should be well organized, dust-free, insulated, well ventilated, dry but not hot, rodent-free, and secure, and provide plenty of storage space. If you can determine exactly how you will use the room, you can estimate the optimal size and begin your floor plans. Within reason, make the room as large as possible. Rooms smaller than 8 by 10 feet (the size of a small box stall) seem crowded as soon as a few saddle racks and tack trunks are moved in. If you have from two to ten horses, try to allow between 100 and 200 square feet for a tack room in your barn plans. Using a piece of graph paper, draw the proposed floor plan to scale, noting the placement of large items such as saddle racks, tack trunks, a worktable, and a sink. A 10-foot by 22-foot (approximate inside dimensions) room can be substituted for two 12-foot by 12-foot box stalls in most barn plans and results in a very useful space. For a larger barn, two or more smaller tack rooms in several locations may be more convenient and efficient.

A tack room should be located very near the grooming and saddling area. Consideration should be given to airflow between the two locations so that dirt, hair, and sweepings from the grooming area are not automatically sucked into the tack room. The

Tack room layouts

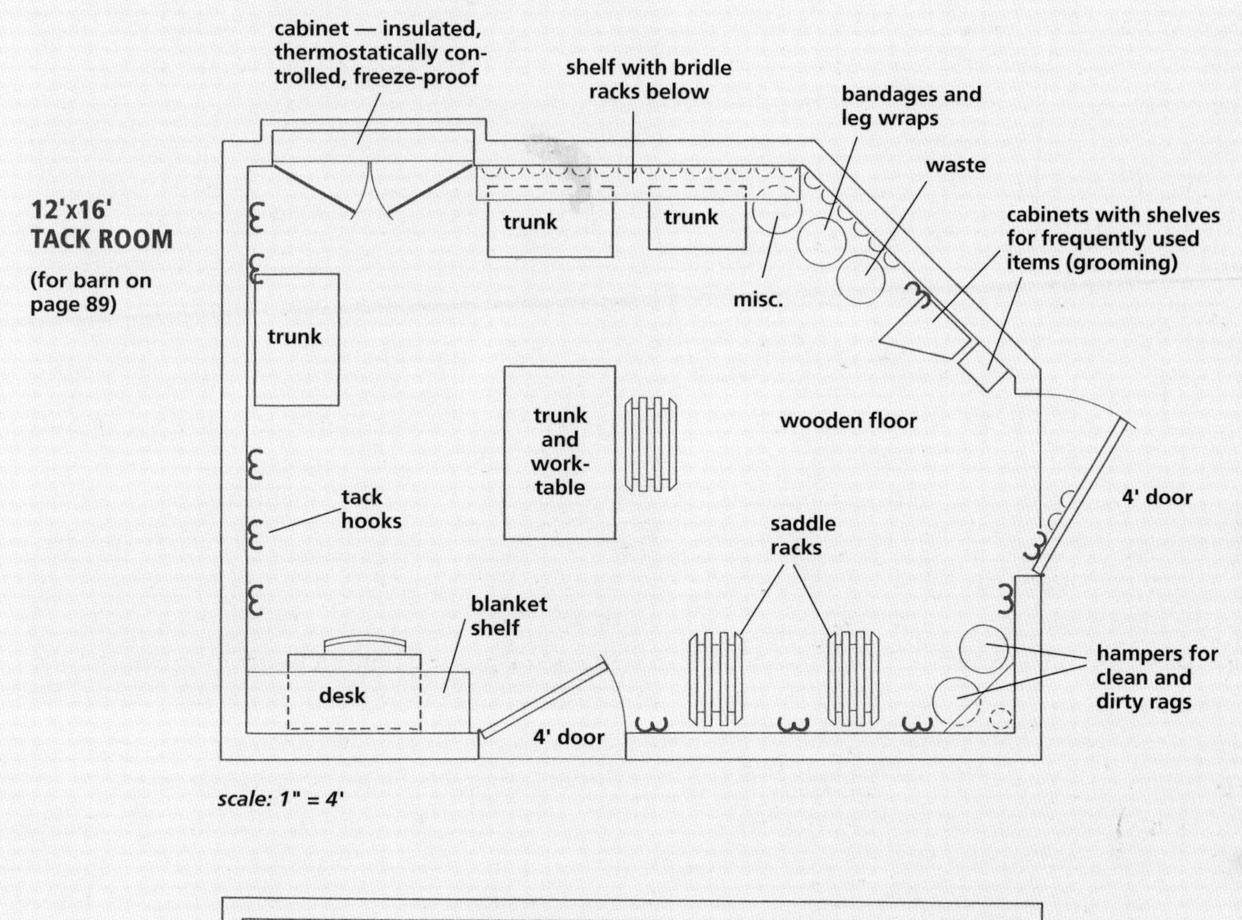

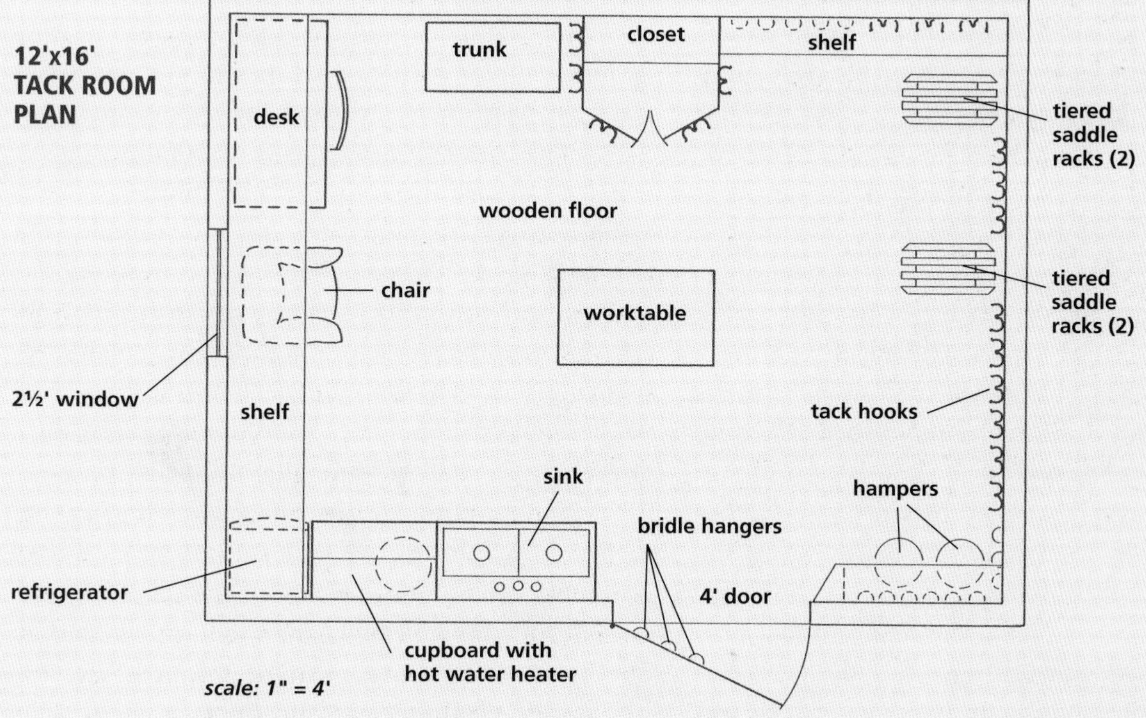

doorway between the grooming area and tack room should be at least 4 feet wide to accommodate a person carrying a Western saddle. All doors should be fitted with strong, durable locks to prevent theft and to satisfy insurance requirements.

Whether you do the construction yourself or hire someone else to do it, the work should be professional and comply with local building codes. After the tack room is framed, and before the walls are covered, the wires for the electrical outlets and lights need to be placed. Outlets should be plentiful, one every 6 feet. Locating some receptacles about 4½ feet from the floor makes their use more convenient when trunks, racks, and boxes line the walls. Adequate overhead light fixtures are necessary to ensure there are no dark corners. One central light in a 120-square-foot room may not be sufficient, as most enclosed ceiling light fixtures are limited to 60-watt bulbs for safety reasons.

Because leather goods are best kept at moderate temperatures and low humidity, the walls, ceiling, and floor of a tack room should be insulated. Even without a heater or air conditioner, insulation has the ability to keep the indoor environment more constant. Some climates may require the use of a dehumidifier to keep mildew from forming on the leather during warm, wet weather. During winter months, a small space heater will keep the chill out of the air and prevent medicines from freezing. Be sure the heater is safe and does not present a fire hazard.

In areas with very cold winters, designing a freeze-proof cabinet may be better than heating the entire room. An insulated, heated cabinet requires far less electricity than is needed to heat the entire tack room and keeps the assortment of veterinary supplies from piling up on the back porch of the house. A small, safe electric heater or large lightbulb with a thermostat control set at about 40°F will prevent ointments, oils, aerosol products, and other items from being destroyed by freezing. Setting the thermostat at slightly higher temperatures will keep salves softer for more convenient application.

Although windows can provide a desirable airflow, the sun's rays shining through glass can be very destructive to leather. If you want windows, those in direct sunlight should be outfitted with shades. Exterior windows, however, unless fortified with bars, decrease the security of the tack room. Most tack rooms are satisfactory without windows.

The corner swinging-blanket-rod system in my tack room holds many saddle pads, coolers, sheets, and blankets.

Just because the room is being built in a barn does not mean the carpentry should be casual. Tight doorways and precise fit of walls to ceiling and floor will prevent infiltration of dirt, bugs, and rodents, all of which can be damaging to tack. The material for the floor should be durable, water-resistant, and easy to sweep. Dirt floors are a constant source of dust, prevent tight wall-to-floor fit, and defeat the purpose of insulating the rest of the room. Cement floors, although easy to keep clean, are very cold. An insulated wood floor is a good choice.

A boot scraper or mat, placed outside the door to remove mud and snow before entering the tack room, will help to preserve the floor.

Above: I like contoured wooden bridle holders because they help bridles keep a good crown shape. *Top right:* These horseshoe halter holders are deep enough to hold four or five halters each. *Right:* A bit hook is a great way to keep extra bits organized. *Far right:* Rope hooks help prevent tangles.

A large amount of wall space is required for hanging equipment, especially bridles and halters. Bridle holders with contoured crown pieces that approximate the configuration and size of a horse's poll area help to keep bridles in good shape. A half-circle of wood about 4 inches in diameter and about 2½ inches thick works well. A ¾-inch lip on the forward edge of the curved surface will keep the bridle from slipping off the bridle holder. Mount the wooden bridle holders 9 inches apart, center to center, to provide ample room for convenient use.

Other gear, such as running martingales, ropes, extra cinches, and nosebands, can be hung on hooks. Many types of commercial hooks are available, and custom hooks can be fashioned from, among other things, old horseshoes. Figure how many hooks you think you will need, then double the number! It is Murphy's Law that the pair of reins you want is always at the bottom of a pile of entangled leather and nylon straps. Outfit your tack room and grooming area with plenty of hooks.

Saddle racks can be freestanding or built-in. The former style is essential for the compulsive rearranger. Freestanding units are useful as saddle-cleaning stands and can be relocated near the heater in January or carried outside to a shady spot in June.

Built-in wall-style saddle racks will allow you to store a stack of saddles in a small space. Mount the racks with ample room in between — more than seems necessary, especially if you use Western saddles. Multiple saddle racks, two or three high, make efficient use of space, but Western saddles are difficult to lift onto a rack 6 feet above the floor.

Wall-mounted saddle racks are great space savers. The slatted design allows the underside of the saddle to air out between rides.

No matter what type of saddle rack is chosen, be sure it satisfies the following requirements. The rack should allow air circulation; the bars or slats should be well spaced so that the sheepskin or panels and stirrup fenders and leathers can dry after hard use. The rack should be designed to approximate the contours of the horse's back. Although some horse owners may think their horse looks like a 55-gallon drum, using one for a saddle rack prevents the saddle from drying and spreads it out in an unnatural shape. Not many horses' backs are shaped like logs, either. A saddle rack that supports the front jockeys and rear portion of the skirts without causing them to become misshapen is ideal.

Wet saddle blankets are often set to dry by turning them upside down on top of the saddle. This might work well during hot, dry weather or if the blanket does not need to be used again soon. But the smell can become rank quickly inside a tack room filled with wet saddle blankets. Faster drying time and a nicer environment can be achieved by locating a blanket-drying rack outside the barn where it can take advantage of the sun and/or air currents.

Tack trunks are handy for storing items that are not used frequently or for seasonal items such as winter blankets. Trunks take up a lot of floor space, however, so it is best if they can double as seats or short-term work areas for simple repair jobs or small cleaning jobs. If you can afford it, plan to put a washer and dryer, sink, and small hot-water heater in your tack room so that you can do your tack cleaning frequently and conveniently. Also, be sure to have a fire extinguisher in your tack room and know how to use it.

Right: **This large wide trunk is ideal for storing winter blankets. When closed, the trunk makes a good seat or workbench.** ***Far right:*** **A washer and dryer come in handy for cleaning sheets, leg wraps, saddle pads, blankets, and towels.**

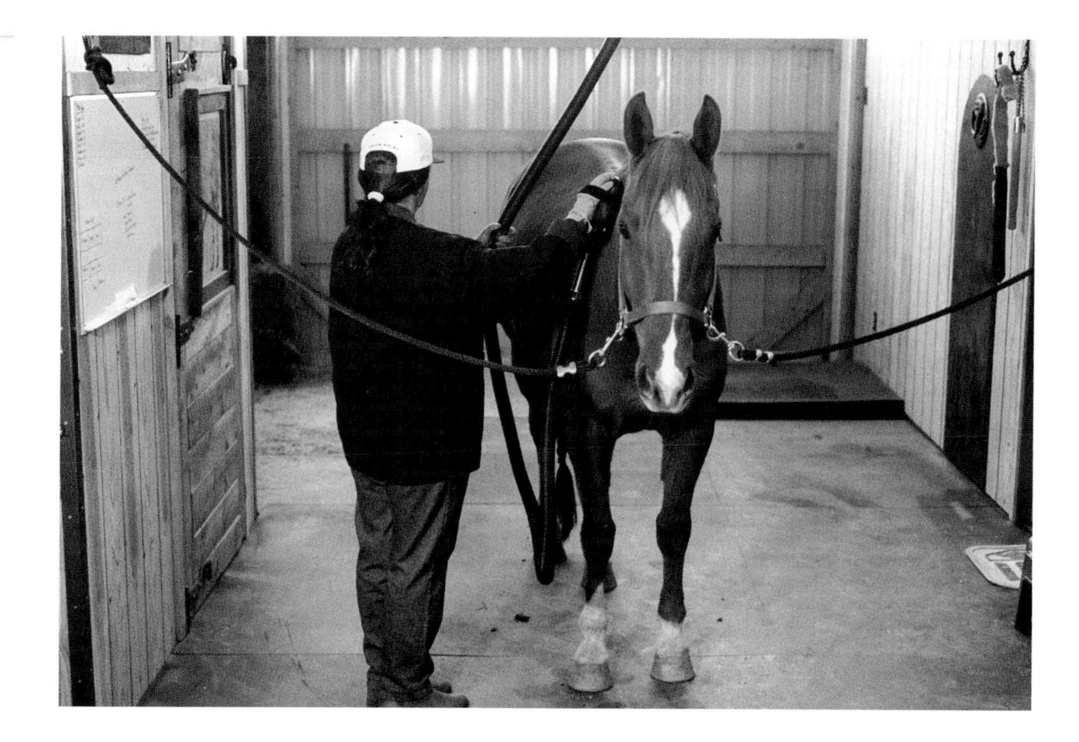

Dickens enjoys being vacuumed. The central vacuum canister is mounted on an exterior barn wall where it is vented outside. The hose is suspended over the grooming area. The aisle is 10½' wide; 5" diameter tie rings are mounted 80" from the floor and installed to wall studs with 3"x⅜" lag bolts. There are two 5' cross-tie ropes.

Shelves, cupboards, and cubbyholes will help to remind you to return an item to its spot. This way gear stays cleaner, lasts longer, and is there when you need it. Bins work well for small supplies such as repair items, bandages, protective boots, spurs, and gloves.

It is convenient to use the tack room for veterinary, farrier, and training records. A corner of the room can serve as an office. Outfit a small desk with a good light, your files, pencils and pens, a calendar, and a coffeepot!

This tack cleaning and repair area is near the sink in my tack room. Supplies are organized in bins with transparent fronts.

Work Areas

Many horsekeeping tasks are performed indoors, so plan for safe, spacious, comfortable work areas for grooming, tacking, veterinary care, and farrier care.

TACKING/GROOMING AREA

Just outside your main tack room door should be an open area specifically set aside for grooming, tacking, and clipping. The floor should be a level, nonslip surface, and there should be a safe place to tie. Whether you use a post or tie ring on a wall or cross ties for tying is a personal preference. Whichever you choose, it should be well designed and sturdy. If you use a tie rail, ring, or post, it should allow you to tie the horse at or above the level of the withers. If you choose a cross tie, locate the rings above the height of the horse's head so the cross-tie ropes angle down to the cheek pieces of the horse's halter where they attach.

There should be plenty of conveniently located receptacles for clippers, vacuums, extra lights, and a

Above: An aisle wide enough for a pickup or tractor but too wide for cross ties. Rubber mats are set flush with the concrete floor down the center of the aisle. *Top right:* A tie ring in the stall mounted 72" from the floor confines Sassy so her foal Sherlock can be handled. *Right:* Zinger stands under the roof of the loose stocks: 68" wide by 90" long with multilevel tie rail and rubber mats.

radio and hooks for gear. Add a shelf or bins for your most commonly used grooming tools and supplies. The shelves should be located where the horse cannot knock down items with his mouth or tail, especially during clipping and shoeing.

AISLES

Aisles should be a minimum of 8 feet wide, but 10 to 12 feet is better. Any wider is a waste of space and can allow a horse to turn around in the cross ties. In the aisle there should be safely designed and well-located hooks for halters and lead ropes, and storage racks for blankets. Be sure not to hang any items where a horse can reach them.

TIE AREAS

Your barn should have plenty of safe places to tie horses: at hitch rails or posts, in cross ties, in loose stocks, and in stalls. An outside tying area is a handy training device, allowing you to teach a horse that he must stand relatively still for several hours at a time. This teaches a horse to accept restraint and be patient, and it will be reflected in his attitude about many other activities. The tie area should be strong and tall enough that you can tie the horse at the level of his withers or higher so that if the horse does pull back, he will not be able to get very good leverage.

Loose stocks limit sideways movement of the horse that is tied or cross-tied at the head of the stocks. Such an arrangement is ideal for young horses or for cooling out any horse after work. The loose stocks are also handy for veterinary work, such as backing the horse into the stocks so that the veterinarian can float the teeth. Ideally, there should be a roof over any tie area for the comfort of the horses and the people working with them.

FARRIER WORK AREA

Be sure there is an area in your barn for your farrier to work that is protected from rain, wind, and direct sun. A farrier works out of his truck, so his work area should be in a place that he can back up to. Often the cross ties of a grooming area or a covered hitch rail work fine. The floor should be level, smooth, and uncluttered; have good traction; and be kept free of gravel and other debris. It's best if the area is lighted from the sides or corners, rather than from overhead, so the farrier's shadow won't shade the hoof that he's working on. Most farriers use electric tools, so access to a 110-volt outlet is essential.

VETERINARIAN WORK AREA

A veterinarian, like a farrier, needs to get his or her truck close to the work space in the barn. Many veterinary procedures require water, so convenient access to hot and cold water is a big plus. Plan a counter or worktable near the vet area for setting down tools and keeping supplies at hand without having to put them on the floor. The floor in the vet area should be solid, have good traction, and be sloped to drain water away from the work area so the vet doesn't have to stand in water. Good lighting is especially important for vet work. Besides ample permanent lights, portable clamp lights or light stands should be available to spotlight certain areas on the horse. Plenty of outlets installed in convenient locations will make it easier for the vet to use clippers, an ultrasound machine, and other equipment.

WASH RACK

If you have space to include an indoor wash rack, you'll find that bathing and other horse care will go more smoothly. Wash rack flooring needs to be solid and have good traction. The floor should slope toward a drain so you won't be standing in water as you bathe your horse. Make the walls waterproof so the wall framing doesn't rot. Place light fixtures so light comes from the sides, not centered overhead, and choose fixtures, outlets, and switches that are weatherproof. Outfit the wash rack with hot and cold water and an overhead infrared heater so you can safely bathe your horse during cool weather. Locate the wash rack where it is protected from breezes so a

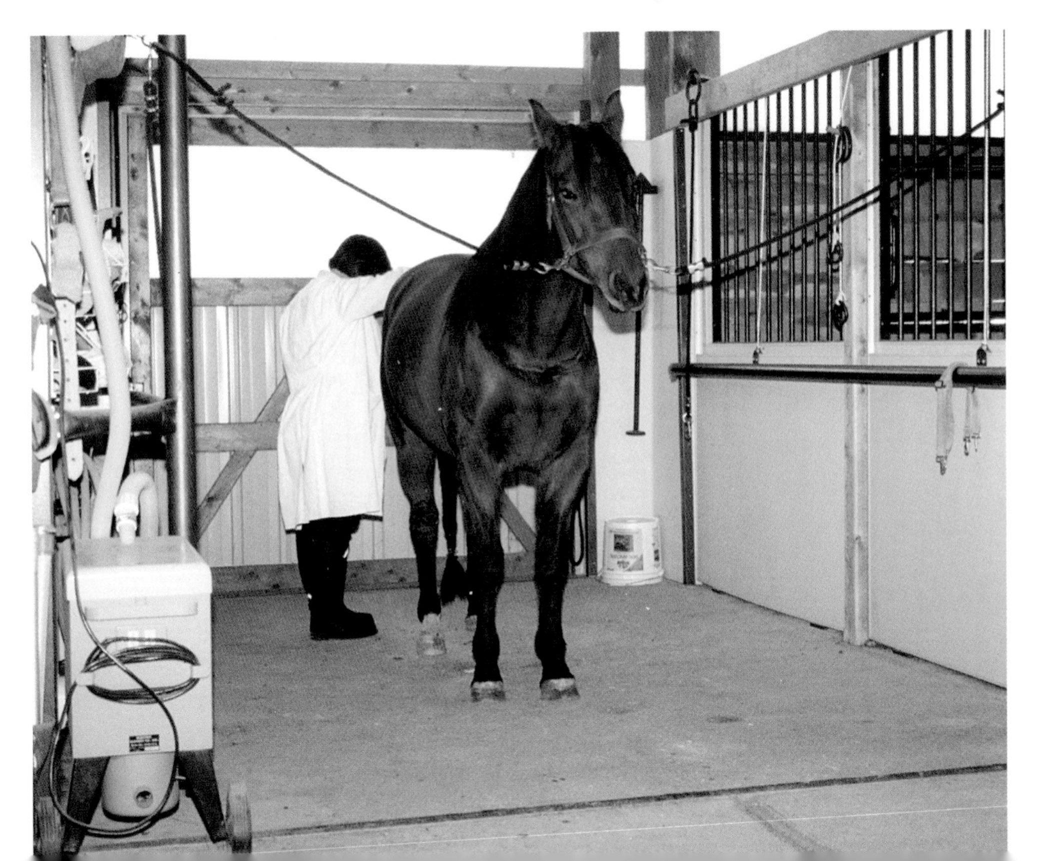

The wash rack doubles as a veterinary work area. Here, Seeker is cross-tied while her temperature is taken.

This wash rack, with its smooth walls and heat lamps, is a safe, comfortable place to bathe a horse.

wet horse doesn't get chilled. Because some horses can move quickly and knock into things when startled by water spray, there should be no protrusions such as water faucets that could injure a horse or be damaged by a horse. If there is no room inside the barn, a wash rack can be located outdoors, but it will be limited to use during warm weather only.

Storage

To keep your barn tidy and the aisle clear, design storage areas when you plan your barn. Allow enough room for a week's supply of grain, hay, and bedding. Decide what kinds of tools and equipment you'll use frequently, and plan a spot for them.

FEED AND BEDDING STORAGE

Although the bulk of your hay and bedding should be stored in a separate building, you should have a small feed room in the barn as well as space for a week's worth of hay and bedding. Be sure the feed room is separated from the stalls and has a horse-proof lock on it. Laminitis (founder), one of the most debilitating diseases of the horse's hoof, often begins with a grain overload from a horse getting into the feed supply.

The feed room should also be rodent-proof, dry, and well ventilated. Concrete floors are ideal here. Grain should be stored in covered containers, such as large garbage cans. A 30-gallon garbage can holds 100 pounds of corn; a 50-gallon garbage can, 100 pounds of rolled oats. Build a work shelf or table to serve as a place to store buckets and measures and to weigh and mix up rations. Outfit your feed areas with both a hay scale and a grain scale.

Bottom left: **Rodent-proof feed bins can be used in lieu of a feed room.** ***Bottom center:*** **Garbage cans with tight-fitting lids will work for feed storage as long as a horse can't get at them.** ***Bottom right:*** **Storing bags of bedding in a stall temporarily is fine but not long term, as it is a fire hazard.**

A wide counter makes mixing rations convenient. A regularly updated feed board helps to ensure that a horse gets the same ration each day, no matter who feeds.

Feed bins in lieu of a feed room can work if they are rodent- and horse-proof, but they are sometimes difficult to work out of. Storing feed in garbage cans in the aisle is risky. I don't know too many horses that couldn't easily remove a garbage can lid. Filling an extra stall with hay or bedding may be convenient, but it does add fire risk.

TOOL STORAGE

Plan a place in the barn that is exclusively for tools and extra equipment. This will help keep the main barn aisle uncluttered. The tool area should be away from traffic areas and out of reach of horses to prevent injury to horses and damage to tools. Here is where you can store brooms, forks, a shovel, rake, cart or wheelbarrow, manure basket, stall freshener, and other stall-cleaning equipment. The tool room is also a good place for a small tool kit, a flashlight, and a fire extinguisher.

Water

Besides the stall waterers and the tack room sink, additional hydrants and faucets in and around the barn are convenient. Hydrants should be self-draining and frost-free and located in a safe place. Pipes leading to the stable should be buried 3 feet or more, depending on the depth of the frost line in your locale. One of the places you will want water, and hot water if you can afford it, is the wash rack and/or laundry area (often located in the tack room). The washing machine can empty into the wash rack drain or into a septic system if you were required to install one for a barn lavatory. If there is no septic system involved in your barn, the drains from sink, wash rack, and washing machine can lead to a dry well, a large hole filled with gravel that is used to dissipate water into the soil. A dry well is much cheaper than a septic system, if your building code will allow it.

Far left: **A place for everything and everything in its place makes for not only a tidy barn but also tools that will be there when you need them.** ***Left:*** **This barn-aisle hydrant has a gravel splash and drain box below it and a hose hanger that helps keep coils tidy.**

9 Outbuildings

Depending on the size and scale of your operation, you may find that you need buildings, in addition to a barn, for your farm or ranch to function properly. As you develop your property, you might want to consider adding a farm equipment shelter, a hay barn, loafing sheds, or other specialized buildings.

Hay Barn

Due to increased risk of fire and the health hazards associated with dust continually falling or blowing into stalls, the traditional hayloft over stalls is no longer recommended. Storing more than a few bales of hay in a horse barn simply isn't safe. That's why it's best to keep housing for horses separate from the storage of feed and bedding.

The hay barn should be located at least 75 feet from the horse barn for fire safety, yet close enough for convenience in transporting weekly supplies of hay, grain, and bedding to the feed room in the barn. To determine your hay storage needs, figure you will need between 3 and 4 tons per horse per year if you feed hay year-round and do not rely on pasture for supplemental feed. A ton of hay, 2000 pounds, is usually composed of about thirty 65-pound bales, thirty-six 55-pound bales, or forty-four 45-pound bales. A ton of hay requires approximately 200 cubic feet of storage space, but fluffy, lighter grass bales will require more space per ton than tight, heavy alfalfa bricks. Two hundred cubic feet of storage space would be a space approximately 10 feet by 10 feet and 2 feet high, or 5 feet by 5 feet and 8 feet high, or 6 feet by 6 feet and 6 feet high.

Storage space

ITEM	SPACE REQUIRED
Baled alfalfa hay	150–200 cubic feet per ton
Baled grass hay	200–300 cubic feet per ton
Straw	300–400 cubic feet per ton
Shavings	300–350 cubic feet per ton

The hay barn or open shed should be located on a well-drained site. Because of the labor involved and the cost of hay, you simply cannot afford to have moisture get into your hay barn. For added insurance, stack your hay on pallets inside the building. You may have to incorporate ditches, berms, or retaining walls to prevent water from collecting around or moving through your hay barn or shed. In addition, the roof must be leak-free, and the building must have adequate ventilation. Build the hay barn large enough so you have room to park your tractor inside too.

Newly baled hay should be allowed to cure outdoors in a stack for a week or two before loading into the barn. (See chapter 13, Land, for more on hay curing.) If you walk into a hay barn and detect a damp or fruity aroma or a moldy or caramel smell, the barn probably contains spoiled or heating hay. Grain and bedding should also be well cured and dry to prevent spoilage or heating. In addition to keeping all grain in rodent-proof containers, you can employ a feline rodent patrol to help reduce the mouse population around the barn.

Opposite: **In arid country, a roof minimizes sun bleaching.** ***Top left:*** **A hay shed designed to receive stacker loads needs to have a doorway at least 17'–6" high.** ***Left:*** **Whenever possible, stack hay on pallets.**

Hay stacked outdoors should be placed on pallets and covered with a waterproof tarp.

Outdoor Hay Storage

Although the best way to store a supply of hay is in a separate building, your budget might dictate that the hay be stored outdoors, at least temporarily until a hay barn is completed. First select a level, well-drained site in a convenient location to distribute the hay for feeding. The stack can be placed to offer some degree of wind protection for animals, but in cold-winter regions, the protected side can end up with deep snow from drifting.

Rather than stacking hay on bare ground, place it on pallets so the bottom layer will remain dry. Used pallets are often available free or very inexpensively from feed mills, lumberyards, or cement plants. In lieu of pallets, 2×4s or 4×4s can be set up side by side and the bales placed to span them.

Stack the bales tightly together, alternating the direction of the bales every two or three layers so that the stack is more stable. If you are planning to cover the hay with a tarp, finish your stacking with a ridge of bales on the top rather than with a flat top. The resulting peak will help water and snow run off the tarp rather than accumulating on top and possibly leaking through.

In some geographical areas, an uncovered stack fares well with very little nutritional loss from sun or precipitation. In places with snowy, freeze-and-thaw, or rainy winters, however, it is best to cover the stack with a good tarp. Still, it is better to leave a stack uncovered than to cover it with a tarp that is full of holes. Water entering a covered stack will make a column of mold from the top bales all the way down to the bottom.

The best cover for a stack is a canvas tarp, as it is waterproof but allows for some air exchange, which minimizes condensation. Although new tarps are expensive, they will last a long time. You might be able to find a used truck or machinery tarp for sale at an auction. Black agricultural plastic is not a good choice because it can tear or be punctured by the hay stems, it is difficult to tie down, and it can result in condensation under the plastic, causing hay spoilage. Blue polyethylene tarps may be initially inexpensive, but they have very low resistance to sunshine and often deteriorate in one season.

Coverings should be secured with twine, rope, or bungee cords at all edges as well as over the top of the stack from side to side and from end to end. Canvas tarps and some polyethylene tarps have sturdy grommets that are useful for tying down the edges of the cover. You can improvise by placing a pebble or marble slightly in from the tarp's edge to create a lump around which you can attach your twine. It is advisable to run several ropes across the length and width of the stack and completely around its circumference to prevent billowing of the tarp by the wind, which can loosen and tear a covering.

Cut up an old inner tube and use the rubber scraps under the rope or twine wherever it looks like it may cut into the covering when you tighten the rope. Holes in most coverings can be patched with a dab of silicone and a scrap of plastic or canvas. If you take a little extra time when you stack and cover your hay, it will have a better chance of retaining its quality throughout the year.

Three-sided shed

This 12'x20' three-sided shed is set on a well-drained site, with an 8' high back wall set to block the prevailing winds. The lower portion of the inside walls is lined with solid wood to prevent kicking damage to the siding. This would be ideal for three ponies, two yearlings, or two small horses.

Pasture Shelter

A shelter can be man-made or natural but it must provide a comfortable place for a horse to get out of the elements. He needs protection from wind, cold precipitation, hot sun, and flies. Trees out in the open provide shelter while still allowing cooling breezes that also take away insects. Thickly wooded areas may provide shelter but are usually insect havens with no breeze, and the undergrowth can be hazardous to a horse's legs.

Man-made pasture shelters are typically rectangles with one of the long sides open. The backside of the shelter is oriented so the prevailing winds (usually from the north, west, or northwest in North America) hit the back wall, and the open front faces the east or south. Although a south or southeast opening usually provides winter sun and protection from the cold north winds, each locale and each spot on your land may have different wind patterns that you must take into consideration before selecting the site and building orientation. Locate the shelter on well-drained, high ground so it doesn't become a

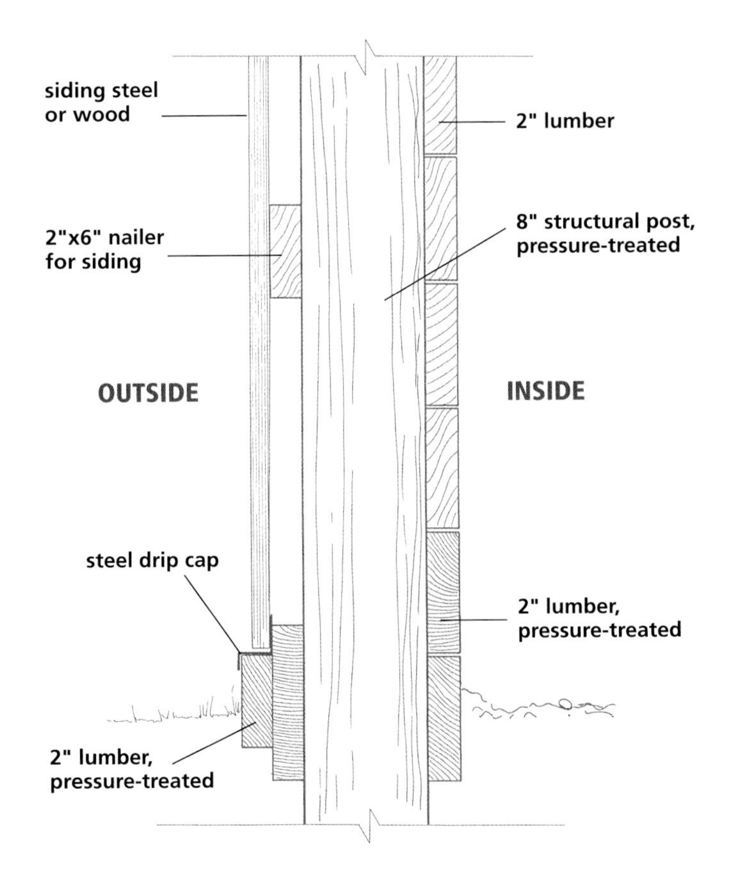

This 12x12 paddock shelter would be ideal for a single horse.

quagmire during wet seasons. Select one of the highest points in the pasture, unless it is the windiest spot. The shed should not be located so remotely that you rarely visit it.

Figure a 12-foot by 12-foot space for each horse. A 12-foot by 36-foot shed should give three horses room enough to cohabitate comfortably.

Pasture shelters typically have shed roofs, those that are higher at the front than the back. The height at the rear of the shed should be a minimum of 8 feet and the front at least 10 feet. Consider a 4- to 6-foot overhang in front to keep out wind and precipitation and to provide more shade for loafing.

The shelter should be constructed of materials that are safe for horses. Wood is usually safe but must be treated regularly to prevent chewing, and nails tend to creep out of wood and so need regular checking. Steel is lower maintenance, but sharp edges must be covered. As with stalls in a barn, a shelter should be lined on the inside with a material durable enough to prevent a horse from kicking through the walls.

Ideally, the footing inside the shed should be a clean, comfortable material that drains well. A deep bed of pea gravel (⅜ inch minus round gravel) works well. Horses readily lie down in pea gravel, stay clean, and don't develop hock sores like they do from bare ground. Avoid using sand if you plan to feed inside the shed during stormy weather (sand colic). If it's in your budget, aim for a shed that's large enough to have a feeding area with a smooth surface like rubber mats and a soft loafing area where a horse can stand comfortably out of the sun or wind or lie down if the pasture is wet or muddy.

Don't bed the shed unless you are using the shed for foals or if you produce bedding on your farm and cost is not a factor. Bedding will invite horses to urinate and defecate in the shed (even though they might have a huge pasture to do so in), so unless you are looking for a daily job of cleaning, skip the bedding. You might want to devise a way to lock horses out of the shed during nice weather to discourage them from using it as a toilet.

Consider a portable shed, one built on skids that can be relocated with a tractor. You can routinely move the shed to a different location or a different pasture depending on the season or when the area around it becomes damaged and needs renovation.

Indoor Riding Arena

If you want to extend your riding season and are considering an indoor riding arena, see chapter 12, Arenas, for preparation of the arena site and footing. As far as the structure itself, the building for an indoor riding arena is often a clear-span metal shell with metal or wooden trusses. Prices start at about $50,000 for a basic 60-foot by 120-foot shell (walls, roof, kickboards), two sliding doors, and one man door; electricity and plumbing cost extra. The building can be a separate structure or an addition onto the horse barn.

Generally, the maximum span possible for wood trusses is 60 feet with a 3/12 slope, so if you want a wider arena or require a steeper roof slope, you will need to have a building constructed with properly engineered metal trusses. Commonly, the sidewalls are 14 feet tall and the door openings are 12 feet tall. You may want to install kickboards along the sidewalls of the arena. Kickboards are added partial walls that begin about 4 feet from the ground and angle inward toward the arena floor. This keeps the horse off the wall, which protects the rider's legs. Lighting is an important consideration for an indoor arena. Daylight illumination can be accomplished by using translucent fiberglass panels at the top portion of the walls, at the gable ends, and in the roof. Auxiliary lighting is best provided by several HID fixtures. (See page 101.)

Footing tends to last longer in an indoor arena than in an outdoor one, but you will need to water and work indoor footing since it won't have the benefit of natural precipitation. Plan for easy access to water and for a tractor and equipment. Overhead watering systems, drop-down sprinklers, and traveling sprinkler systems are costly but convenient

Above: **This raised-center-aisle barn had an arena added across the back, making for a convenient training center.**
Below: **Fabric arenas take advantage of ambient light.**

Cover-All Building Systems

alternatives to dragging out hoses and setting up a sprinkler or using a water wagon. Infrared heaters can help prevent the footing from freezing. (See chapter 10, Equipment, for more information on machinery.)

Ventilation can include doors, windows, ceiling vents, cupolas, and fans. Other extras to consider would be a tacking area, restroom, and small lounge. A recent innovation for indoor riding arenas is a steel frame with truss arches covered by a translucent fabric, which results in an arena illuminated by natural lighting. Whether such an arena would be suitable for your acreage will depend on temperature, precipitation, and the seasons you would most likely want to ride indoors.

Breeding Facilities

If you have more than two or three broodmares, you may wish to construct a separate barn or outdoor facilities for foaling and housing mares and foals. Your broodmare barn could have comfortable quarters for observers adjacent to the foaling stall. Or you could have a closed-circuit TV in the foaling stall, which you could monitor from the tack room, lounge, or house.

If you plan to handle a large number of young horses, you might want to have specially designed weanling and yearling barns and facilities. Features should include smaller spaces between fence rails and ample, safe room for the explosive outbursts of youngsters.

If you will be keeping a stallion, his facilities must be extra high and strong. His stall should be a minimum of 14 feet by 14 feet with 6-foot-high solid walls. His paddock should be located away and upwind from mares but close enough that he can see other horses. Stallions need very experienced, competent handling and extra attention. They cannot be cooped up without exercise for very long before their energy gets potentially uncontrollable. Be sure to check zoning ordinances and local laws regarding the keeping of stallions.

If you will be standing a stallion for breeding, then you will want to develop some breeding facilities, which may include a breeding shed containing a tease rail, a mare preparation stall, a foal-restraint stall, a phantom mare for collecting semen from the stallion, and a laboratory to process the semen for artificial insemination. See the recommended reading list for additional help.

Equipment

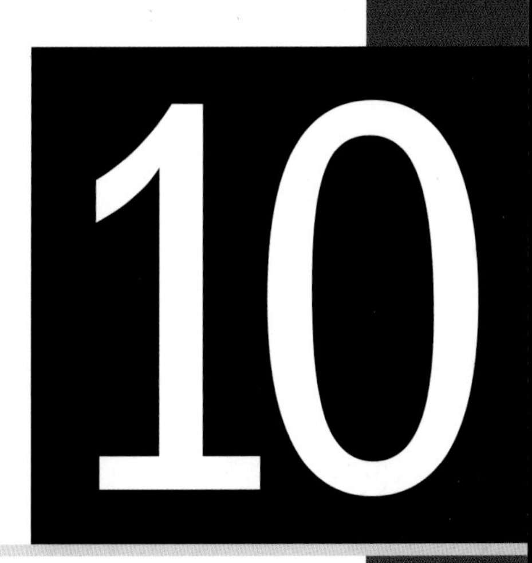

Although you will probably never eliminate the need for a wheelbarrow and a fork, the size of your horse operation and its associated tasks may require some specialized machinery. Such equipment usually falls into one of two categories: a truck with a trailer and a tractor with its attachments.

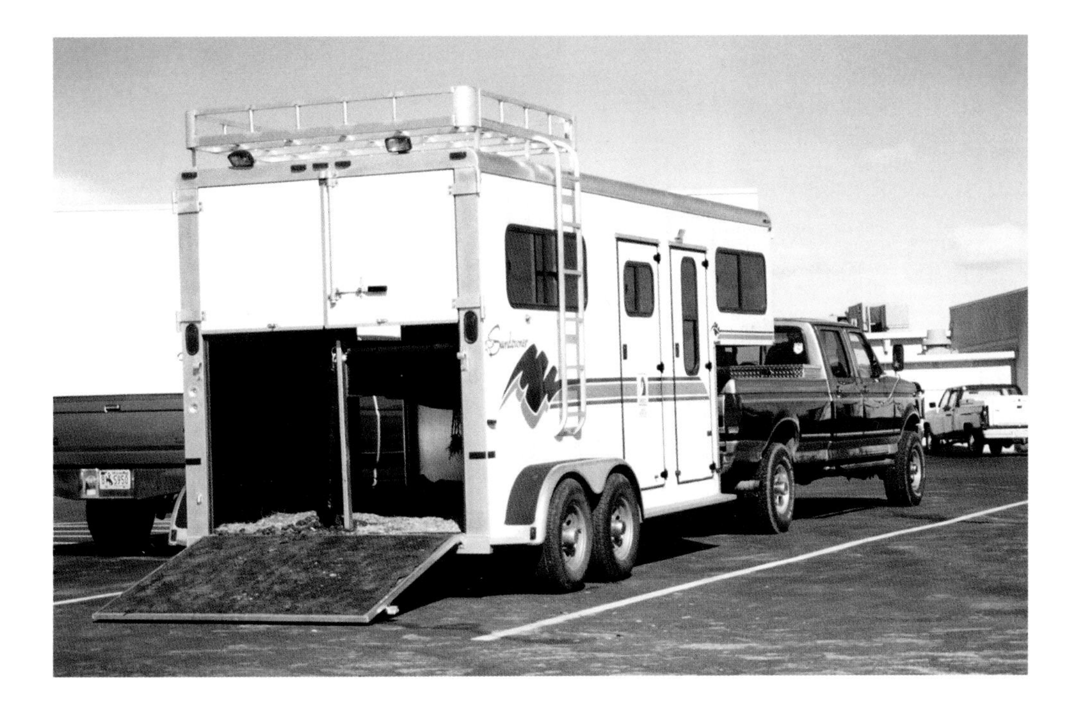

Truck and Trailer

A truck and horse trailer is a must for an acreage horse owner. In case of fire or flood, you may need to evacuate your horses. And you should have the means to transport a horse to a veterinary clinic in the event of an emergency. Even for routine veterinary care, you can often avoid paying a farm-call charge by taking your horse to the clinic. You will also need a trailer if you plan to show your horse or haul him to trail-riding areas. (For a more detailed discussion of trucks and trailers, see Hill, *Trailering* [Storey, 2000].)

THE TOWING VEHICLE

Evaluate your present vehicle to see if it is suitable for pulling a horse trailer. What is its towing capacity?

Especially for interstate driving, the weight of a loaded towing vehicle should be at least 75 percent of the weight of the loaded trailer. Is the distance from the front axle to the back axle (wheelbase) of your towing vehicle at least 115 inches? The longer the wheelbase, the less likely the trailer is to sway while in motion. Does your truck or car have a powerful enough engine to haul the extra weight of a loaded trailer? What you save in gas mileage with a smaller engine may result in greater repair costs due to excess strain on the engine and transmission. Most often, a ½-ton or ¾-ton pickup truck is appropriate for pulling a two- or four-horse trailer, respectively. If you plan to pull a larger trailer, or if you plan to use your truck to carry heavy loads of hay or grain, you may want to consider purchasing a 1-ton truck.

Does the towing vehicle have heavy-duty springs and shocks (suspension) and good brakes and steering (power preferred)? Are the tires dependable, and do you have a good spare, a lug wrench, and a jack? Are they easy to get to, and do you know how to use them? Tires should be inflated with equal, optimum pressure and regularly rotated and balanced.

Ideally, your truck should have a special towing package, which includes a heavy-duty radiator, a heavy-duty transmission with a special cooler and a low gear ratio, and heavy-duty springs and shocks. Look in the owner's manual for the specifications on your vehicle or ask your dealer to help you with the answers to these questions.

Routine servicing of the engine, cooling systems, suspension, tires, wheel bearings, brakes, and other mechanical components not only prolongs the life of a vehicle but also allows you or your mechanic to uncover problems before they become emergencies. Some common problems encountered with a hauling vehicle are overheated engines and transmissions, flat tires, and brake or hitch failures.

THE TRAILER

Maintaining the safety and comfort of your horse is the main function of a horse trailer.

Trailer Selection

You must consider many factors when investing in a horse trailer. New trailers will cost from $2500 for a bare-bones two-horse model to $10,000 and higher

This enclosed trailer has a breast bar (rather than a manger) that allows a horse to lower his head to blow en route.

Left: **A stock trailer is open and airy, which makes some horses feel more comfortable.** ***Right:*** **The truck chassis of a horse van delivers a comfortable ride in a stall-like environment.**

for a deluxe four-horse model with dressing room and tack room. Many good new trailers can be purchased for $6000, and used trailers can be found for substantially less. Whether your budget dictates new or used, you must make some basic decisions.

There are three types of horse-hauling vehicles: the enclosed trailer, the stock trailer, and the horse van. Enclosed trailers are typically two-horse and four-horse models and are the most common trailer seen on the road. The height inside trailers ranges from 72 to 90 inches, with most near 76; length of standing room ranges from 66 to 88 inches (depending on style), with the average around 70; the width of one stall is from 26 to 32 inches, with most toward the low end. A 16-hand horse can fit into a standard trailer, provided he is levelheaded about loading and unloading. Better suited for large horses are the 7-foot-high Thoroughbred and Warmblood trailers, which allow ample space for the tall and long-bodied breeds.

Choose a trailer that allows your horse to hold his head at a comfortable, natural level without his ears touching the ceiling. The trailer should be long enough to keep the horse from being tightly sandwiched between the chest bar and the butt bar and should provide sufficient room for him to lower his head and blow. If you are planning to haul a very large horse, you may need to look into a custom trailer or van.

Stock trailers are usually the equivalent of a four-horse trailer in length and basic style, but the sides are slatted rather than enclosed and are designed to haul horses somewhat loose. This comes in handy when hauling mares and foals and young horses or when hauling saddled horses. Horses have less of an enclosed feeling because they can see out of the stock trailer, but because of the slatted sides, horses can get dusty, cold, and wet.

A van is a horse stall on a truck chassis. It is more comfortable for the horse than a conventional trailer and is the most expensive of the three types of transport.

Trailer Construction

Materials and workmanship dictate, to a large degree, the cost of a trailer. Materials commonly used include steel, aluminum, and fiberglass. A trailer with a frame and skin of steel is generally very sturdy. Substituting an aluminum skin while retaining the steel frame will decrease weight and rusting. Fiberglass is often used for roofs and fenders, as it is cool, lightweight, and easy to repair. Quality workmanship will be evident in the straightness of the frame, the fitting of seams, the finishing of edges, and the paint job.

Good-quality trailer suspension should be sturdy but not stiff. Whether you decide on leaf springs or rubber torsion suspension depends on the quality of each, but the latter gives a less stiff ride. Be sure the suspension is independent: That is, when one wheel hits a bump or a hole, it will absorb the shock independent of the rest of the trailer. The trailer must have its own brake system (usually electric) and adequate operating lights and clearance lights for night driving.

Trailer Design

You must decide how you want your horses to ride in the trailer — facing forward, backward, sideways, or loose. Most trailers are designed for rear loading and place the horses side by side, facing forward. These side-by-side, or straight-load, trailers are typically two stalls but can be for four or six horses; the stalls can be any length and thus can accommodate large horses. Other options, such as a slant-load or reverse-load, are available if your horse is a difficult

Left: **A straight-load trailer consists of side-by-side stalls that face forward.** ***Top:*** **A ramp can help or hinder loading.** ***Above:*** **A slant-load trailer has diagonal stall dividers; horses ride at an angle.**

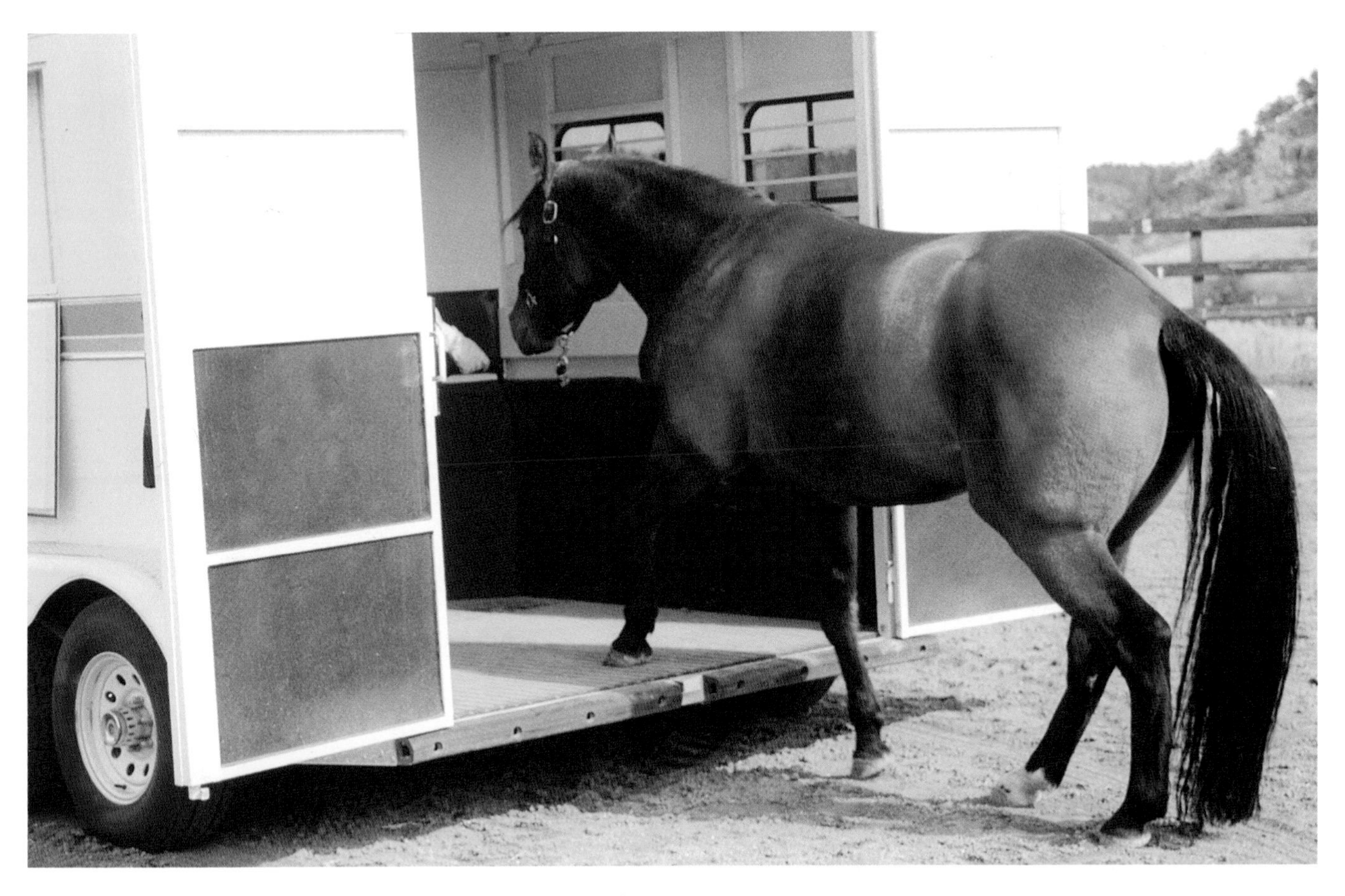

Seeker demonstrates that a step-up trailer is safe and straightforward. It is the most common option for both straight- and slant-load trailers.

traveler. In a slant-load trailer, horses stand diagonally across the width of the trailer; slant-loads are limited by maximum allowable width on the road, so may not the best choice for large horses. With reverse-load trailers, the horses load in the front and face the rear of the trailer. In many trailers, the center divider is removable to accommodate a large or difficult traveler or a mare and foal. If the center divider does not go all the way to the floor, your horse will likely travel more comfortably, as he will be able to move his feet farther sideways to balance.

Horses load into a hauling vehicle by stepping up into the trailer or by walking up a ramp. The step-up style is less expensive, straightforward, and most common. The ramp style is more expensive, can be difficult to use on uneven ground, and can spook horses when they step on the ramp and it moves.

Trailer Features

Are you looking for a tagalong trailer or a gooseneck style? A tagalong (also called a *straight pull* and a *bumper pull*) trailer attaches to a hitch that is mounted to the truck's frame. A gooseneck trailer attaches to a ball that must be installed in the bed of your pickup truck. What type of tack room do you need — a small compartment for just a saddle and bridle or a larger one that can also be used as a dressing room?

Consider the following options, realizing that for every one you add, the price will increase. How many vents, windows, and interior lights do you require? A minimum of one bus-style window per horse on each side of the trailer is suggested. Does the center divider of the trailer need to be removable? Does the center divider go all the way to the floor? Do you want padding on the sides of the stalls, on the center divider, and at the chest? What type of flooring is available? Oak and pine are both fine as long as the quality of the wood is good — no warping or knots. Pressure-treated wood will withstand manure and

Top: **Straight-pull trailer hitches are most commonly seen on two-horse trailers.** ***Bottom:*** **A gooseneck hitch should be used for any trailer hauling more than three horses.**

urine longer than untreated wood. What type of mats come with the trailer? Removable rubber mats are preferred. What type of release bars are at the chest or head, the tail, and at the center divider? Check to be sure all releases work easily. Some are very difficult to operate if the trailer is on less than 100 percent level ground.

Don't assume that just because a trailer looks good it is safe. For example, a well-known manufacturer overlooked one small detail in rear-door design that let the doors lift easily off their hinges. When a horse leaned against the rear door, the door came off. The horse fell out of the trailer and was dragged behind it. Get competent help when you select a trailer.

With the trailer hitched to its towing vehicle, assess its balance. The truck hitch should be at the right height so that the trailer floor is level. Most of a trailer's weight should be borne by its wheels. At a standstill on level ground, only about 10 percent of the weight of a tagalong trailer should be transferred by the tongue to the towing vehicle. A trailer that is tongue-heavy will overstress the rear end of the towing vehicle and cause excess wear on the ball of the hitch. A trailer that is rear-heavy, by contrast, can cause dangerous swaying (fishtailing) when the vehicle is in motion.

Trailering can be very hard on a horse's muscles, bones, joints, ligaments, and tendons. Rough roads, long miles, inexperienced or inconsiderate drivers, inadequate trailer suspension, poor floor mats, and improperly maintained tires can all cause unnecessary wear and tear on your horse. To help ensure that a horse arrives at his destination refreshed rather than fatigued, make the trailer as safe and comfortable as possible. Organize a maintenance plan for your horse trailer as you do for your towing vehicle.

Trailer Maintenance

Store your trailer on level ground with the hitch jack adjusted so that the trailer's weight is balanced between the tongue and the tires. Park the trailer out of the weather to preserve the paint, and on pavement whenever possible to protect against tire rot. When storing it for several months, jack up each side and place blocks under the axles where the springs attach to take the weight off the tires.

Evaluate the following major items at least once a year and repair or replace worn or broken parts.

Wheel bearings. The grease that lubricates wheel bearings accumulates dirt and dries out. Have the wheel bearings cleaned and repacked with grease annually. The seals will be replaced at the same time.

Brakes. The wheels should be removed so that accumulated dust and dirt can be cleaned from the brakes. Also, have the pads checked for wear and replaced if necessary. Several times each year, perform a brake check and adjustment. A gravel roadway makes a good area for this test. Enlist the aid of a knowledgeable observer to tell you whether a particular wheel is locking up or rolling free when the others are stopping properly. In either case, the brakes of that particular wheel will need to be adjusted.

Once all of the brakes of the trailer are stopping evenly, you can proceed to the next part of the brake test. Preferably on level ground, accelerate to about 30 miles per hour. Then, without using your truck's brake pedal, bring the rig to a stop by using the manual electric brake controller mounted on or under your dashboard. If the trailer brakes grab too quickly, adjust the controller down. If the trailer can't stop the rig, adjust the controller up. Once the trailer brakes alone, stop the rig properly and use your brake pedal, which activates both the towing vehicle brakes and the trailer brakes.

Tires. Rotate and balance tires at least once a year to equalize wear. Check for bare patches, bulges, and other defects.

Suspension. At least once a year, grease springs and shackles. Check the bushings for wear where the spring ends are pinned to the shackles. Check shock absorbers, if present on your trailer, and replace when necessary.

Floor. Check the floorboards at the beginning of the season for rotting, splintering, shrinking, or warping. Replace any boards that are remotely suspicious. Use clear (no knots) planks that match the dimensions of the rest of the floor. You may wish to treat the floor with a preservative to combat the effects of manure and urine. Use resilient mats with "life" to help absorb road vibrations and shock. Replace mats when they have become excessively chewed up by shod hooves or compressed from long use.

Sideboards. The bottom 2 to 3 feet of the sidewalls of your trailer get abuse from the inexperienced or scrambling horse. If the walls are metal, check for rust. Consider installing ¾-inch plywood or rubber kickboards over the metal walls for added protection for both the horse and the trailer.

General. Be sure that the hitch, safety chains, chest bars, tail bars, dividers, doors, and windows work properly. Each time you use your trailer:

- Sweep out the stalls of the trailer and remove the mats so the wooden floor can dry out. The mats can be hung over the stall divider.
- Clean mangers of old hay and grain to prevent mold (danger to the horse) and rust.
- Check tire pressure. (See the specification on the sidewall of your tires.)
- Check to see that wheel lug nuts are tight.
- Check tires for irregular tread wear, bulges, defects, or weather checking (cracks caused by sun, moisture, and freeze and thaw).
- Check the spare tire and jack and know how to use them.
- Oil any hinges, latches, and other moving parts that do not function freely.
- Check running lights, turn signals, brake lights, emergency flashers, and brakes.
- Check the inside of the trailer for things such as hornet and mouse nests.
- Add fresh bedding (sawdust) if desired.
- Wash the trailer as needed and wax twice a year.
- Check the hitch and safety chains.
- Make sure the trailer registration is in the towing vehicle.

Clean out your trailer after each use. Remove unused feed, sweep the floor, and lift up the mats to allow the floor to dry.

Tractors

If the activities on your acreage justify investment in a tractor and attachments, take plenty of time to research the market so that you buy the best equipment to fit your needs. You might find that it would be more economical to hire a neighbor or local farmer to do large jobs or field work for you. But if you find that your needs warrant the purchase of a tractor, realize that the tractor is just half the equation. There are many implements to consider as well. Here's where working with a knowledgeable, reputable dealer can really save you time and money.

TRACTOR SAFETY

Operating farm equipment can be dangerous. A tractor is not a toy for children to play with or for adults to operate carelessly. Every person who will be operating equipment on your farm should be well informed about safety practices and agree to follow safe operating procedures.

In 1984, the Occupational Safety and Health Association (OSHA) instituted safety regulations for tractors manufactured in the United States. Since that time, U.S. tractors must have a rigid, arched rollover bar called a rollover protection structure (ROPS); seat belts; and a protective power takeoff (PTO) cover. Most Japanese tractors have safety features comparable to U.S. tractors, but some other foreign-made tractors do not. And many used small-farm tractors, because of their age, do not have safety gear. If you purchase an older tractor or a foreign tractor, it is up to you to update your machinery and to maintain it and operate it in a responsible manner. All major equipment manufacturers provide tractor safety programs, including information about retrofitting older tractors with ROPS.

Safety features are put there for a reason. Keep all protective equipment in place and be sure all guards, shields, and safety signs are installed properly. If you remove, modify, or disengage safety features, you increase your risk of injury or death.

Don't alter or remove the ROPS. If a ROPS has been damaged, it should be replaced.

Use a seat belt with the ROPS. If your tractor overturns, the recommendation is to hold the steering wheel firmly and don't leave the seat until the tractor has come to rest. A seat belt will help ensure that if the tractor overturns, your body will be held in place and protected by the ROPS.

Never use a seat belt when operating a tractor not equipped with a ROPS. Not being restrained by a seat belt will make it possible for you to jump or be thrown clear if the tractor overturns instead of being trapped beneath it.

The following are the most common tractor accidents: falling off a tractor when mounting or dismounting or when hitting a bump while speeding; becoming entangled in the PTO shaft while it is running; rolling over as a result of an unbalanced load, unlevel ground, or a sharp turn; and flipping over backward due to improper hitching.

What tractors can do

With a tractor and implements, you can:

- Collect manure
- Spread manure
- Harrow fields
- Mow weeds
- Clear brush
- Haul bedding
- Haul and move feed
- Maintain arena
- Plow snow
- Grade and maintain roadways
- Maintain pens
- Move portable shelters and large feeders
- Dig postholes
- Pull out old posts
- Power a water pump or generator
- Spray, fertilize, and seed pastures
- Cut, bale, and stack hay
- Cultivate soil for crops or gardens

Safe tractoring rules

- ☐ Know your tractor and its capacity.
- ☐ Know your implements.
- ☐ Keep manuals handy for easy reference.
- ☐ Avoid running a tractor at high RPMs when it is cold.
- ☐ Before starting your tractor, look for children, animals, and other bystanders in the vicinity.
- ☐ Do not carry passengers on the tractor or implements.
- ☐ When you stop the tractor and plan to leave the driver's seat:
  - Disengage the PTO.
  - Lower implements and attachments to the ground.
  - Shut off the engine.
  - Apply the parking brake.
  - Put the transmission in neutral.
  - Remove the key if the tractor will be unattended.
- ☐ Dress safely. Loose clothing and long hair can be caught on moving parts.
- ☐ Stay clear of all rotating PTO implements and be sure other people and animals are clear of them too.
- ☐ Never attempt to service your tractor or implements with them running.
- ☐ Keep brake pedals latched together at all times.
- ☐ Know the terrain you will be driving on. Walk unfamiliar terrain to identify ditches, large rocks, debris, or bogs.
- ☐ A tractor is a workhorse, not a racehorse. Drive at safe speeds; reduce speed before turning to prevent overturning.
- ☐ When driving in hilly terrain, keep tractor in gear — never coast downhill, and never depress clutch while going uphill or downhill. Drive straight down all slopes, never diagonally. If you must turn on a slope, turn downhill. When you must drive up a slope, consider backing up for added safety.
- ☐ Always carry the front-end loader as low to the ground as possible, and try to center materials in the bucket.
- ☐ When driving your tractor on a public highway, if legal, use flashing amber warning (hazard) lights and a slow-moving vehicle (SMV) identification emblem. (Note that in some areas, use of hazard lights when driving is illegal.)
- ☐ Hitch only to the drawbar or three-point hitch when pulling a load. There is no place on the rear of a tractor that is safer or more effective for pulling than the drawbar. Hitching to the axle housing, seat base, or top link of the three-point hitch reduces the pulling capacity of the tractor and increases the chance of the tractor rotating over the rear axle and flipping over backward. Be sure no loose chains, ropes, or cables are dangling or dragging from either the tractor or an implement. They can catch under a wheel or on a stump or rock and cause a serious or fatal accident.

BUYING A TRACTOR

Generally, you can buy farm equipment in three ways: through a dealer, by private contract, and at auction. Working with a reputable dealer is ideal because you will get experienced advice and have a much better chance of finding what you really need. A good dealer will specifically ask what type of work you plan to do with your tractor, the size of your acreage, and how much you expect you will use the tractor. Most dealers offer warranties on new equipment, some offer a limited warranty on reconditioned equipment, most offer repair service, and many, if you become a regular customer of theirs, will provide advice and answers after purchase. A dealer may also offer low-rate financing from the manufacturer.

If you already have experience with machinery and equipment, buying privately or at auction can be a viable option, but if you have not driven, hitched, and pulled a fair bit, get a professional opinion before you buy or bid.

Is the tractor you want mechanically functional now? If not, what needs to be repaired and what would be the cost of the parts and labor? Are parts readily available? What is the item's value? (See the next section, Buying a Used Tractor, for specifics.) If you ask a tractor dealer or mechanic to evaluate a used tractor or other equipment for you, be prepared to pay an appraisal or consultation fee.

Some of the most common mistakes made in purchasing a tractor include being uninformed, using price alone as a buying criterion, buying a fix-up, and buying too small.

One single feature on a tractor could make it unsuitable for your use. If you are not aware of these types of things, you could end up with a tractor that seemed 100 percent right for you but was actually 10 percent wrong, and that 10 percent made the tractor unusable. For example, "tricycle" tractors with narrow front ends (a single front tire or front tires 6 to 8 inches apart) are highly maneuverable but very unstable. They have a reputation for tipping over, one of the main causes of injury and death on the farm. A tractor with a narrow front end is a risk in

Top: **Become familiar with all the features of a tractor before you buy.** ***Middle:*** **Buying from a dealer often includes financing, warranty, service, and advice.** ***Bottom:*** **If you buy privately and are not experienced, bring a mechanic with you to evaluate the tractor.**

Top: **Buying at auction can yield bargains or disasters. Inspect any equipment thoroughly before you bid.** ***Bottom:*** **Narrow-front-end tractors are maneuverable but unstable, and can easily tip over.**

any situation and should be avoided, but is an accident looking for a place to happen, particularly on hilly or mountainous terrain.

Other poor purchases would be a tractor with no PTO, no three-point hitch (referred to as a "bareback" tractor), or no rear hydraulics. These three features are a must.

If you approach the matter of buying a tractor solely with price in mind, you may pay a low initial price but end up adding, replacing, or repairing many costly items on the tractor. You should not consider purchasing a fixer-upper unless you are very experienced with tractor repair and have good knowledge of parts availability.

And when you buy, be sure you are comparing apples to apples and oranges to oranges. For example, a listed tractor may state that it has a PTO, but not all PTOs are equal. Read about the difference between a live and a two-stage PTO later in this chapter. Similarly, there is great variability in how transmissions operate, which will affect the ease and efficiency of your work.

Within reason, it is usually better to buy a more powerful tractor than your current needs require, because you will almost always find more demanding work to perform with your machine. It is a costly mistake to buy a small tractor and then try to accomplish work more suitable for a big tractor. The mechanical workings of the small tractor simply will not hold up, and eventually you will be faced with extensive repair bills. No single tractor will do all jobs for you perfectly, but try to choose the type of tractor that is best suited for your needs.

BUYING A USED TRACTOR

Just as when buying a horse, when looking at used tractors there are many things to consider, and it's up to you to discover a tractor's strengths and weaknesses. If you lack experience, have the tractor evaluated by a professional.

Does the radiator show signs of leakage? If there are stains near the radiator cap or overflow tube, or mineral deposits inside the radiator (remove cap), the radiator may require flushing or repair.

Does it start easily? A tractor that starts cold tells you that it has a good battery, compression, and ignition and a functioning fuel system. If it doesn't start easily, it may be a good tractor in need of a battery or tune-up, or it may have serious problems.

Is the thermostat functional? A thermostat should prevent liquid from circulating through the radiator until the engine is warm. Right after cold starting and with the engine at a fast idle, remove the radiator cap. If you see bubbling liquid, it means the thermostat is stuck open or has been removed. Question this.

Does it run well when hot? Run the tractor for half an hour, keeping an eye out for oil or antifreeze leaks. Shut off the tractor, then see if it will start hot.

Does it smoke? Diesel tractors puff a little black smoke on start-up but soon should run clean. Blue smoke indicates oil burning, which could mean problems with rings, pistons, or valve guides. White or black smoke could indicate the need for a tune-up (carburetion or ignition).

What kind of noises do you hear? Ticking might mean a valve adjustment is needed. Loud knocking, clunks, or thuds are usually serious, involving crankshaft, bearings, or pistons.

Is there leakage in the middle of the tractor? A rear main seal that is leaking can get the clutch oily and ruin the clutch lining. If the rear main seal needs to be replaced, the tractor will usually need to be split in two; this is a costly repair.

What does the oil look like? After the tractor runs for half an hour, turn it off and pull the dipstick. Water or foam could indicate a serious problem.

Has the air cleaner been serviced? On an older tractor with an oil-bath air cleaner, look at the oil level and sediment to determine if the cleaner has been regularly serviced. Dry filters should be free of dust and dirt; if clogged, the engine has been taking in dirt due to poor maintenance.

When buying a used tractor, bring a checklist with you and make notes about potential problems that require repair.

Do you see oil leakage around the head gaskets? The gaskets might need to be replaced (not a big deal), or the heads might be warped or cracked (a big deal).

What is the oil pressure reading on the gauge on start-up and when hot? If the gauge does not work, have a mechanic test the oil pressure for you. Oil pressure is higher when the oil is cold and thick and drops as the oil warms up and thins down. A gas engine tractor should carry at least 15 pounds of oil pressure when warm; a diesel, 20 pounds.

What is the engine compression test result? To get an idea of the condition of the engine, have a compression test performed on each cylinder. The results, when interpreted by a competent mechanic, can indicate the condition of the cylinders, pistons, and rings, as well as valves and gaskets.

Does the clutch work well? There are two ways to test the clutch. First, with the tractor running, push in the clutch and put the tractor in gear. If it grinds, this could mean the clutch is on its way out. Next, put the tractor in low gear, and with your foot on the brakes, let out the clutch. The engine should lug down and die. If, instead, the clutch slips and the engine keeps running, that means the clutch is shot or close to it.

Do the hydraulics work without leakage or letdown? Extend the hydraulics, paying attention to whether they operate smoothly or chatter. Lift an implement with the hydraulics and see if there is any loss of lift power over a few minutes. If the implement starts lowering on its own, the hydraulics require attention.

Is the tractor structurally sound? Look at the frame and all steel and cast components for obvious cracks or hairline cracks.

Does it look like all moving parts took grease during the life of the tractor? Look for the presence of excess grease around moving parts.

Do the zerks (fittings) still take grease? Or are they clogged with dirt, and are the joints "frozen"? Try to add grease to the zerks with a grease gun. Manually or with tractor power, manipulate joints to see if they move freely.

Do the tires have deep cracks or fissures in them? If so, beware: the tires could come apart at any minute.

Does the age match the hours and the serial number? If the number that appears on the hour meter doesn't jibe with the stated age and serial number, ask the seller to verify.

What is horsepower?

Horsepower is defined as the amount of energy or work required to raise a weight of 33,000 pounds a height of 1 foot in 1 minute of time or to overcome or create a force that is equivalent to doing that amount of work. Therefore, in simplified terms, horsepower is 33,000 foot-pounds of work done in 1 minute.

Tractor manufacturers represent horsepower in various ways. Some rate their equipment with the horsepower rating of the engine, but this doesn't take into account that the energy must be transferred through the tractor's drive train to the wheels or PTO in order to be used. Other manufacturers use laboratory tests to take horsepower measurements, but often the tests are not "real world," so the value can be inaccurate.

A useful measure of how much work a tractor can perform over a long period of time is sustained or continuous horsepower, measured by a dynamometer. To test a tractor's pulling ability, its wheels are rested on the dynamometer platform and the tractor is operated at varying speeds in all gear ranges against standardized resistances. The performance is then averaged, and the resulting pulling capacity is known as its drawbar horsepower, or how much power the tractor can continuously deliver to the drive wheels. In another test, the dynamometer is connected to the tractor PTO and the engine is operated at various speeds against standard resistances. The resulting averaged value is known as the PTO horsepower, or how much power the tractor can continuously deliver to an auxiliary piece of equipment through the PTO.

PTO horsepower = 75–85 percent of engine horsepower

When buying a new tractor, be sure you are comparing the same measure of horsepower among models. The specifications are readily available in sales brochures.

TRACTOR SIZES

Tractors can be loosely grouped into four categories as dictated by size, weight, horsepower (hp), and suitability for purpose: lawn and garden tractor, compact tractor, utility tractor, and farm tractor.

Lawn and garden tractor (up to 20 hp). A lawn and garden tractor is handy for driving through barn alleyways and pens or for pulling a small manure or feed cart. It would be unsuitable, however, for routine field work, arena work, or large-scale feeding or manure handling. Lawn tractors can be an expensive option for horsemen — by the time you buy a tractor, a cart, and other attachments, you might have reached the price range for a compact tractor. You might end up with half the tractor for the same price. Therefore, consider purchasing a lawn tractor for light-duty work if you are planning to buy two tractors.

Compact tractor (approximately 20–50 hp). Also called a *midsize* or *acreage tractor,* this is a convenient, easy-to-operate tractor. Because these are not very tall, they are pretty easy to mount. The hitch is low to the ground, making attachment of implements convenient. Generally, these are good tractors for teenagers to learn on. (*Note:* Some manufacturers divide this category into subcompact [approximately 20–30 hp] and compact [approximately 31–50 hp].)

Tractors in this category are the equivalent of the 8N Ford, which was known as the estate tractor when it was manufactured (1939–1952). As a rule of thumb, you can probably find a fairly decent used tractor in this category in running order for approximately $2000. Not many midsize tractors were manufactured from the late 1950s to the 1970s, so used tractors in this size are either at least 50 years old or

Top: **Lawn and garden tractors are handy for small jobs.** ***Middle:*** **Because of their size, compact tractors are easy to mount.** ***Bottom:*** **A utility tractor is powerful and can operate heavy-duty equipment.**

25 years old or less. Since the late 1970s, most compact tractors are Japanese-made, many of them four-wheel drive. They go for $10,000 new and for $5000 to $6000 used.

Compact tractors are good for all-around small acreage chores, but they are limited to the size of the attachments that can be used with them. They can pull about an 8-foot pull-type disc or a 6-foot three-point disc. They work well with a small manure spreader, especially the friction-drive type. (See Manure Spreaders on page 154.) With a front-end loader or a 6-foot blade on the back, a midsize tractor works well for cleaning out pens and runs.

Utility tractor (approximately 50–90 hp). This is a taller, more powerful tractor, able to operate more heavy-duty equipment. If you have a large arena and want to use a disc with two 8-foot sections, you will want to consider a utility tractor. If you need to handle large amounts of manure and you use a heavy-duty PTO-driven spreader, you will need a tractor of this size. Depending on economics at the time you are ready to purchase, you may well find a utility tractor for the same price that you would pay for a compact tractor. All other things being equal, if you think you might need the extra power, buy the larger tractor.

Farm tractor (over 90 hp). Generally, this type of tractor is designed for commercial farming and is not necessary or suitable for small acreage farms or ranches.

What about a utility vehicle?

Utility vehicles are handy for chores if you have to carry feed a fair distance or need to check on pasture horses. There are several manufacturers, and each offers different options and accessories. Most are four-wheel drive, and some feature a cargo box like a small pickup bed that has room for hay and grain and can haul and dump up to 1000 pounds on some models.

TRACTOR OPTIONS

Certain features on a tractor are a must, others can be added according to personal preference, and some may be unnecessary for your particular needs. Learn all you can about tractor options before you go shopping.

Gas or Diesel

Since 1974, all tractors (except for lawn and garden tractors) manufactured in the United States have had diesel engines. That is because a diesel tractor engine is two to three times more economical to run and produces more torque per cubic inch than a gas engine tractor. As far as the environmental benefits, diesels have lower greenhouse gas emissions, and they emit less carbon monoxide and unburned hydrocarbons than do gas engines. However, there is concern that a diesel engine's higher emission of nitrogen oxides and diesel particulates pollutes the environment. Using new particulate filters and low-sulfur diesel fuel may correct these problems so that diesel engines can meet the U.S. Environmental Protection Agency's standards for particulate matter (PM) and hydrocarbons (HC). Keep updated on these options by talking with your tractor dealer.

Make

Some types of tractors should be avoided because they are so rare that it is difficult, if not impossible, to find parts for them. Do not buy any make of narrow-front-end tractor. Dry-land tractors, those without a three-point hitch, are very impractical. Buying an old Minneapolis-Moline for $200 may seem like a great deal, but when you realize that the drawbar on the back limits you to using a pull-type disc, it might not seem so wonderful. And when you try to find parts to repair it, they will be difficult to locate.

Generally, most models of Ford, Massey Ferguson, John Deere, International Harvester, and Allis Chalmers still have parts available and would likely make good choices. Because of a renewed interest in restoring old tractors, after-market companies now offer parts and manuals for older tractors. These parts are not original manufacturer's equipment (OME), however, and because the market is for collectors and restorers, the price of the parts is sometimes high. Still, the parts are more readily available than they have been in the past.

Size

Choose a tractor for the work. Be sure it will fit into the areas you need to work in, such as in a loafing shed or under a barn overhang, and that it will fit into the storage shed where you will park it. Make sure you feel comfortable mounting, dismounting, and operating a tractor before you buy it.

Horsepower and Power Takeoff

Buy a tractor with enough horsepower to run the implements you require. It is better to have more horsepower and not need it than to need more horsepower and not have it. The power takeoff (PTO) is an extension of the drive train that allows power to be mechanically transferred to other machinery or implements via a removable drive shaft with splined couplings. Live or continuous PTOs are independent; that is, they rotate independently of the tractor drive train. With a live PTO, you can power an implement with the engine declutched and the tractor standing still. Two-stage, semi-live PTOs are engaged and disengaged in conjunction with the tractor drive train by operation of a two-stage clutch. Therefore, the tractor must be in motion for a two-stage PTO to be operable.

The PTO transfers power from the tractor to the gearbox of implements such as a manure spreader, posthole digger, rotary mower, and brush hog. Because the PTO is a revolving splined shaft on the back of the tractor that turns at 540 (standard) revolutions per minute, if an operator's clothing or hair gets caught in the mechanism, serious injury or death could occur. Modern tractors and equipment have mandatory shields or guards around dangerous parts of farm machinery like the PTO, but such safety requirements did not exist when the older models were manufactured. If you use a PTO, even if it is shielded, be sure to keep children away from the equipment and use caution when working around it yourself. If you have an unshielded PTO, contact the manufacturer of your tractor for information on retrofitting the PTO with a shield, or have a blacksmith or welder fabricate guards for your machinery.

Transmission

A transmission is a device (full of gears) that uses gearing and torque conversion to effect a change in the ratio between engine rpm and driving wheel rpm. A transmission can make a tractor user-friendly or user-angry. A tractor's transfer case is made up of at least two transmission shift levers. Commonly, the range shift lever provides four major speed changes, such as slowest, slow, medium, and high range. The gearshift lever provides four smaller gear changes with each range. This results in sixteen forward and sixteen reverse gear speeds.

Manual shift (standard) transmissions are durable and ideal for pulling heavy loads, but they are not handy. Manual transmissions are best shifted when the tractor is not in motion. Shift while the tractor is moving and you'll get grinding, because the two gears that are trying to mesh are rotating at different speeds.

A *synchro shift transmission* is a standard transmission with synchronizers that minimize grinding when you shift between specified gears when the tractor is in motion. The clutch pedal must be depressed when you shift, but the tractor motion does not have to be stopped.

A *shuttle shift transmission* synchronizes forward and reverse and allows you to shift back-and-forth between the two with one movement. Shuttle shift is handy for repetitive back-and-forth operations like picking up a bucket of manure, backing up, then moving forward to load it into a spreader.

A *power shift transmission* allows you to shift between different speeds while the tractor is in motion without using the clutch pedal, much like an automatic transmission in a truck.

Hydrostatic transmissions are controlled by a foot pedal that allows you indefinite speeds. When you take your foot off the pedal, the tractor stops, which is a good safety feature. These types of transmissions are common in utility vehicles.

Hydraulics

Hydraulics is a system of pressurized hydraulic oil that transfers power via its hoses and fittings to operate implements that have hydraulic pistons and cylinders. It also raises and lowers the three-point hitch and powers the front-end loader.

Bar Hitch

Most tractors have a straight bar hitch: a frame-mounted bar of steel at the rear of the tractor where implements can be attached. The straight bar hitch is used for pulling trailers, wagons, and some discs and harrows.

Three-Point Hitch

A three-point hitch is the linkage on the back of the tractor that is hydraulically powered to raise and lower equipment such as discs, plows, posthole diggers, and so on. The two side arms, also called *lift arms* or *draft links,* serve to both lift and pull. The top arm, called the *top link* or *center link,* dictates the angle of the implement to the ground and serves as a pivot point to raise the implement. When the implement is raised, the added weight on the tractor's rear wheels increases traction.

Most modern compact tractors have either a Category 1 or 2 three-point hitch (see chart). Some Category 1 implements can be adapted to Category 2 hitches, but Category 2 implements cannot fit a Category 1 hitch. (*Category* refers to the size of the connecting pins and the strength of the components.)

Appearance

Don't be overly influenced by a shiny appearance, as fresh paint can often cover up significant problems. Ninety percent of all tractors are stored outdoors. This means that many tractors that are functionally sound may look weather-beaten, with oxidized paint and tattered seat covers. One item of appearance that you should be concerned with is weather-checked tires.

Three-point-hitch classifications

CATEGORY 0

Tractors with up to 20 hp

Top-link pins are ⅝ inch in diameter

Lift-arm pins are ⅝ inch in diameter

Width of arm spread is 19 inches when implement is centered

CATEGORY 1

Tractors with 20–45 hp

Top-link pins are ¾ inch in diameter

Lift-arm pins are ⅞ inch in diameter

Width of arm spread is 26 inches when implement is centered

CATEGORY 2

Tractors with 55–90 hp

Top-link pins are 1 inch in diameter

Lift-arm pins are 1⅛ inches in diameter

Width of arm spread is 32 inches when implement is centered

CATEGORY 3

Tractors with 95–100+ hp

Top-link pins are 1¼ inches in diameter

Lift-arm pins are 1⁷⁄₁₆ inches in diameter

Width of arm spread is 31–33 inches when implement is centered

Top: **New tires for a tractor might cost $800, but if well cared for can last many years.** ***Bottom:*** **Beware of deep weather checking on used tires — they may be unsafe.**

Tires

Some weather checking is normal for tractor tires because they are often stored outside and exposed to the deteriorating effects of sun, mud, and water; on used tractors, the tires are often quite old. If the weather checking has become so severe that it results in deep fissures, however, beware, as the tires could come apart at any time.

Within reason, tread depth is not as important on tractor tires as it is on your car or truck. However, try to find lugs with at least 50 percent of their tread depth left. New 28-inch tires have 1-inch tread depth, and new 38-inch tires have 1½-inch tread depth.

A type of tire well suited for acreage tractors is the R1 agricultural tread, which has diagonal bars that run from the inside to the outside edge of the tire. R4 industrial tires are also popular. They are more like backhoe tires or a cross between a turf tire and an agricultural tire. They have bars that are wider and not as deep (aggressive) as those on agricultural tires. They support weight without digging in as deeply. Chains can be added to any tire to increase traction.

Be aware of rear tires that have calcium chloride added for weight, as the fluid can rust out the wheels quickly. Rust around the valve stem could indicate that the wheel is rusting and might need to be replaced. If you feel your tractor tires need more weight for traction or stability, there are some new nonsalt products available that your tractor dealer can order for you. Weight can also be added with wheel weights.

Front tractor tires can be purchased in the price range of truck tires, approximately $50 for two-wheel-drive tires and $80 to $90 for four-wheel-drive tires. However, rear tractor tires will be somewhere around $125 per tire used or $295 new. When considering a tractor's price, therefore, don't overlook the condition of the tires, as it may greatly affect whether you are getting a good deal.

Two-Wheel Drive or Four-Wheel Drive

Although a substantially priced option, four-wheel-drive (4WD; also called 4×4) tractors are becoming more popular because they provide added traction in ice, snow, and mud and during rough or hilly work, and they tend to hold their resale value better than two-wheel-drive (2WD) models. The extra weight of the front axle mechanism in 4WD tractors helps to even out weight distribution, which increases balance and stability, increases pulling power, and adds traction for the use of a front-end loader. If you opt for a 2WD model, look for one with a differential lock that locks the rear wheels together (like PosiTraction), which will help keep you from getting stuck.

Power Steering

While manual steering is fine for long straight lines, if you are maneuvering in tight spaces, as you would when cleaning pens, power steering is very handy.

Age

Don't be overly concerned about the age of a tractor. Because we have become accustomed to buying a new car or truck every so many years, we often look at other vehicles that way too, but with tractors, that is false economy. If it runs and is paid for, the extra age of a tractor just adds to its character. And the quality of workmanship is really fine on most older tractors. There are a lot of very usable, good tractors from the 1950s and 1960s. Just be sure that parts are available for repair.

Good equipment retains its value and in some cases actually appreciates. In 1948 a brand-new 8N Ford was sold right off the showroom floor for $900. That same tractor is worth $2000 today.

Wear on tractors is measured by hours of engine operation rather than calendar age. If you're buying a used tractor, try to purchase one with fewer than 5000 hours on it.

Warranty

New tractors usually have a full 2-year or 2000-hour warranty for parts and labor. Used tractors might have a 90-day parts-and-labor warranty if reconditioned and sold through a dealer, but otherwise used tractors are usually sold "as is"; that is, "Buyer beware." Although you will pay more for a reconditioned tractor than an "as is" tractor, you will have the peace of mind and financial assurance that if anything goes wrong, it will be fixed. Most things that are going to require fixing seem to turn up within the first few weeks of operation. An "as is" 8N Ford might cost $1500 while the same one reconditioned, from a dealer, with a warranty, might cost between $2000 and $2200. Dealers have to add at least a 25 to 30 percent margin to the price for their risk. A dealer may also offer a fifty-fifty engine drive train warranty for 90 to 120 days, wherein the cost of any repairs will be split between the dealer and the customer.

TRACTOR MAINTENANCE

For every $1 spent in maintenance, $10 is saved in repairs. Completely service your tractor every 150 hours of real working time. This can be monitored by the engine hour gauge or by using a logbook. Each time you service your tractor, change the oil, replace the oil filter, clean or replace the fuel filter and the air cleaner, and grease all zerks. Check hydraulic fluid levels when servicing and regularly in between servicing. Keep radiator free of all bugs, dirt, and chaff.

If you are in a temperate climate, every fall be sure the antifreeze in the tractor's radiator is good for –30°F. This maintenance item is not to be taken lightly; the damage that occurs when an engine block freezes and breaks is very costly to repair.

Tractor repairs can be very expensive. It is best to spot little problems before they become big ones. Just like you give your horse or your tack a daily check, on a regular basis walk around your tractor to identify any loose, broken, or missing parts.

Before operating the tractor, visually check for oil leaks on the engine, transmission, and axles. Correct leaks as soon as possible. Hydraulic and diesel fuel leaks can occur under high pressure; these require special diagnosis and repair. Place a piece of cardboard under your tractor when you park it to help you locate suspected leaks.

While operating your tractor, keep your eyes and ears open for odd sounds or change in operation.

Maintenance checks

- ☐ Using the dipsticks, check fluid levels regularly for engine oil, power steering fluid, and transmission oil. On 4WD tractors, check the front axle oil level according to your manual.
- ☐ Check radiator level regularly, especially in hot weather. Verify in the fall that the antifreeze will protect the tractor to –30°F or lower if you live in a temperate climate.
- ☐ Periodically check all hoses, belts, and clamps.
- ☐ Perform all routine maintenance as prescribed by your owner's manual or by the tractor manufacturer.
- ☐ Grease all fittings every 50 hours or as specified in your operator's manual.
- ☐ Change the engine oil and filter approximately every 150 hours or as specified in your operator's manual.
- ☐ Change transmission oil and power steering oil and filters approximately every 300 hours or as specified for your tractor. (Use the same interval for front axle oil in 4WD tractors.)
- ☐ Drain and flush the radiator annually.

Regularly check the following:

- ☐ Fuel filter sediment bowl — empty and clean.
- ☐ Radiator and air screens — blow off debris.
- ☐ Air cleaner — remove and blow out; consider replacing every 50 hours if warranted.
- ☐ Battery — check cables and electrolyte level.
- ☐ Brakes — check and adjust the brakes regularly and ensure that they are evenly adjusted.
- ☐ Tires — monitor for checks, cuts, bulges, and correct pressure. Service or replace tires as needed.

Tractor Implements

Avoid package deals (tractor plus certain implements) unless they contain what you need. Don't buy an attachment just because the person selling it says it comes with the tractor. If you don't need a particular implement, ask what the price of the tractor would be without it. Choose only those items that are in very good working condition. When in doubt, get professional advice.

LOADER

A front-end loader is a large scoop bucket mounted on the front of the tractor. Pipe-frame trip-bucket loaders, common in the 8N Ford era, raise and lower hydraulically, but operation of the bucket itself is an all-or-none situation. Because of this lack of finer control, trip buckets are notorious for skimming over the top of what you are trying to load or digging in too deep and gouging the earth underneath.

Double-action hydraulic buckets are much more versatile than trip buckets. They allow you to adjust the position of the bucket to a much greater degree. One advantage of this is that you can sprinkle the material you are unloading rather than just dumping it in one pile, as is the case with a trip bucket.

With any loader, look for signs of stress: bowed arms, welded repair spots, worn-out pins and bushings, and leaking hydraulic cylinders.

A quick-attach option for loader buckets is an extra cost, but handy if the bucket obscures your view and you want to remove it when not needed or you require different bucket sizes for different jobs: for example, large bucket for sawdust, small bucket for gravel. Also, with a quick-attach option, you can easily remove a bucket and attach a front-mounted posthole digger, a hay spear or tines, or a fork for moving pallets or posts.

Top right: **Deep snow is difficult to plow; a front-end loader helps Richard scoop and move the snow elsewhere.** ***Right:*** **When it's time to harvest the humus from the compost pile, a front-end loader makes Richard's job loading the manure spreader easier.**

DISC

A disc is a plow that uses rows of evenly spaced circular discs to turn over soil. A disc is handy for lightly working the soil in an arena or for aerating a very compacted pasture. Pull-type discs are common and inexpensive because they are very difficult to transport. They cannot be raised and lowered like hydraulic or three-point discs. To be moved, pull-type discs have to be either lifted with a loader onto a trailer or dragged behind the tractor, discing everything along the way, including driveways, road surfaces, and grassy areas! In addition, if you use a pull-type disc to work an arena, you will not be able to back deep into the corners and will end up with an oval area of worked soil.

Hydraulic discs are raised either by the tractor's three-point hitch or by a hydraulic ram on the disc. Smaller hydraulic discs (6-foot) are operated off the three-point. They are raised in the air to be carried from point A to point B. Larger, heavier discs have tires that are lowered to carry the weight of the disc for transport. Hydraulically raised discs prevent damage to areas that you do not want worked. They also allow you to position the disc in tight spots. Using reverse gear, you can raise the disc; back it into the corners of your arena, field, or pasture; then set it down and work the earth right up to the fence line.

Top: The steel discs of this hydraulic disc cut into the soil and turn it over. *Bottom:* Discing breaks up the soil in a hard-packed arena to prepare it for harrowing.

An average 6-foot three-point disc will cost about three times as much as a pull-type disc. Both, however, do an equally good job of discing. Old discs with box bearings should be avoided because box bearings are very difficult to find today. It is best to stick with discs that have sealed bearings.

HARROW

Harrows break up and smooth plowed or clumped soil in preparation for creating a seedbed or a smooth working surface. Harrows are useful for smoothing an arena after discing, breaking up and spreading manure in pastures, and aerating compacted soil.

There are basically five types of harrows, or drags, as they are sometimes called: the chain (or English), the spike tooth, the spring tooth, the rotary, and the combination drag.

English harrows are made of heavy rods that crisscross each other in a diamond-shaped configuration and have protrusions called teeth on the bottom side. They are very heavy and expensive but do a wonderful job of smoothing rough spots. They are good for leveling manure in a pasture as well as for aerating the soil without ripping it up. Homemade drags, simulating the English style, have been made with chain-link fence, but the lack of teeth and their light weight make them bounce on top of the soil and result in little smoothing and leveling. Like pull-type discs, English harrows are difficult to load, and when you move them by dragging them behind the tractor, they harrow everything in their path.

The teeth of the *spike-tooth harrow* look something like old railroad spikes but are adjustable for

Top: **A rotary harrow smooths and levels the arena footing as it rotates behind the tractor.** ***Bottom:*** **Once a year, we dismantle all the pens and level and resurface the footing; that's when I find that a blade comes in handy.**

work or transport. You can use them in a full upright position to rip up pastures in the spring or you can set the teeth in a flat position so you can use the harrow more as a leveler and smoother. After long or hard use, the teeth will become rounded and/or short, but most spike-tooth harrows have replaceable teeth.

The *spring-tooth harrow* has half-moon spring steel bars, making it a cross between a ripper, a mild plow, and a harrow. Its configuration allows it to rough up the earth, but it does not do a very good job of smoothing or leveling the soil.

The *rotary harrow* rotates as it is pulled behind a tractor. The welded tubular circle with crossbars has teeth that aerate and smooth the soil as the harrow rotates.

The *combination drag* is several implements in one. It could include a chisel plow, leveler (float bar or paddle wheel), scraper box, pulverizer, and roller. It usually is a freestanding implement with two tires.

BLADE

A blade is a cutting edge and moldboard mounted on the rear of the tractor for pulling or rearranging dirt, manure, bedding, or snow. The typical horse farm blade is a 6-foot, rear-mounted, three-point blade. The blade can be mounted at an angle or swiveled around so that it can be used either to push or to pull. A blade is handy for scraping pens, leveling driveways, and moving light snows.

MANURE SPREADER

Manure spreaders are wagons with a mechanical apparatus designed to distribute manure as the tractor is driven through a pasture or field. The smaller, older spreaders are friction-driven; a PTO powers the bigger, newer spreaders.

Friction-drive spreaders are ground-driven, the power for the mechanics of the spreader being generated by the tires rolling on the ground. There are two levers, one to control the speed of the apron chain, which moves the load toward the rear of the spreader, and the other to control the beater bar at the back of the spreader, which flings the manure into the air. This type of spreader can be operated behind a pickup or a team of horses, as it is a self-unloader.

Spreaders powered by a PTO are usually bigger, heavy-duty spreaders suitable for a commercial farm. They can be more difficult to hook up but have many advantages. They have larger capacity, and because the ground speed and the spreader speed can be controlled separately, the manure can be spread heavier or lighter or unloaded in one spot if desired.

MOWER

Mowers come in various styles and are designed to cut hay, weeds, or brush. *Rotary mowers* are built as either pull-type or three-point hitch and usually come in widths from 5 to 20 feet. You can raise and lower rotary mowers to adjust mowing height. *Sickle-bar mowers* use a bar of cutting blades pulled back and forth with a pitman arm that slides across a fixed-position bar to cut grass crops on relatively smooth surfaces. The sickle bars range in width from 6 to 8 feet and can be used for trimming banks as well as mowing fields. Although they cut cleaner than a rotary mower, they are more costly and don't have as wide a mowing height range as rotary mowers and tend to plug. *Disc mowers* are a cross between a sickle-bar and a rotary mower and result in less plugging than the former and higher ground speed than the latter, but with height adjustment similar to a sickle mower. If your pasture is uneven, a rotary mower would probably be most appropriate. For cutting hay, a sickle-bar or disc mower would be best. If you have brush or heavy weed areas that need to be removed, consider a brush hog, which is a heavy-duty rotary cutter/shredder.

Top: **A small friction-drive spreader, suitable for a two- or three-horse operation, could be pulled behind a pickup or small tractor. The PTO spreader on page 139 is appropriate for a farm with more than four horses.** ***Right:*** **This 6' wide rotary mower can be adjusted to cut at various heights.**

A flatbed trailer like this one can haul building materials up to 18' long and up to 7000 pounds of hay.

POSTHOLE DIGGER

A posthole digger is a large auger that is positioned vertically and powered by the tractor to drill holes. The digger can attach to the tractor in several ways; the most common is to the three-point hitch at the rear of the tractor. This is less than ideal, however, because the only downward pressure that is applied is from the weight of the auger and the three-point hitch, which can make drilling in hard ground difficult. A more recent popular option is an auger that attaches to the bucket mounts or to the bucket itself to provide greater downward pressure from the tractor's hydraulic system. Used diggers are hard to find in good shape because either the augers are worn or the gearboxes are shot. You can't rent the type of posthole digger that is used with a tractor, although two-man gas-powered ones are usually available for rent. If you are just beginning to put in your facilities and see lots of fence posts in your future, it would pay to purchase a posthole digger for your tractor ($350–$500). Otherwise, it would be more economically sound to hire someone to dig the holes for you ($1–$5 per hole). Or you can dig the holes by hand using a hand posthole digger. You'll need one of these hand tools anyway to clean loose dirt out of the drilled holes.

FLATBED TRAILER

Choose a utility trailer that you can use with your tractor and your pickup. A common size is 16 feet long and 6½ feet wide with tandem axles. The weight limit on such a trailer is 7000 pounds, so if you wish to use it to haul hay, you can transport 3 to 4 tons at a time. A flatbed trailer is handy to move panels, gates, and posts around the farm with a tractor.

HAY BALER

A baler is an implement that can collect, compress, and tie wire or string around grasses, alfalfa, or straw. Balers have a pickup trough to bring up the grasses into an auger that moves them into position to be pressed by a piston. When the pressed grass reaches a configurable length, a mechanism is tripped to cause the string to be tied and cut. The bale is then ejected and a new one is started.

A hay baler, along with a mower, rake, field wagon, and other hay-making equipment, could run you $20,000. Add to this the cost of gas, baling twine, and your time. Unless you are putting up hay on at least 40 acres and enjoy doing it, it might be better to hire a custom baler to work your fields for you. You might get a quote of something like $1.50 per bale, or the farmer might offer to bale your hay on shares.

Unless you are experienced with the mechanics of farm equipment, think twice about buying a used baler — they can be quirky.

Sharecropping usually involves you supplying the field of hay; the custom farmer supplying the equipment, supplies, and labor; and the two of you splitting the crop (percentage varies widely).

Balers have a deserved reputation for mechanical quirks. They are amazing and intricate machines, and many things can go wrong. If a piece of equipment is going to break down on your farm, it will be the baler. They are difficult to understand and troubleshoot, expensive to work on, and will frustrate you, especially if you don't know anything about farm equipment! So unless you can justify baling your own hay economically, stay away from balers.

Machinery storage

Most compact tractors measure 7'3" at the top of the roll bar. Many new compacts offer a fold-down roll bar that breaks in the center to accommodate standard garage door heights of 7 feet. Most utility tractors measure between 8'2" and 8'6" to the top of the roll bar or factory cabs. When figuring storage space, measure your tractor or implement and add 1 to 2 feet on all sides for a buffer zone to make parking easier and allow you to walk around the machinery.

EQUIPMENT	APPROX. SQ. FEET
Utility tractor with loader (50–90 horsepower)	150–250
Plow	25
Disc (14 feet)	140
Harrow (14 feet)	100
Mower (5 feet wide x 7 feet long)	40
Manure spreader	130
Blade	25
Posthole digger	20
Cart	20
Trailer	160
Baler	130

CARTS

Many small tasks around your horse farm are best accomplished with a cart.

During chore time, a two-wheeled cart can handily haul feed behind a garden tractor.

Every horse farm needs one or two handcarts, depending on the particular use. Two-wheeled carts are more stable than wheelbarrows when it comes to moving heaping loads of manure or bedding.

- Lightweight garden carts are ideal for gathering and feeding hay dregs.
- For daily stall and pen cleaning, choose a medium-sized, lightweight but strong aluminum-frame manure cart.
- A heavyweight cart with tires is best for moving gravel and heavy loads.
- A large-capacity wooden cart with bicycle wheels is ideal for moving hay and bedding.

Fencing

11

Good fencing serves many purposes. Fences keep horses separated and in a particular place, away from the residence, lawns, crops, vehicles, buildings, and roads. Fences maintain boundaries and property lines and thus promote good relationships between neighbors. Good fencing is designed to keep horses from getting hurt. And attractive fencing can set off acreage and add to the value of the property.

Fences decrease liability because they lessen the chance of a horse doing damage to others' property or getting on the road and causing an accident, and fences keep people (especially children) and animals (especially dogs and other horses) off the property.

When it comes time to choose fencing, get ready to do some research and comparison. In the last 10 years, a great many new types of fencing for horses have been developed that increase safety, enhance appearance, and minimize maintenance. But it is important to choose the fencing that fits your situation and budget. It is likely that no single type of fence will be suitable for all of your plans. The ideal fencing will be different for particular needs, such as pens, paddocks, runs, pastures, round pens, and arenas. You may find it perfectly logical to have five or more types of fencing on your horse acreage.

As you choose fencing materials, be aware that for horses, the risk of injury due to fencing is greater than for other livestock. Because a horse's main attribute is movement, leg injuries, frequently associated with fence accidents, can put a horse temporarily or permanently out of service. Besides being safe, good horse fencing is sturdy, low maintenance, highly visible, attractive, and affordable.

No barbed wire!

Barbed wire is not a suitable horse fence because the sharp points, coupled with a horse's natural tendency to fight when tangled and flee when frightened, lead to horrible injuries. Barbed wire should not be used for horses, so it won't be discussed. *Never* use barbed wire for a horse fence.

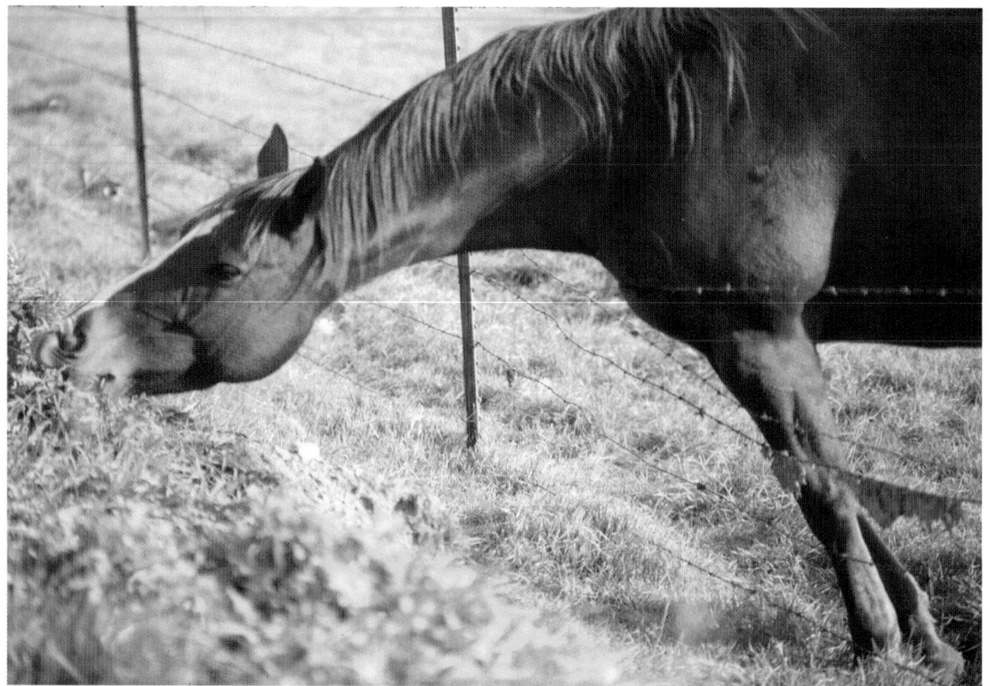

Above: **The risk of injury from poorly maintained fencing is high with horses, especially when crowded.** ***Right:*** **The grass is always greener on the other side, and this horse endures barbed wire on his chest and legs to reach for better grazing.**

Planning

To make a fence plan for your acreage, draw a scale map of your land on graph paper. Design a complete perimeter fence with entrance gates that can be easily closed and locked so loose horses are kept off your property and your horses are kept on your property. Consider a buffer zone (double fence) for perimeter fences to keep your horses away from neighboring property, buildings, or animals. Properly planned, a buffer zone can be used as a riding trail.

A property boundary line should be determined exactly by survey. Check your local and state fence laws to determine whether your perimeter fence should be set on the property line and the installation and maintenance costs should be shared by you and your neighbor(s), or whether your fence should be set inside the property line on your land and paid for and maintained entirely by you. This is an important distinction that should be made before you dig your first posthole. Errors can be costly and disputes can get ugly.

Draw on the map all permanent structures and objects, including things such as trees, water, large rocks, and buildings. Draw current and proposed traffic patterns. Locate places where gateways would be convenient and decide if they should be man gates, horse gates, or equipment gates. Then draw in fence lines on your map, including cross-fencing that you may plan for the future.

How much fence do you need?

NO. OF ACRES	FENCING (IN FT.)
20	3960 (or ¾ mile)
10	2640
5	1980
2½	1320
1¼	990

Note: Fencing is sold in 20- or 80-rod rolls and 100-foot rolls; 10 rods = 165 feet. (See diagram on page 60.)

When laying out fence lines, avoid acute angles where a horse can become cornered by other members of the herd, even if only in play. When running, whether from fright or exuberance, horses can go through or over fences.

Once you've drawn your fence map, use rocks, boards, or stakes and flags as markers to translate the proposed plan from paper to the land. Make adjustments. When you find a layout that will work, mark your postholes (inverted marking paint works well). Finally, calculate the number of feet and amount of materials necessary: corner posts, brace material, line posts, gates, fencing material (such as planks, rails, rolls of wire), and miscellaneous supplies such as bracing wire, staples, insulators, and electric wire.

To discourage horses from jumping, 4½ feet is the absolute minimum fence height that you should consider. Aim for a finished fence height of 5 to 6 feet (wither height), except for stallions, larger breeds, or those specifically bred and trained for jumping, which need an eye-level fence. For smaller horses and ponies, a fence that is just a bit higher than the withers is usually safe.

Fence height refers to the height of the top strand or board and includes the height of all horizontal elements of the fence, as well as gates. Use five or six rails or strands spaced 9 inches apart as the optimum, and adjust according to the size of your horses and budget. The bottom rail should be 8 to 12 inches off the ground.

Purchasing fencing materials in large quantities often saves you money.

5-acre fence plan

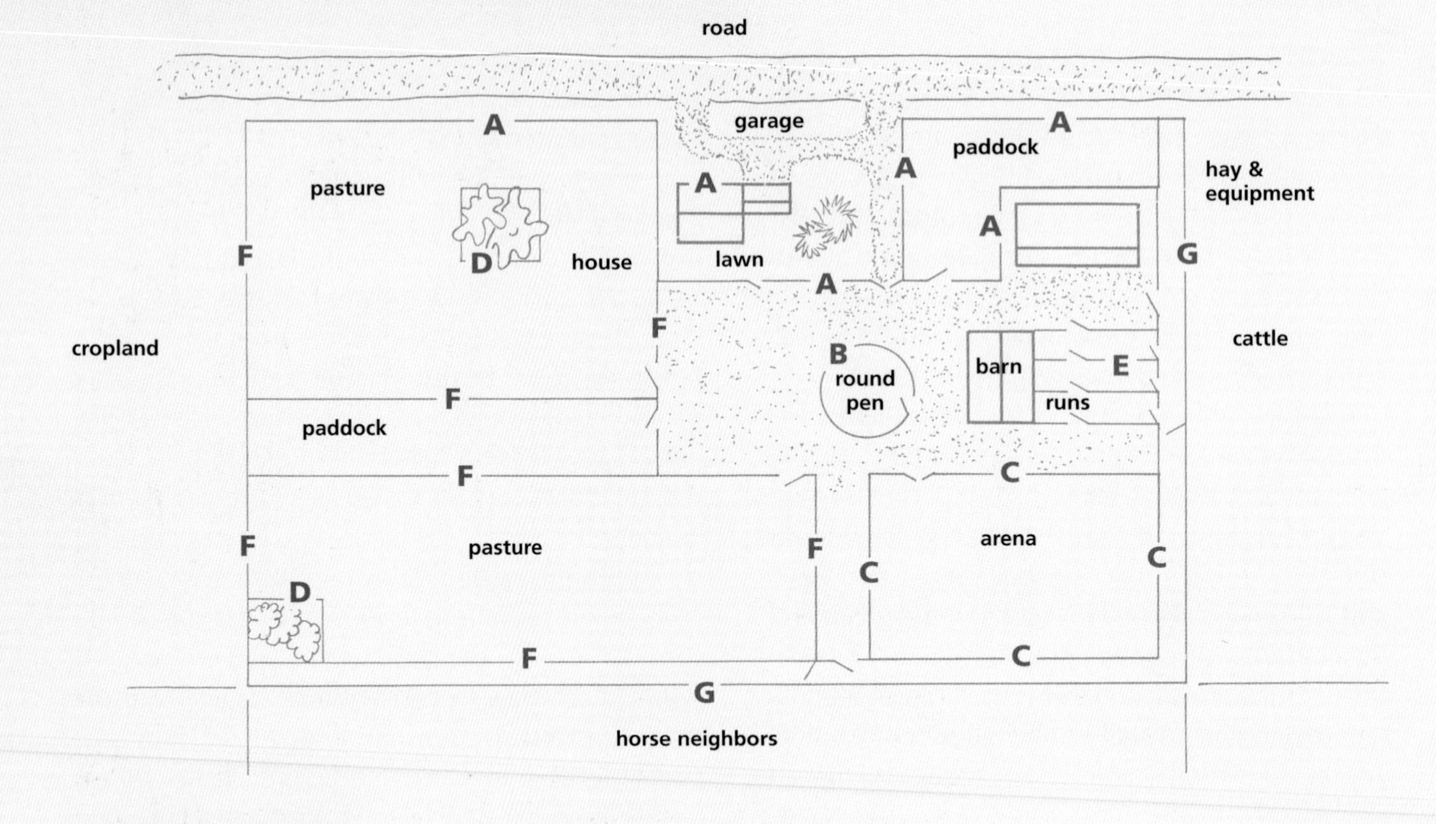

KEY

A: FOUR-RAIL POLY
For the front of the property and around the yard, choose an attractive, low-maintenance fence such as PVC post and board.

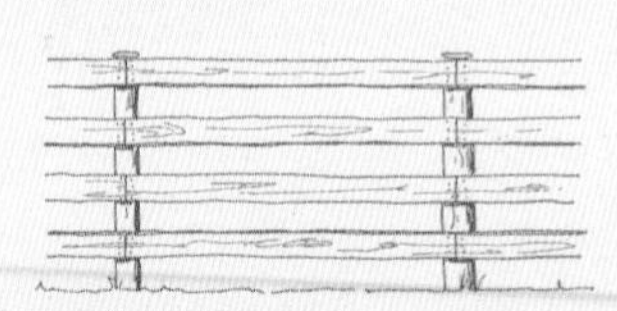

B: ROUGH-SAWN FOUR BOARD
Rough-sawn boards set on the inside provide strength and a good visual barrier for training pens.

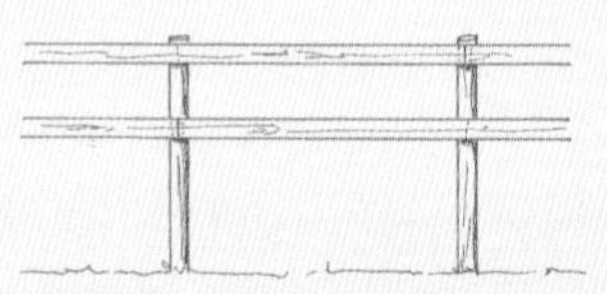

C: TWO-RAIL SMOOTH BOARD
An arena used only for riding, not for turnout, mainly needs a visual barrier provided by one or two smooth board rails set on the inside.

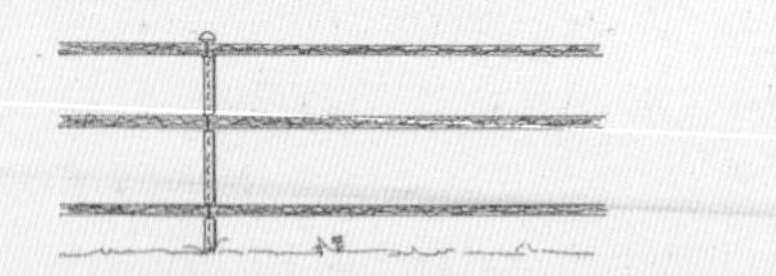

D: THREE STRANDS OF ELECTRIC TAPE Electric tape should not be trusted by itself for boundary fence, but it is ideal for temporary fence and for keeping horses away from trees.

E: 6' METAL PANELS It is hard to beat sturdy metal panels for fencing small areas such as runs that need to withstand a lot of horse contact.

F: CONTINUOUS PIPE ON WOOD POSTS WITH ONE ELECTRIC ON TOP
Continuous pipe fence on treated posts with an electric fence wire on top makes a very secure perimeter fence that a horse will not go through or lean over.

G: FIVE STRANDS ELECTRIC ROPE Where animals are on both sides of a fence, nothing commands fence respect like five strands of electric rope. Horses, cattle, sheep, pigs, dogs, and emus will look, but not touch more than once. And even without electricity, it is an effective physical barrier.

Posts

Although the type of posts that you use will depend on the style of fencing you choose, some information about posts relates to nearly all fencing systems.

Wooden posts are used with many other types of fencing besides plank and rail. Wooden posts are 6 to 10 feet long and are set 2 to 4 feet in the ground. Postholes are dug or drilled and the posts are set with tamped native soil, gravel, and often cement. If the ground is not too rocky, wooden posts can be driven into the ground using a mechanical post driver.

Preservatives are applied according to standards for use aboveground, with ground contact, or as a foundation (below the ground). It is important that fence posts are treated adequately for use below the ground. Four-by-four landscape timbers, although less expensive, are treated for ground contact, *not* for below-the-ground (foundation) applications. Each post or piece of treated lumber should have an ink stamp or plastic tag stapled to it that specifies the level of treatment, the intended use, the treating plant and date, and a corporation guarantee or the logo of an accredited inspection agency.

Posts are generally available in 2- to 8-inch diameter; the size refers to the diameter of the small end. Corner posts and gateposts should be 8 inches in diameter; brace posts at least 4 inches; and line posts at least 3 inches.

Steel posts cost less, weigh less, and don't burn in case of grass fire. They don't require a hole to be dug or drilled; they are set by driving them into the ground with a manual or tractor-driven post driver. They can help ground a fence against lightning, especially when the soil is moist. Steel T-posts are available in 6-inch increments from 5 to 8 feet. Generally, they are set about 1½ feet into the ground, just so the flange is below the surface of the ground. Metal fence posts are not as strong or stable as wooden posts, so it is best to add wooden posts to the line fence every 60 feet or so (closer in soft, wet, or sandy soil) to maintain the integrity of the fence.

Posts for electric fencing can be wood or steel (which require insulators) or PVC or fiberglass (which do not require insulators).

When setting posts, be very certain that the line for the fence is absolutely straight; if not, it will cause you repeated problems both in the attachment of the fencing and with future maintenance. Set the end posts first and then stretch a carpenter's building line between them to set up the line for the intermediate posts. If you are using roll fencing, such as wire or tape, unroll it between the end posts to make a

Strength of 7-foot posts set 3 feet deep

SIZE AND TYPE OF POST	BREAKING FORCE REQUIRED IN LBS.*
Wood 3" diameter	400–700
Wood 4" diameter	800–1500
Wood 5" diameter	1400–2700
Concrete 4" diameter	295–310
Metal 1" T-post	152–160 *(results in permanent bend, not a break)*

*Force exerted at the top.

How long will wood posts last?

TYPE OF POST	NO. OF YEARS
Untreated black locust and Osage orange	20–25
Treated cedar	20–25
Most other wood, treated*	15–30
Untreated cedar	15–20
Most other wood, untreated	2–7

*Treated pine is the most common wooden fence post.

Left: Notching the vertical post to accept the horizontal brace post makes a more secure corner. *Above:* A spike is driven through the vertical post into the end of the horizontal brace (see H-brace drawing on page 167). The corner brace wire is set in a notch and stapled to hold it in place.

Post length and depth

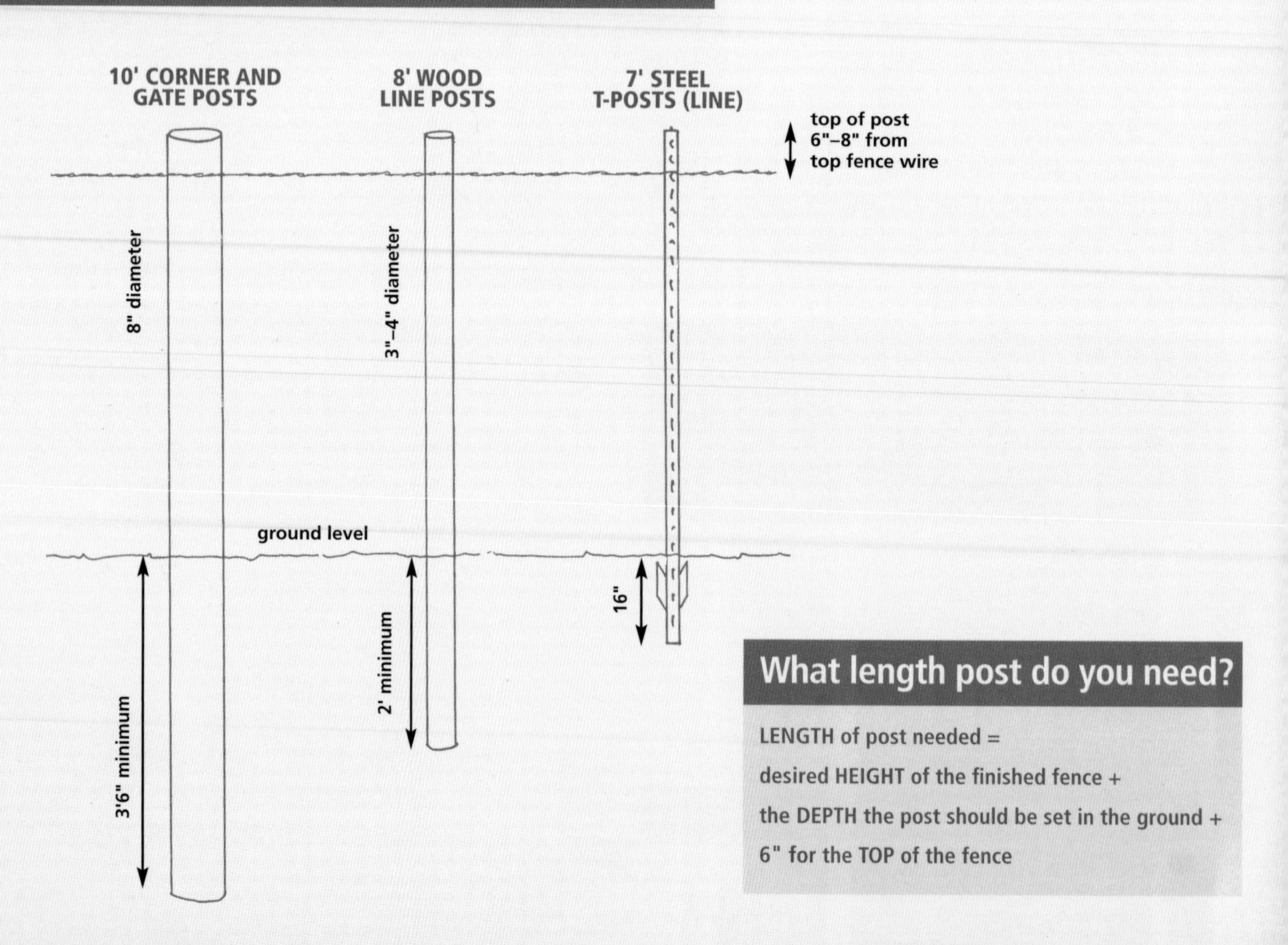

What length post do you need?

LENGTH of post needed =

desired HEIGHT of the finished fence +

the DEPTH the post should be set in the ground +

6" for the TOP of the fence

straight line. You can make an exception to straight lines with electric fencing, which can be installed in a curve.

Posts are usually set 6 to 15 feet apart, depending on the fencing material. Some material comes in continuous lengths, so the distance between posts is not as critical. But if you are using lumber or other materials that come in precut lengths, the distance between your posts must be accurate. It is better to make the distance between posts a few inches less than the lumber lengths so you can trim off any split ends of boards.

Top: The flange of metal T-posts should be set below ground level for safety. _Left:_ If ground is rocky and prevents you from setting T-posts deep enough, cover the exposed flanges with rocks or concrete. _Bottom:_ If you want curved fence lines, consider electric fencing that doesn't require corner posts or high tension.

Gates

Gates should be free swinging, light to handle, and roomy enough for the intended purpose. All gates should be the same height as the fence. Locate gates in the line fence, not in a corner where a horse or person could get trapped. Man/horse gates should be 4 to 6 feet wide, machinery gates between 12 and 16 feet wide. Walk-throughs are small spaces in the fence where a person can slip through sideways. These can be handy, but may also be a hazard for a horse if improperly designed. Gates used with rigid metal corral panels are usually a panel, a bow gate, or a gate in an arch frame.

Gate sag can be remedied by setting the gate on a rock, a cement block, or a wood block when the gate is open and closed. If your gate is angling down, you can lift it off its pinties (the pins on which it swivels) and adjust them. Screw in the top pintie and unscrew the bottom pintie to attain a more level gate. The fence on either side of a gate might need bracing; see the illustrations on pages 162 and 167.

Materials for gates vary as much as for the fences themselves. A wire stretch gate is inexpensive and lightweight and can be made to look just like the wire fence it is a part of. However, wire gates aren't handy to open, so would be most suitable for an occasionally used field gate. Wooden gates work as 4-foot man gates, but longer ones are often very heavy. Unless properly engineered, it would be difficult to keep a 16-foot wooden gate from sagging. And horses tend to chew wooden gates. Heavy gates should always have a support block at both the open and closed positions to prevent sagging and to extend the life of the hinges. Some tubular metal gates are light, sturdy, and safe. Flat metal farm gates with sharp edges should not be used in horse facilities.

Right: A rock set under the latch end of a gate prevents sag at the hinges. _Top right:_ Make machinery gates 12' to 16' wide, depending on the size of your tractor and implements. _Bottom right:_ A two-way, horse-proof slam latch is convenient to operate on foot or on horseback.

Walk-through

Gates

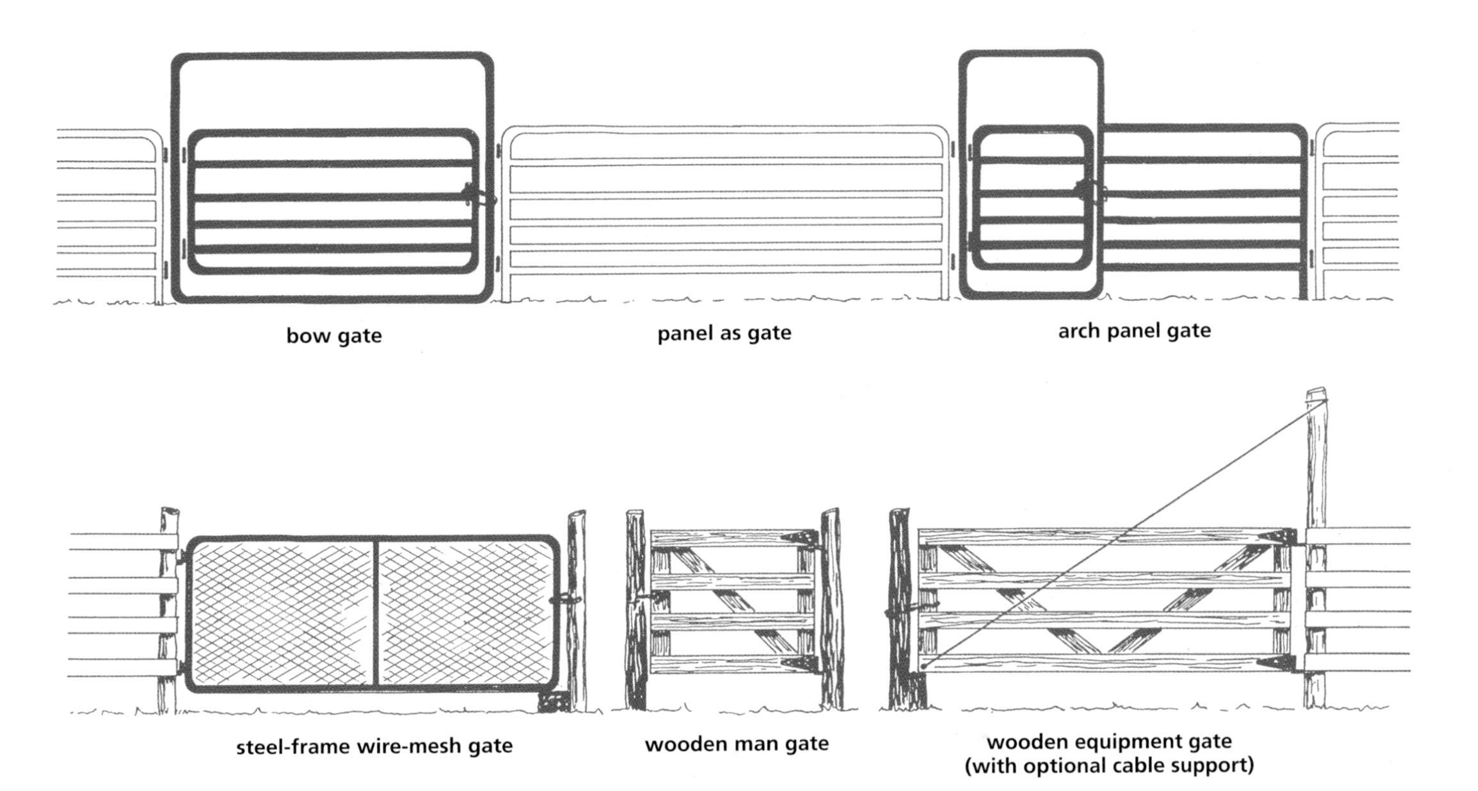

Gate and fencing tips

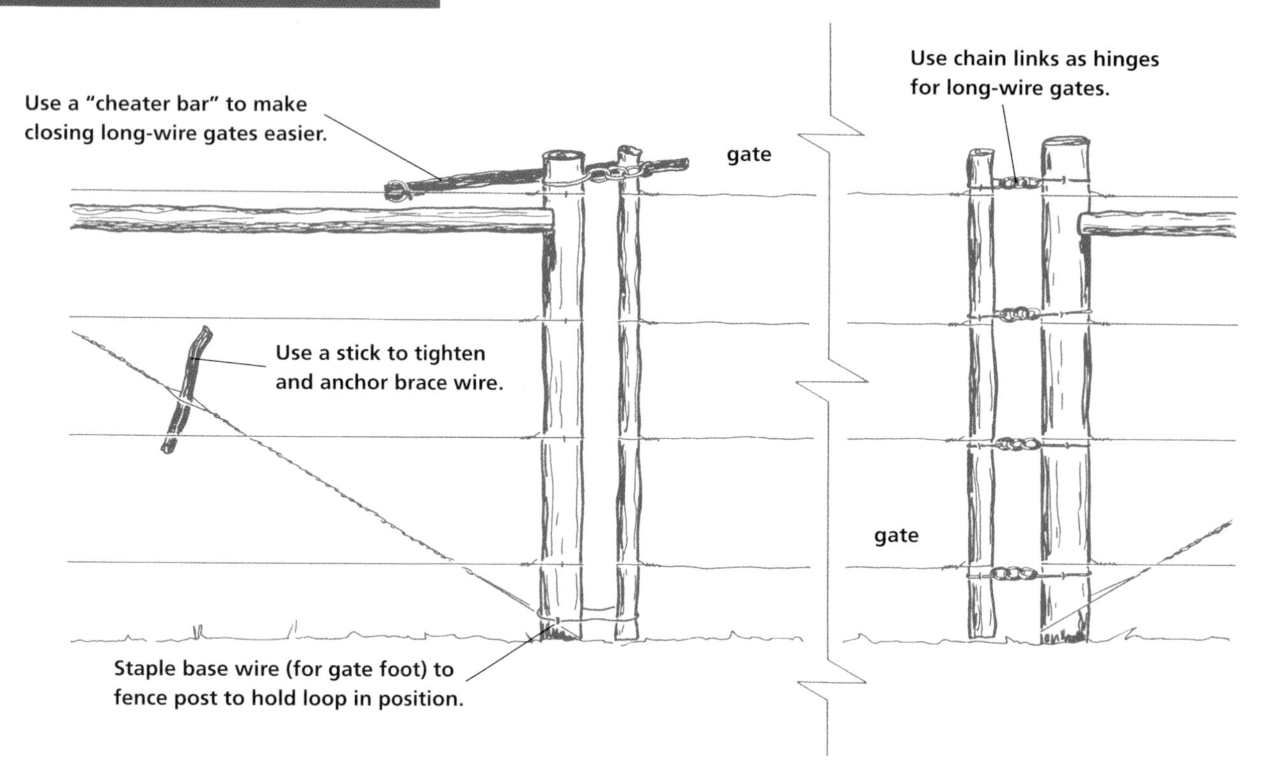

Gate support

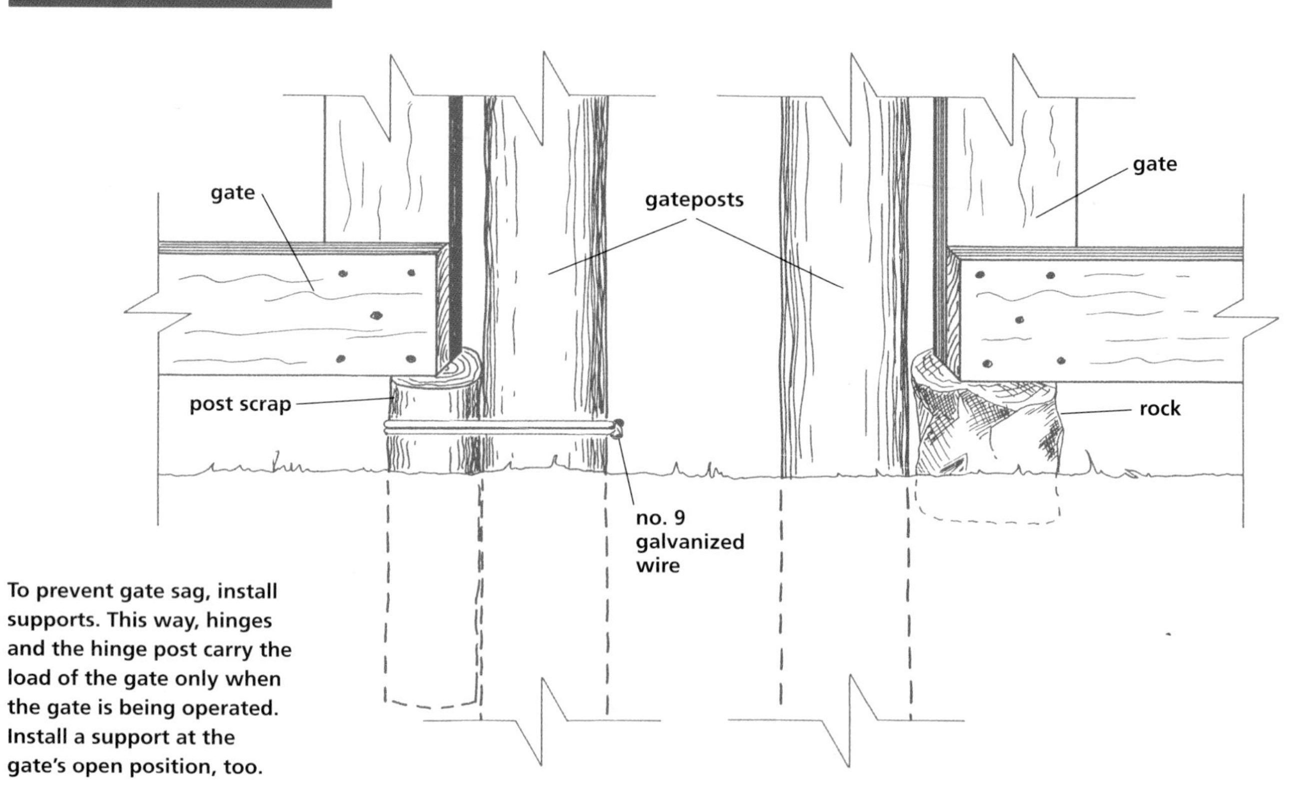

To prevent gate sag, install supports. This way, hinges and the hinge post carry the load of the gate only when the gate is being operated. Install a support at the gate's open position, too.

Fence construction overview

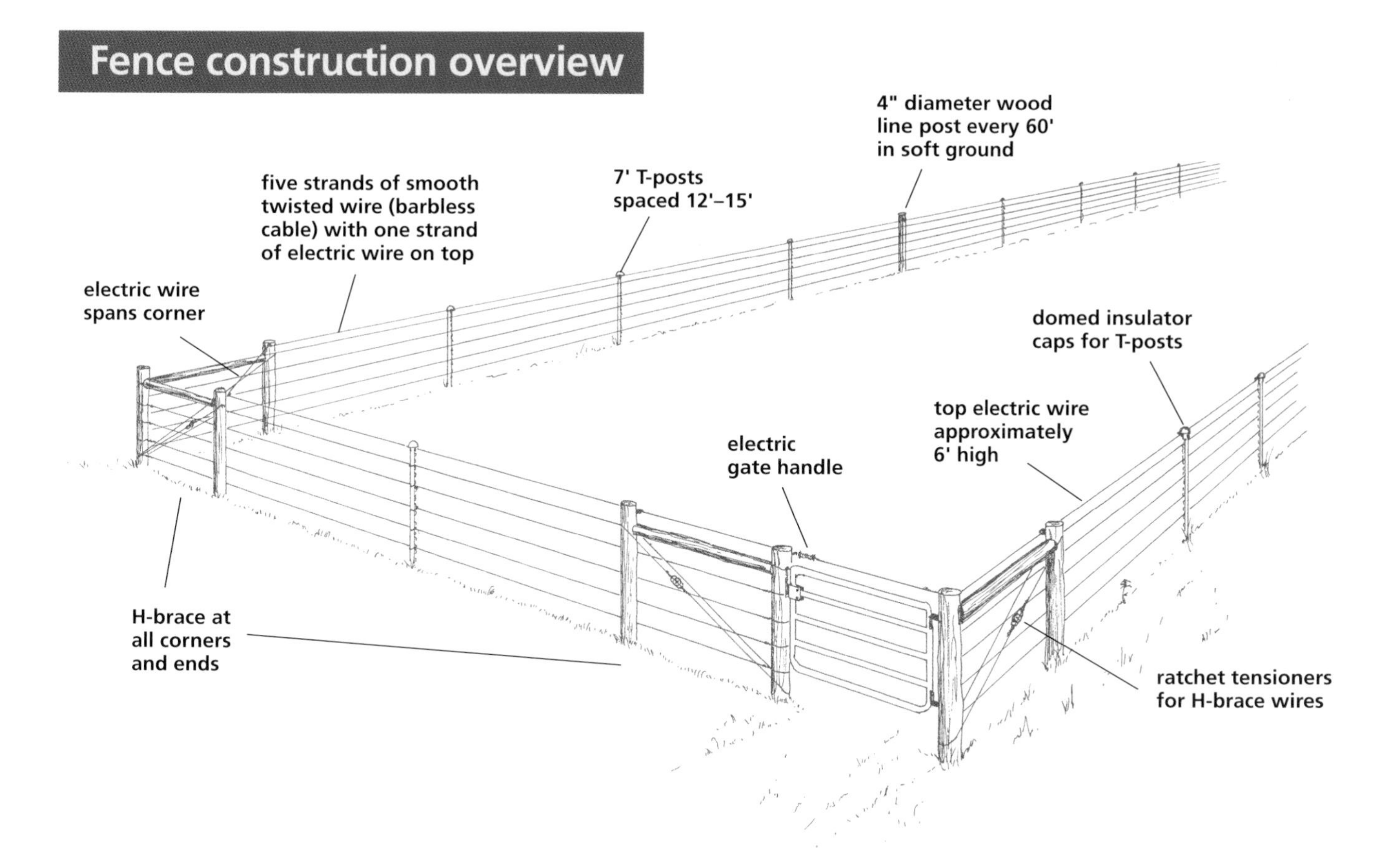

H-brace for gate or corner

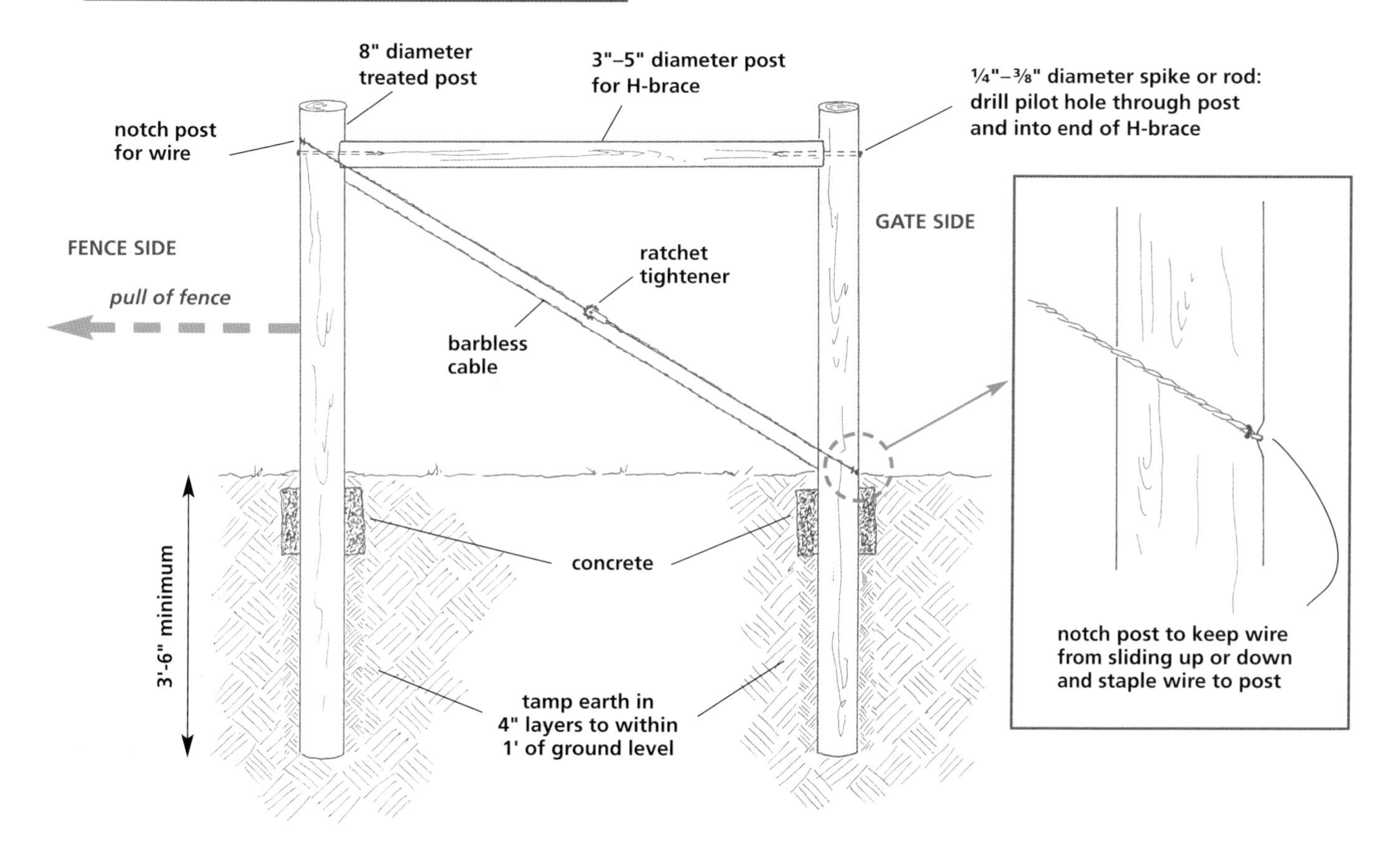

Fence Types

Most fences have either a continuous sheet of fencing (such as woven wire) or three to six rails, pipes, or wires, the bottom element being at least 1 foot from the ground. Any closer and a horse is more likely to get a leg caught between the fence and the ground when lying next to the fence. Continuous fencing can be stretched with a fence stretcher, a come-along, a truck, or a tractor. Fencing material should usually be put on the inside of the pen — that is, toward the horses — to prevent horses from contacting the posts or pushing the fencing off. The temptation is to put the railing on the outside of the posts for aesthetics, but with most fencing it is less secure that way.

WOOD FENCE

Wood fences are traditional and have a certain aesthetic appeal for a horse farm. If they are well installed and maintained, wooden fences increase property value. The three most common types of wood fences are board (plank), post and rail, and buck fence. Plank fences are strong if put up correctly, with boards nailed to the inside of the fence posts and the joints of the boards lined up evenly. Visibility, safety, and security are good with wooden fences, except when boards are nailed on the outside of the posts or when broken or splintered boards and exposed nails are not regularly repaired.

Wood should be kiln-dried and treated with a nontoxic wood preservative. Wood is used as posts, boards, rails, and as the poles in buck fence. Posts should be a minimum of 4 inches in diameter and up to 8 inches in areas requiring extra strength. Wooden corner posts and gateposts should be 8 to 10 feet long, set 3 to 4 feet in the ground. Board

Wood fence has a traditional look, and if well maintained is safe and attractive.

fences often have three or four rails and are made of 2-inch by 8-inch pine boards, as pine tends not to splinter. Pole fences, post and rail, and buck fences use 8-foot to 16-foot poles of varying diameters, peeled or unpeeled, treated or untreated, with three or four rails. Sometimes pole and split-rail fences are made of cedar, which weathers to a gray color and is resistant to decay.

Wooden fences require regular maintenance. They should be checked yearly to assess the need for paint or preservative. Broken, splintered, chewed, or rotten boards need to be replaced, but it is often easier to replace a board or two in a wood fence than to repair other fence systems. Expansion and contraction of the wood from changes in weather and moisture causes nails to creep out of wood, sometimes protruding an inch or more in a matter of a few days. All protruding nails should be regularly hammered to a flush position; using screws instead of nails is also an option. Wood fences with rails that fit into slots (mortises) in the post have a unique appearance, but unless engineered and installed properly, they are not well suited for horses. Rails can fall or pop out of the mortises in posts, and the post can split where the mortise has weakened it.

Wooden fences invite chewing *(top)*, require frequent maintenance *(middle)*, and can warp and twist *(bottom)*.

Left: Mortise-and-tenon fence is attractive and easy to construct, but the holes weaken the posts. *Right:* Pressure-treated wood does not prevent chewing; apply an antichew product.

Wood Preservatives

Wooden fences must be protected from chewing horses with special antichew products that are painted or sprayed on the wood, by metal strips fastened to the edges of the wood, or by the addition of electric fence wires to keep horses away from the wood.

A preservative is needed for wood that will be in contact with the ground, where conditions are ideal for rotting, termite damage, and fungus, and for wood that will be subjected to extreme conditions aboveground. Wood deteriorates not just from moisture, but also because an organism is eating it. Preservatives work by making the food source poisonous to these organisms.

Most preservatives will *not* prevent horses from chewing the wood. Most preservatives will not protect wood from exposure to sunlight; you still have to use paint, stain, or sealers. Not all preservatives provide protection against termites.

Preservatives can be forced into the wood using pressure (pressure-treated wood) or applied to the wood by dipping, spray, brush, or roller.

The main wood preservatives that have been used in the United States include creosote, carbolineum, pentachlorophenol (penta), and chromated copper arsenate (CCA). There is a host of new-generation preservatives based on borates and copper quat (ACQ) that are more environmentally friendly, including Preserve, NatureWood, and Wolmanized Natural Select.

Coal tar creosote. Coal tar creosote (a distillate of coal tar) has been used as a wood preservative since 1889 in the United States and is still used for treating railroad ties. It is a dark brown or black, thick, oily substance with a pungent smoky odor. Although it is not suitable for application inside a barn, creosote-treated railroad ties are ideal for the base of a round pen or arena. Because creosote contains strong acids and can burn the skin, care must be taken when handling creosote-treated wood. Due to the dangerous nature of the product, the U.S. government banned the use of creosote without a license in 1986. Creosote-treated wood is not paintable. It does resist termites and chewing horses.

Carbolineum. This thick, brown, oily blend is derived from coal tar creosotes and is available for application without a license. It can be used safely on trees that are at least 2 inches in diameter, fences, and other exterior applications. It is marketed as an antichew product as well as a preservative.

Pentachlorophenol. Penta is a manufactured crystalline organic compound developed specifically for the wood preservation industry. Penta is normally dissolved in petroleum oil and forced into the wood with pressure or applied topically, giving the wood the dark brown color that you see on telephone poles. Penta chemicals can slowly volatize into the surrounding air and so would not be appropriate for indoor application. Purchase and use have been limited to certified applicators since 1986. Penta-treated wood is not paintable. It should not be used where a horse can lick or chew it.

Chromated copper arsenate. CCA is a waterborne chemical preservative that contains arsenic, chromium, and copper, and until recently was the most widely used preservative for pressure-treating posts and fence lumber. The copper in CCA usually gives the wood a greenish color, and treating wood with it leaves wood paintable. The chemicals in CCA are bonded tightly to wood, so when used properly, leaching should be minimal. CCA-treated wood was approved by the U.S. Environmental Protection Agency for use around people, pets, and plants until its voluntary withdrawal for residential use; it continues to be widely used in agricultural, marine, highway, and industrial applications. CCA-treated wood emits no vapors, so is suitable for use inside the barn as well as for fences. Arsenic is a confirmed human carcinogen; anyone working with CCA-treated wood should take appropriate precautions.

Borates. Borates occur naturally in soil, water, plants, and animals. Borate wood treatments are used for protection against termites, beetles, carpenter ants, rot, and fungi. Borate pressure treatment penetrates to heartwood, so end cuts on treated wood do not need to be re-treated. Borates are water-soluble, however, and will leach out if used in contact with ground or water, reducing their effectiveness. Borate wood preservatives are odorless, nonirritating to skin and eyes, and considered safe to use around people and horses.

Although there may not be an EPA requirement to wear gloves when handling a particular treated wood, it is always a good idea to protect your hands from splinters when handling treated or untreated wood of any kind. When cutting treated wood, avoid inhaling sawdust or getting dust or wood chips in your eyes. Wash your exposed body parts well after working with treated wood. *Never* burn treated or other manufactured wood. Not only are the vapors and ash harmful to your health, but they will also pollute the environment. Keep treated wood away from waterways and out of the groundwater. Whether you can haul treated wood to your local landfill will depend on your area regulations and whether the landfill is lined. How to properly dispose of treated wood is a growing concern, and there is no simple answer. Here at Long Tail Ranch, we use scraps of treated posts and lumber as gate props, as building blocks for small retaining walls, and for edging between gravel and grass in nontraffic areas.

Use treated-wood scraps as a gate prop to prevent sag.

Buck fence rests on top of the land and so is ideally suited for rocky terrain.

BUCK FENCE

A buck fence is ideal for rocky or very irregular terrain where postholes would be difficult if not impossible to engineer. A buck fence is a series of triangles that sit on top of the ground and are connected by rails. Because it is usually made of long, thin poles (bark on or off), a buck fence looks very rustic. It would be an open invitation for wood chewing if used for a small confinement area. It is most suitable for large mountain pastures.

PVC FENCE

Usually referred to as PVC, polyvinyl chloride fence is available in several styles: post and board, post and rail, and post and ribbon. Used in traditional white, it makes a very handsome fence that never needs painting; PVC is also available in other colors. While expensive, PVC fencing has elastic action and is highly visible, so tends to result in fewer injuries and veterinary bills than many other types of fencing. Another advantage is that horses don't generally chew PVC. Today's PVC is highly resistant to ultraviolet rays. In humid climates it might be necessary to periodically wash it with mildew-removing agents.

POLYETHYLENE-, PVC-, OR VINYL-COATED WOOD FENCE

Planks and square wooden posts covered with synthetic sheathing are available in the traditional horse-farm white as well as brown and black. The ends of the boards and posts have protective caps to prevent moisture from getting into the wood. This fence requires very little maintenance; is attractive, durable, safe, and highly visible; and deters chewing. It is also quite expensive.

PIPE FENCE

Steel pipe makes a very strong, safe horse fence, particularly good for pens and runs, and needs little maintenance, especially if rustproof metal or paint is used. Pipe posts must be set in concrete. Often 2-inch drill-stem pipe is used for the top and bottom rails with three or four strands of cable or sucker rod in between. Pipe rails can be 20 feet long with posts set 8 to 10 feet apart.

Pipe fence offers little flexibility, however. It looks very businesslike, but some feel it lacks the aesthetically pleasing appearance of other types of horse fencing. Materials can be expensive, unless you are located near an oil field, where the 2-inch drill-stem pipe and sucker rod are available as salvage. A cutting torch, a portable welding unit, and someone to operate them are also needed.

CONTINUOUS FENCE

Continuous fence is made of tubular steel, which makes a tidy, safe, low-maintenance horse fence; it is most similar to the portable horse panels that are discussed at the end of this chapter. You can purchase components in various pipe diameters and choose the number of rails and spacing. Generally, the posts are set about 10 feet apart, and the pipes or bars are attached to the posts via metal clips. Depending on the manufacturer, you can choose from various types of pipe, posts, and connectors, and the materials can come painted or you can paint them after installation. Sometimes called *estate fencing,* continuous fence is custom fence that requires no welding.

Top: PVC fence is low maintenance, highly visible, and expensive; wood is high maintenance, traditional looking, and moderate in price. *Middle:* Steel pipe and rod fence is strong and safe and requires little maintenance. *Bottom:* It's hard to beat continuous tubular steel fence; it's tidy and can be customized to suit your needs.

Priefert

Wire gauges

9 gauge
10 gauge
11 gauge
12.5 gauge
14 gauge
14.5 gauge
16 gauge
17 gauge
18 gauge
20 gauge

Top: **Fence tightener.** ***Bottom:*** **Roll of two-strand smooth-wire (aka barbless) cable.**

WIRE FENCE

Wire fences include smooth-wire and mesh fencing. Wire fences have a galvanized (zinc) coating to delay rusting. The thickness of the coating determines the length of time the fence is able to resist the corrosive effects of the weather. Class 1 indicates the lightest coating of zinc, Class 3 the heaviest. Class 3, with two to three times the zinc coating as Class 1, can last 5 to 10 years longer, a feature especially important in humid climates where rusting is a problem.

Smooth-Wire Fence

Smooth wire is also called twisted barbless cable because it is like barbed wire but without the barbs. Smooth wire is economical, especially for large pastures, and when properly installed and maintained, it is a safe horse fence.

Mesh Fence

Mesh fence consists of crisscrossed wires fastened together to make a large screen. It is usually 4 to 5 feet high and comes in rolls 20 rods (330 feet) long.

There is an important difference between welded wire and woven wire. Welded wire is spot-welded together at junctions; woven wire is actually tied together with special knots. When a horse crashes into, kicks, or rubs on a welded-wire fence, the welds often break or the wire is permanently deformed. With woven wire the joints are more secure and the mesh is more flexible, so it is less easily deformed and is much easier to install over hilly terrain.

All mesh fencing needs to be stretched tight to prevent sagging and bulging. Especially during shedding season, horses love to rub on this type of fence, which adds to the sagging and bulging problem. Often, to maintain fence shape, electric wire, heavy-gauge smooth wire, pipe, or boards are used at the top of the fence to prevent horses from leaning over it. If boards are used, they will likely need to be protected by regular application of an antichew product or by an electric fence wire. An electric "stand-off" wire is sometimes installed at the height of a horse's hip to prevent rubbing and pushing on the fence.

Sassy waits patiently while I repair smooth-wire fence that had been stretched by deer.

Mesh fence is often set 6 to 12 inches off the ground to get a higher fence for horses, but this leaves a space underneath where a rolling horse could get a leg caught. Setting the fence a maximum of 2 inches from the ground will make it safer for mature horses, but a foal could still get a foot caught underneath. One solution is to bury the bottom of the fence a few inches, but this leads to early fence corrosion. A better fix is to attach the bottom of the fence to a treated board, like a 2×8, at ground level. The mesh should be securely stapled to the board on the inside to prevent chewing. This will also prevent horses from pushing out the bottom of the fence in search of grass and prevent small animals and children from crawling under the fence. But it will make the fence a better "trash collector" on its windward side and will make trimming grass along the bottom of the fence more difficult.

This well-designed V-mesh fence incorporates a top board to prevent fence sag when a horse leans over it. At just 5 feet tall, this fence is a bit short for the Warmblood shown.

Field Fence

The type of woven-mesh fence that has uniform rectangular spaces, from 6 to 12 inches, all the way up the fence is called *field fence* or *stock fence.* Because the openings are large enough that a horse could put a hoof through and get caught, it is not a good fence for horses.

Hog Panels

Hog panels are made of very heavy (4-gauge) welded wire and are from 3 to 5 feet high and 8 feet long. They have smaller, 2-by-8-inch openings at the bottom; 4-by-8-inch openings in the middle; and 8-by-8-inch openings at the top. Because they are so sturdy, it is tempting to use them for horses, but again, the large openings at the top could trap a hoof, so they are unsafe.

Chain-Link Fence

Chain link is a flexible mesh fence with 2-inch openings and comes in rolls from 3 to 6 feet high. It typically has 2-inch-square spaces and is stretched within a framework of horizontal and vertical pipes. It is relatively expensive and most appropriate along borders with high-traffic areas because it will keep horses in and keep people and all animals larger than a gopher out. Chain-link fencing has sharp ends at the top and bottom that must be protected by pipe or boards or buried. Some chain-link fencing is galvanized after it is woven, and the zinc coating can easily flake off. Once chain-link fencing is deformed from horse contact, it cannot be stretched back into shape. It is generally not appropriate for horse fence.

Nonclimb Fence

One of the safest mesh fencings for horses is "nonclimb," or 2-inch by 4-inch fencing, so named because it has spaces 2 inches wide by 4 inches high, small enough to prevent even a young horse's hoof from going through. It comes in rolls 2 to 6 feet high, typically has heavy 10-gauge top and bottom wires with 12.5-gauge filler wires, and is tough enough to withstand horse kicks. It is a woven mesh that is relatively flexible and easy to install. "Utility" fencing looks like nonclimb but is welded, not woven, and is not appropriate for horses.

V-Mesh Fence

V-mesh is another excellent mesh horse fence. It typically consists of 12.5-gauge horizontal wires spaced 4 inches with 14-gauge wires laced between the horizontal wires in a zigzag or V pattern. The resulting small spaces make it very safe for horses. Often, bulges in V-mesh fencing can be smoothed out by crimping the horizontal wires using fencing pliers or a claw hammer.

High-Tensile Wire Fence

High-tensile wire fences are single strands of smooth or twisted barbless (12.5-gauge high-tensile galvanized) wire held between end posts and with posts and battens or droppers between to keep the wires evenly spaced. End posts must be securely placed, and wires must be attached to intermediate posts in such a way that the wires can move sideways for retensioning (tightening), which is done once or twice a year.

Properly installed with in-line stretchers and tension springs, it can be stretched to withstand 200,000 pounds of pressure per square inch. This means it is strong, but also that if a horse collides with it or gets tangled in it (if the fence was not maintained), the fence will likely not break and the horse will be injured. Not a highly visible fence, it is most suitable for large pastures or rangeland. Fence flags or ribbons can increase visibility. One or more of the wires can be installed with insulators and be electrified to prevent rubbing. Closer post spacing increases visibility and decreases risk from entanglement.

Coated-Wire Fence

Compared to standard high-tensile fence wire, coated 12.5-, 14.5-, and 16-gauge plastic coated wire is more attractive, safer, and more visible. Cost can be up to ten times greater than for standard high-tensile wire, but post spacing can be farther apart due to increased visibility.

High-Tension Wire-Rail Fence

This plastic fence is one of the safest, most attractive, and most expensive fencing options.

It is from 2 to 5 inches wide and looks like a wooden board fence from a distance. It has two or three 12.5-gauge wires embedded in the plastic, so the fence can be stretched under high tension. It requires no paint. It flexes rather than breaks on impact and therefore tends to have a lower risk of injury. Post spacing of up to 12 feet is common because of the high visibility of the rails. This type of fence requires well-braced ends that take time to build properly.

Above: **Every other strand of this seven-strand, high-tensile wire fence is electrified. (Notice the insulating sheaths on the corner post.)** *Below:* **Wire-rail fencing is safe and attractive.**

RAMM Fence

ELECTRIC FENCE

Electric fencing falls into a category all its own and requires some specific comments. It can be used as a temporary or permanent fence and by itself or with other fencing. A temporary electric fence is most suitable for use when camping or when grazing is limited to a particular area within an already fenced pasture. Such fences are portable and easy to put up and are usually battery-powered. Electric fences can train horses to not chew, lean over, push, or rub on other types of fences. An electric fence system consists of wire, insulators, posts, and a grounded power source.

Insulators

The traditional insulator used with steel wire is ceramic. With the advent of plastic electric fence wire came plastic insulators. Insulators snap onto steel T-posts and are nailed or screwed onto wooden posts. Fiberglass posts require no insulators — the wires pass through holes in the posts. Generally, posts are set 8 to 15 feet apart.

Power Source

The power source is called a controller, a charger, a "fencer," a transformer, or an energizer. It is usually one of three types: plug-in, battery, or solar. Battery-powered transformers are handy for temporary or remote situations, but the need to replace or recharge the batteries every 2 to 6 weeks adds more cost and time. And as a battery begins to wane, it no longer has enough charge to do its job; thus it often leaves the fence insecure. Solar-powered chargers have photovoltaic cells that convert sunlight into electricity to charge the battery continuously. It takes only 3 to 4 hours of sunlight per day to maintain a charge, and a fully charged battery can last 3 weeks without a sunny day. A solar charger should be mounted facing south in a place where horses can't reach it. Plug-in electric transformers (also known as *mains*) that run off 110/120-volt household current are effective and economical.

The charger converts electricity (from a battery or a 110-volt service) into shock pulses that are transmitted to the fence wire. Some models emit a

Electric fencing

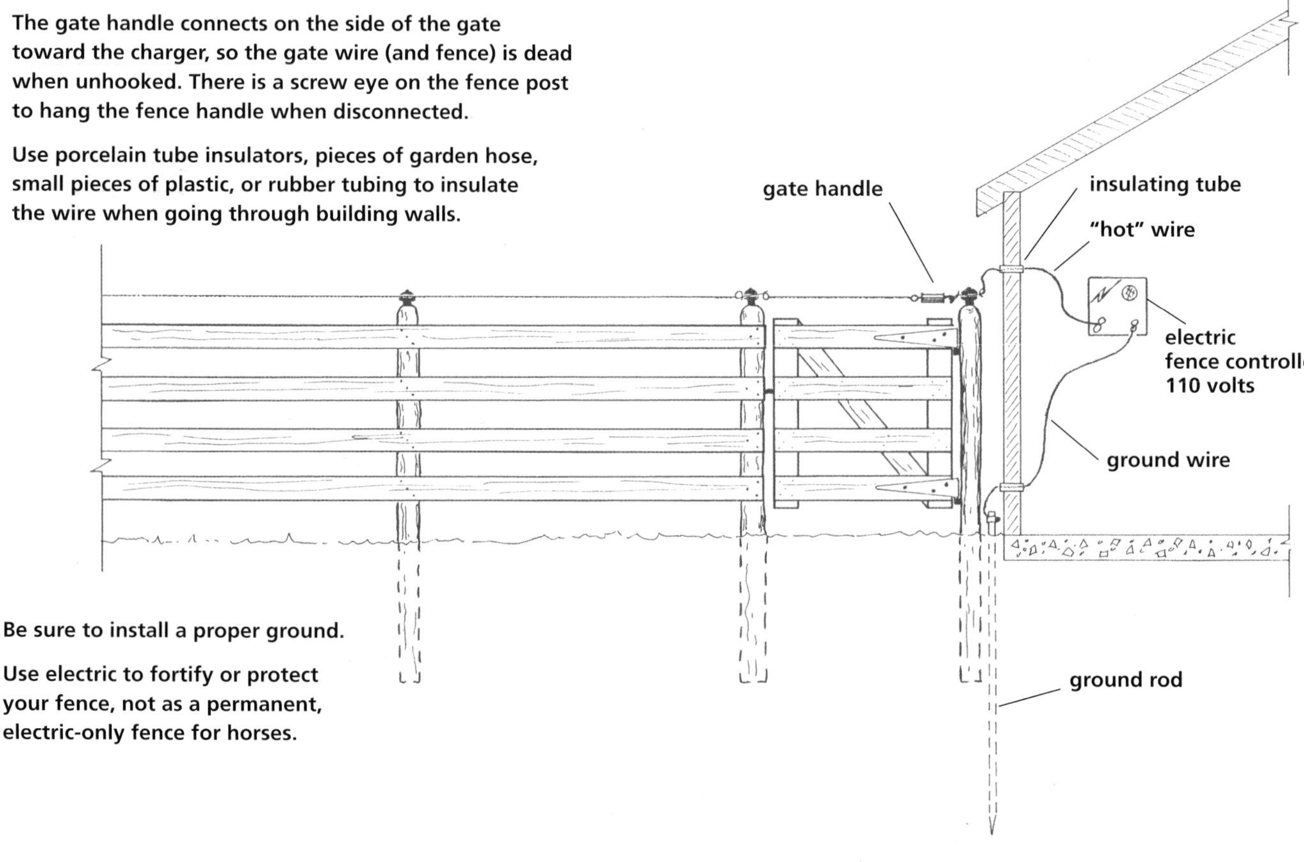
The gate handle connects on the side of the gate toward the charger, so the gate wire (and fence) is dead when unhooked. There is a screw eye on the fence post to hang the fence handle when disconnected.
Use porcelain tube insulators, pieces of garden hose, small pieces of plastic, or rubber tubing to insulate the wire when going through building walls.
gate handle
insulating tube
"hot" wire
electric fence controller, 110 volts
ground wire
ground rod
Be sure to install a proper ground.
Use electric to fortify or protect your fence, not as a permanent, electric-only fence for horses.

Above: **Electric fencing is great, but two strands as a perimeter fence along a highway is risky at best.** ***Top right:*** **The stand-off insulators and electric wire keep Zinger's band and Sassy's band from interacting over the fence.**

pulsating charge, others a steady one, and some can be switched from one to the other. Pulsating charges result in less chance of electrocution.

Choose shock power of at least 1 joule per mile or 2000 to 3000 volts, especially necessary for horses with long winter coats. The higher the joule rating, the stronger the shock and the farther the shock will travel down the fence. The farther down the fence a horse touches, the less of a shock he receives; the more weeds that touch the fence, the less of a shock it delivers. High-voltage units are highly resistant to being grounded out by tall grass and weeds.

In semiarid climates with dry soil conditions, an alternative is a wide-impedance charger, which increases the amount of power to the charger.

The power source (controller) must be properly installed. Follow the directions for your particular model, but here are some general guidelines. Some models require a clean, dry location where moisture cannot drip or blow onto the unit. Others are designed to be weather-resistant and can be mounted right on a fence post; however, it is still advisable to protect them by a weatherproof box. The exception

On this below-zero morning, Zipper waits patiently while I clear snow off the solar panel of a controller.

Electric fence wire choices

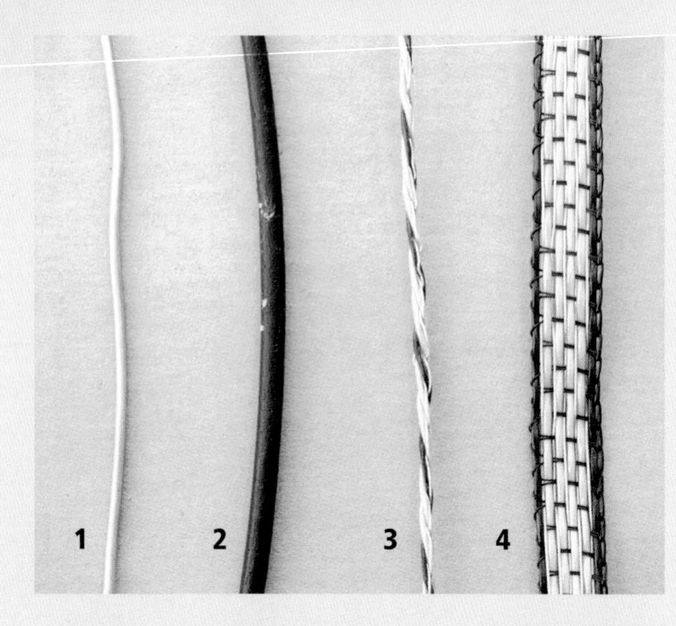

VOLTMETER

- *Solid steel wire* (photo 1). Use galvanized 12-gauge smooth or barbless twisted galvanized cable. Lighter gauges restrict the flow of the current and limit the efficiency of the fence.
- *Insulated wire* (photo 2). Use insulated wire to run an electric fence underneath a gate (see bottom illustration on page 183).
- *Polymer rope* (photo 3). 1/8- to 3/8-inch polymer rope or cord with stainless steel or copper woven into the rope; can be somewhat to very stretchy.
- *Polymer tape* (photo 4). 1/2- to 2-inch polymer tape or ribbon has high visibility and, if properly installed and maintained, can be safe. Not as easy to repair as rope or cord, and it is susceptible to wind damage. Good for temporary fence in pasture or on trail ride.

For horses and managers, polymer electric fencing has advantages over steel. Due to the colors of polymer, it is more visible than steel wire. And if a horse gets tangled, polymer will break more easily, whereas steel will often cut the horse. Polymer fencing can be put up or repaired with just a pair of scissors and a simple knot. It is light and flexible and does not need to be stretched tight, as does steel wire. However, polymer wire and tape require a greater power source to deliver the same jolt as through steel wire. Check the power with a voltmeter.

Fences by Premier
POLYMER ROPE

Fences by Premier
POLYMER TAPE

to this is the solar fence charger, which must be open to the sun.

If your power unit is inside a building, as most plug-in types will be, you can install a tube insulator through the wall so you can run the "hot" wire from the unit outside to your fence without it being grounded by the wall. A piece of small-diameter plastic pipe or rubber tubing works well.

Grounding

The controller unit itself must be grounded. In order for a horse to feel a shock, the current must travel from the controller to the wire to the horse to the ground (earth) and back to the controller. (That's why a bird can sit on an electric fence and not be shocked; it hasn't completed the circuit by touching the ground.)

Improper grounding causes many fence problems. The ground rods or tubes should be a solid copper rod and should be driven 6 to 8 feet into the ground, where they reach permanently moist earth. If you are in a very dry climate, you may need three or more ground rods driven deeper than 8 feet. This will help the controller work steadily and efficiently. Connect a copper ground rod to the controller using 14-gauge copper wire. A hose clamp works well to attach the wire to the rod. You can solder the attachment for a positive connection.

Using the Fence

You will be able to power from 3 to 25 miles of electric fence with your controller, depending on the rating of the model and the type of fencing you are charging. The rating indicates capacity for a single strand of steel wire and is based on using quality insulators and having no weeds touching the wire. If your model is rated for 20 miles and you run two strands of steel wire, the charger will power 10 miles of fence. Polymer wires and tapes add resistance to the flow of electricity and decrease the distance capacity to about 10 percent, so a charger rated for 25 miles would power 2.5 miles of one polymer rope. Low-impedance chargers are specially designed to maintain high power where polymer rope or tape is used or where weeds are a problem; they are also less likely to cause a fire.

The fact that the transformer is making a clicking sound does not necessarily mean the fence is working. Although horses seem to be able to sense when a fence is working and when it is not without actually touching it, few humans have that ability. You should purchase a fence tester so that you can check to be sure that ample current is flowing through the fence without having to touch it! A handheld voltmeter provides a specific reading and is easier to read in bright sunlight than a tester that lights a bulb.

You should formally train your horses to respect an electric fence by putting feed across the fence from them. Once they realize what this new fence means, they will be less likely to run through it. You may wish to or be required to post warning signs for people so they will be informed that a fence is electrified. (See top photo on page 177).

This solar charger has a gauge that indicates the level of power being transmitted to the fence.

Fences by Premier

Although these horses are very interested in something on the other side of the electric fence, they respect it and keep their distance.

Gates

Gates can be protected and reinforced with a single hot wire and an insulated gate handle. Design the electric gate wire to be dead when opened by having the handle disconnect from the side of the gate that is toward the charger. If the gate doesn't require protection and you don't want to unhook a wire every time you go through, you can bypass the gate by running an insulated cable under the ground. For this, use type UF cable (direct bury). Twelve- and 14-gauge two-wire or two-wire with ground are the most readily available, but you'll only need to use one wire. You could also use extra-tall posts on both sides of a gate and run the wire across the top of the gate opening. In such a case, the wire should be at least 8 feet high. A single electric fence wire can also be placed diagonally across the corners of fenced areas to keep horses from congregating there and chewing on wooden corner posts.

Short Circuit

A short circuit can significantly weaken the shock of an electric fence or ground it out and make it totally "dead." A short occurs when the electric fence wire touches a stationary object, most frequently a metal post or steel fence wire, and the electricity is carried to the ground. Weeds and grass can also ground out an electric fence, especially when they are wet. If the wire comes close to a grounding object without touching it, the electricity can arc and spark across the gap. Arcing can also occur between the ends of a broken metal wire strand in polymer wire or ribbon. This arcing can melt and weaken the plastic, causing it to break, and can even cause a fire if highly combustible material is close by.

Continuous electric gateways

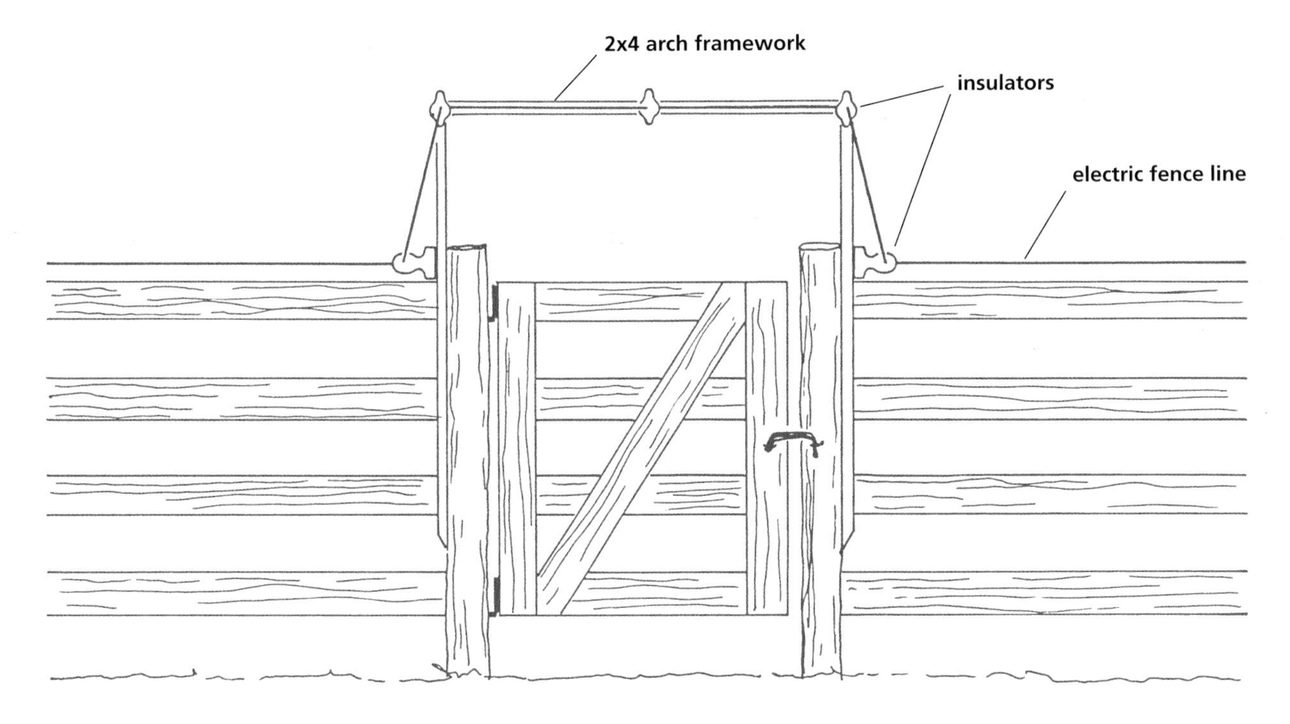

OVERHEAD

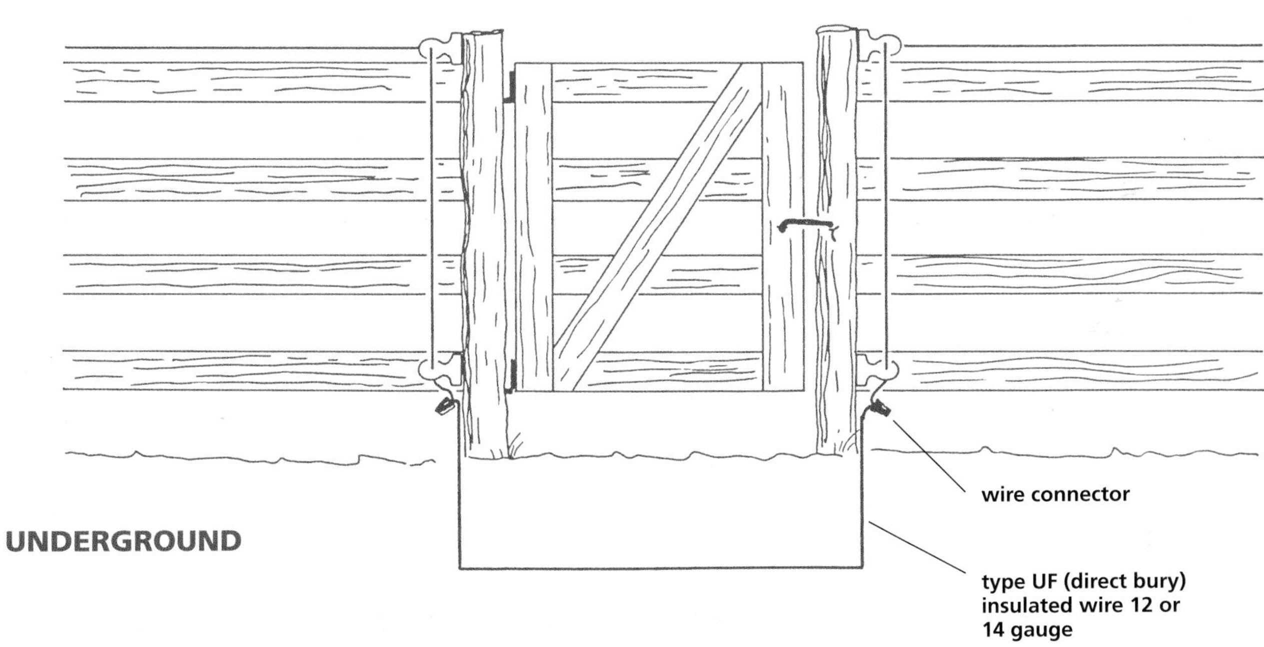

UNDERGROUND

Sassy and I patrol the perimeter fence looking for deer damage. Riding the fence is especially important if you run electric fence.

Riding Fence

On the large cattle ranches of the West, one ongoing task is checking the fence. Although you may not *need* to saddle up for this job, take advantage of the opportunity and include a fence check in your daily routine. Making small repairs often avoids larger problems later. When using electric fencing, find shorts or breaks before batteries are drained or the horses find out the fence is not working. In deer and elk country, this means twice a day, because these animals frequently break electric fencing when going over or through fences. Checking electric fences at dawn or dusk might enable you to hear and see sparks caused by arcing short circuits. During snowy weather, electric fences can become inoperable until the ice and snow melt or are brushed off the wires and charger.

Make a simple kit you can carry with you. Include a hammer, fence pliers, some spare insulators, staples, and fence wire.

Look for worn or weathered materials, protruding nails, loose wires, rails that have been broken or dislodged from the poles, leaning or loose posts, places where the fence might be sagging and need to be tightened, or signs of chewing. Check along road fences for litter such as cans, bottles, and items blown out of vehicles. Walking or riding the boundaries of your paddocks and pastures every day will also allow you to assess whether the land is being overgrazed or if it needs other attention.

MISCELLANEOUS FENCING

Some fences that appear on horse farms aren't designed specifically to contain horses. They are included here for the sake of completeness.

Stone Walls

Stone walls make very attractive fences reminiscent of the British countryside. They are virtually indestructible and also offer a windbreak. They can be used over rocky terrain where posts are difficult to set. The cost of labor to construct them, however, is incredibly high, and unless you have access to stones or rocks on your property, the materials and hauling costs can be prohibitive.

Hedges

Hedgerows make an attractive, natural fence, but in order to be effective they must be about 10 to 15 feet wide and 8 to 10 feet high; otherwise, horses jump or push through them. Hedges take up a lot of pasture space, so are suitable only for very large fields. They require good soil, climate, and moisture and take 3 years or more to reach fence size. They may also require periodic watering, pruning, and fertilizing. Horses do eat bushes, so yew should be avoided because of its toxicity.

Snow Fence

To keep a road or other passageway free from snow, place a snow fence 140 feet from the area you want to protect and perpendicular to the prevailing winter wind. Extend the fence 60 feet on both ends of the protected area. A traditional snow fence of red wooden pickets is half the price of plastic snow fencing and just as effective. Attach the snow fence to 6-foot T-posts spaced 8 feet apart, or to 5-inch by 8-foot wood posts spaced 12 feet. Leave a 6-inch gap between the fence and the ground.

Comparative cost of fencing

This cost comparison relates to initial cost of materials only; installation and labor costs will vary by location. No maintenance costs are considered in this comparison.

MOST ECONOMICAL

- Electric
- Smooth wire

MIDRANGE

- Woven wire
- V-mesh
- High-tensile wire
- Buck fence

MOST EXPENSIVE

- Pipe
- Wood board
- Wire rail
- PVC (polyvinyl chloride)
- Synthetic post and rail
- Polymer-coated wood

Not only is a stone fence attractive, but it also serves as a windbreak.

Fencing Turnout Areas

Although all fencing must be safe, be sure to choose the very safest fencing you can for small enclosures such as pens, runs, and paddocks. The smaller the enclosure, the greater the "pressure" on the fencing, the greater the likelihood the horse will get hurt, and the safer and stronger the fence should be. Make sure that corners are safe and that waterers and feeders do not protrude with sharp edges or create dangerous spaces where a horse could get caught. Be certain that no bolt ends protrude; use round-headed bolts (carriage bolts) on areas exposed to horses whenever possible. Design all gates to be the same height as and flush with the fence when closed. Low roof edges and the corners and bottom edges of metal sheds are particularly dangerous, and turned-out horses should not have access to them.

Keep absolutely no junk, garbage, or machinery in any area that horses frequent. Guy wires for telephone poles, power lines, or antennas located in pastures should not be accessible to horses. If you have power poles running through the middle of your pasture, ask your power company to provide plastic sleeves for the guy wires and/or tie something on the guy wires so they are more visible. Or you can set two 8-inch-diameter posts with a crossbar at the base of the guy wire to keep horses from running into it.

The areas where horses are turned out vary in size, footing, and amount and kind of vegetation present. *Pens* are at least the size of a generous box stall (16 feet by 16 feet) and are meant to be a horse's outdoor living quarters.

A *run* is usually a long, narrow pen specifically designed for exercise. A 20-foot by 80-foot run will allow a horse to trot; if you want your horse to be able to canter, provide 150 feet and make sure there is enough width for him to turn around safely at high speed at the end of the run.

A *paddock* can be thought of as a large grassy pen or a small pasture, ranging from one-half to several acres.

Pastures are large, improved, well-maintained grazing areas provided mainly for their nutritional value, with the added bonus of exercise. (See chapter 13, Land, for more information.)

PANELS

Metal panels can be used to make "instant fence" — enclosures of various shapes and sizes to house or train horses — round pens, turnout pens, runs, or outdoor covered "stalls." The great benefit of metal panels is their portability, making them ideal for a temporary situation as well as for long-term use. With panels, it is not necessary to dig postholes, as it is with permanent fencing.

Many livestock panels on the market today are designed for general farm animal use; they are usually not suitable for horses. Those designed for cattle (or sheep or hogs) are often too short, and many have leg-trapping connections or dangerously wide

These 5-foot-tall, general-purpose farm panels were too short for Sherlock as a yearling.

spaces between the rails. Some panels are too lightweight to use with horses; rubbing horses can easily move them, and normal horse activity can dent and bend them. If panels do not have design features that suit horses, they can result in serious injury and should be avoided.

HOW TO CHOOSE PANELS

When choosing panels, whether you are buying new or used, the first priority should always be safety. Select panels by reviewing features rather than just choosing a manufacturer's name, because most manufacturers make several grades of panels, from utility to premium to heavy duty. Horse panels should be of strong construction and need to be weather-resistant. It's a plus if they are easily portable, attractive, and affordable too.

Connectors

The way panels connect to each other affects their safety, ease of setting up, and stability. Certain types of panel connectors require that the ground be absolutely level for secure attachment. The gap between panels should be minimal so a hoof or leg cannot get caught. If there is a gap, the top of it should be blocked off so a hoof or stirrup cannot get caught. If you need to make old-style panels horse-safe, consider using *panel caps,* which are a simple, easy to install, and tidy way to block the gap on unsafe panels. They are tough, durable plastic connectors that are hinged to conform to any angle up to 90 degrees. They are adaptable to fit various sizes and styles of panels and gates and quickly attach to the panels with nylon zip ties.

It is ideal if panel connectors allow three-way and four-way corners, so that when you want to set up a group of pens, two pens can share a common panel and be connected properly at the end. Some connecting systems make this difficult or impossible.

A popular type of panel connector consists of chains that wrap around the adjacent panel, slip into a slot, and drop through a hole. There is essentially no gap between two chain panels when the chains are properly fastened and the pen is on level ground.

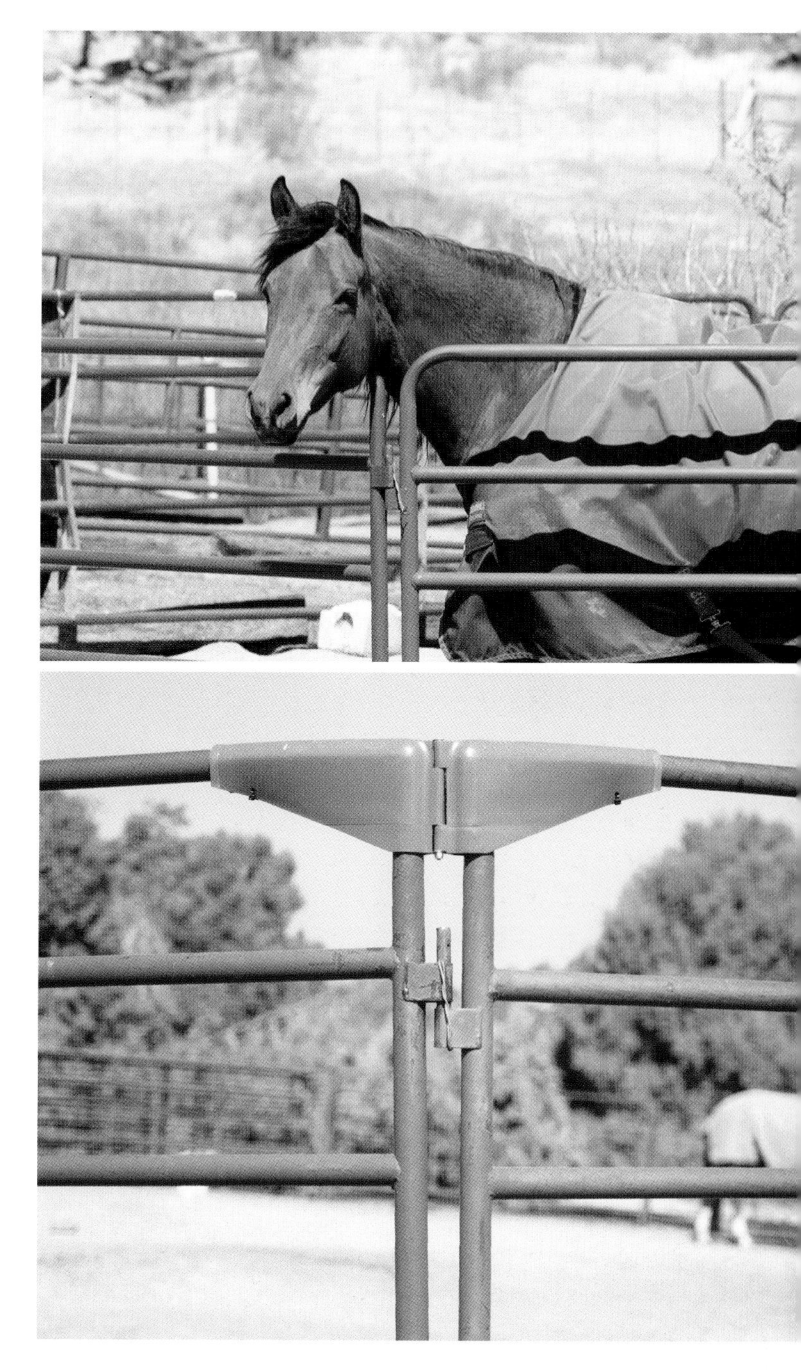

Top: **Seeker slides her neck into the dangerous gap between the farm panels, probably getting ready for a rub. Even more dangerous would be a rearing horse coming down with a leg in the gap.** ***Above:*** **To prevent leg-trapping accidents, cover the gaps; panel caps (by Panel Caps) work well.**

CHAIN-AND-SLOT CONNECTOR

RUBBER PANEL CONNECTOR

CLAMP CONNECTOR

The type of panel connectors you choose will affect ease of pen setup.

Another connector is a high-density molecular rubber strap that wraps around two panel verticals and fastens to a special hook. The gap between the panels is 1¼ inches. Because of the rubber connectors, the pen is somewhat flexible and quiet on impact. However, a rubbing horse *can* cause these connectors to come undone, and the rubber degrades after several years in the sunlight.

The "butterfly" or W-shaped connector consists of two two-piece clamp sets with one carriage bolt and nut per set. A wrench is required to connect the panels. When using the panels for a single enclosure like a round pen, the smooth bolt heads can be located on the inside for safety, but if installing the panels side by side as with runs, the sharper nut ends of the bolts might present a possible injury site or a place for rubbing for the horse whose pen they project into.

Panel connections can be loose or tight, depending on the design.

Some horse panels use the pin-and-bracket type of connectors. Steel pins drop into strap or pipe brackets that are welded onto the panel ends. Such connectors are easy to set up, and some are adaptable to uneven terrain. However, the gap between two panels can range from ½ inch to 3 inches, depending on the type of strap bracket, the levelness of the terrain, and the shape of the pen.

Panel Height

When choosing panel height, consider the height of the horses you have and the use of the panels. You need to decide whether you want to prevent horses from putting their heads over the top rail. Playing or fighting between horses over a top rail can be tough on both the facilities and the horses. And if you have to lead horses past a short rail pen, you run the risk of having the pen resident reach over and lunge at or bite the horse you are leading. Pen rails that are too short can also be a hazard if a horse rears and bucks when playing.

When used for training pens, panels with low rails allow a horse to evade the trainer by putting his head over the rail, and a short panel can seem like a viable escape route to a horse that is pressured during ground training.

In general, the taller the panel, the better. Several 6-foot-tall panels and a few 74-inch panels are on the market. Panels shorter than this would be more suitable for ponies and horses under 14.2 hands.

Rails

The number of rails, the spacing between the rails, and the distance between the bottom rail and the ground can make for a safer or more dangerous pen. Exactly what spacing is right will depend on whether you are housing ponies, mares and foals, or large adult horses or using the panels for training facilities.

Panels usually have five to seven rails (depending on the panel height), and the rails are 8 to 12 inches apart. An 8-inch space is ideal for housing adult horses, since a 1250-pound horse would need at least a 10- or 11-inch space to put his head through. However, weanlings and yearlings can easily put their heads through an 8- or 9-inch space. Panels with 12-inch spacing between rails are more suitable for riding pens and arenas than for housing horses.

The space between the last rail and the ground is an area of contention. While a generous space (16 to 20 inches) would allow a horse that rolled under a panel to get his legs free more easily, it would also allow a foal to roll completely out of his pen or a

The wide spacing of these panel rails invites Dickens to put his head through. He'll have a problem if he turns his head and tries to pull it out.

yearling to possibly get his barrel wedged under the bottom rail. Yet a tight space (8 inches or less of clearance) would make it difficult for any horse to get untangled without hurting himself. Choose the larger spacing if you have only adult horses. If you have young horses, choose the smaller space and check the horses frequently.

Feet

Loop-style skids, the most popular feet on horse panels, prevent the panel from sinking into mud, snow, soft earth, or gravel. They also make it easier for a person or horse to slide the panel across the ground. Loops can be square, rectangular, or triangular (J-leg). Depending on the length and depth of the loop, the skids can be another place for a horse to trap his legs or hooves. Loops that are 12 inches long and 12 inches deep are standard. Smaller loops than this should be avoided; larger loops create less of a trap.

Another style for panel feet consists of a separate flat steel plate with a vertical steel bar welded onto it. The vertical bar slips into the open bottom of the pipe panel end. Panels with feet like this can be a hassle to move because the feet either drop down and snag the ground or fall out and have to be reinstalled by a helper, making setting up or moving these panels a two-person job.

Length

Depending on the manufacturer, panels are usually available in lengths from 6 to 20 feet in 2-foot increments. Twelve-foot panels are the most popular size horse panel because they are fairly movable by one person. To make a custom pen, run, or corral setup, you might need panels of various lengths to fit your spaces. Some manufacturers have a wide variety of panel lengths and gates to choose from, while other companies offer limited sizes, colors, and accessories.

Verticals

Depending on the length of the panel and the manufacturer, each panel will have vertical reinforcements ranging from none to three. For example, some 12-foot panels have one vertical reinforcement; others have two vertical reinforcements.

Some vertical supports are made of 1- or 2-inch round pipe that has been cut to fit between the horizontal rails and welded in place. This results in a safe reinforcement that is flush on both sides. Others

Three examples of vertical reinforcements for panels: a horizontal cut to fit the vertical and welded *(left)*, continuous horizontal flattened where it crosses vertical *(center)*, and straps on either side of horizontal *(right)*.

use a single pipe that is flattened where it lies over one side of the rails. Another style of vertical uses straps (for example, two 14-gauge 1¾-inch straps with rounded edges), one set on either side of the rails.

The tops of the end posts of each panel vary in the way they are finished. Ideally, the sharp edge of the pipe should be covered and moisture prevented from entering the panel. The drier the bare steel interior of the pipes remains, the longer the panels will last. Some are capped with metal domes or plastic plugs. In others, the end posts are drawn into a distinctive arc and welded to the top rail. In another design, the corners are cut on a 45-degree angle and welded to make a weatherproof joint. The effectiveness of this style depends on the integrity of the welds.

Material

Most horse panels are made from either square, round, or oval steel tubing, usually 1⅝- or 2-inch stock. Oval steel tubing is round tubing that has been flattened. Oval tubing panels require less storage space when stacked (such as on the side of a trailer), so would be a good choice for a portable show or trail-riding pen. However, oval pipe panels can bend more easily than round pipe panels of an equivalent dimension and gauge.

As with fence wire, the thickness of steel is described by gauge. Fifteen- or 16-gauge steel is most commonly used for horse panels because it is the best compromise between strength and weight. Fourteen-gauge steel is thicker and weighs more, so it is more suitable for crowding cattle or housing rough stock. Eighteen-gauge steel is thinner and lighter and can bend and dent fairly easily from normal horse activities.

Weight

The weight of panels is determined by design, gauge of the steel, and the height and length of the panel. A horse panel should be substantial enough to stay in place and withstand normal horse activity, yet should be easy to move around. There is a great variety in weights of 12-foot horse panels on the market

Heavy-duty, round steel-tubing panels and gates such as these are used for confinement pens.

today, from 53 to 107 pounds. If you need a truly portable pen, opt for lighter weight but realize that you are most likely sacrificing durability and perhaps safety. Especially if you are using panels to make fairly permanent pens or runs, choose heavy panels that will withstand rubbing and general horsing around.

Finish

Horse panels are either galvanized or painted. Common panel colors are gray, brown, green, and silver (galvanized). The finish must be able to withstand horse rubbing, banging, and chewing as well as the effects of sun and moisture. One of the main factors that creates a durable finish is the paint and the way it is applied. A good finish deflects chewing attempts and discourages rust.

"Powder-coated" panels are electrostatically treated with a high-quality enamel or polyester/urethane paint that can result in a durable, long-lasting panel finish that will stand up to horse and weather abuse.

Galvanized panels are coated with zinc and are nearly impervious to chewing and rust. The best panels are galvanized *after* they are welded together. Panels that are welded after the pieces have been galvanized have painted welds that are more susceptible to rust.

Panel Gates and Latches

Most manufacturers offer more than one style of gate that will work with their panels. Most gate styles are available in 2-foot increments. Gates can be a separate 4-foot or wider bow (arch) gate; a bow gate incorporated in a 10-, 12-, or 16-foot combination panel; or a 4-foot or wider corral or panel gate.

If you are incorporating a pen or run off a building that has posts or off solid fence posts, consider using corral gates (much less expensive than bow gates) that hang on the post from hinge pins. Corral gates are usually available in 2-foot increments from 4 feet to 16 feet.

Latches should be horse-proof, safe when open or closed, and easy to use one-handed while leading a horse. When the gate is open, the latch should not have a portion (such as a protruding pin) that could snag a horse blanket or coat. Latches that have a secondary keeper or a feature (such as a hole for a snap or padlock) that allows you to secure the latch are useful for mouthy horses. When in doubt, secure gate latches with a short piece of chain and a double-ended snap or another means appropriate for the specific latch.

When cold, wet weather hits, many panel gate latches freeze. To prevent this, prior to winter weather, oil all latches with WD-40 or light machine

Left: A bored and curious horse can easily learn how to open a latch, so a safety chain is a must. *Right:* This "horse-proof" latch with keeper is easily opened by a rubbing horse; a safety chain prevents loose horses.

oil. If a latch does freeze, use a small propane torch to thaw it out.

Price

When it comes to price, it is important to compare similar heights, lengths, and quality of materials. Be sure you are not comparing the economy panel from one manufacturer with the premium panel of another. The durability will vary between them as well as the price.

The final price you pay will also be affected by tax and delivery charges. If there is no dealer near you, you might incur some out-of-state taxes and a considerable delivery or shipping charge.

Caution

Often the actual opening of the bow gates incorporated into panels is much narrower than indicated. The opening of a so-called 4-foot gate in reality could be as narrow as 40 inches, a space pretty cramped for leading a hefty, blanketed, or saddled Quarter Horse through. And even though many wheelbarrows or manure carts might fit through a 40-inch opening, a 48-inch space would make it easier to do so without bumping wheel hubs.

A 4-foot gate (with full 48-inch opening) with at least a 7-foot head clearance provides a safe passage for blanketed horses and manure carts. A 6-foot-wide gate with 9-foot head clearance is necessary for leading a saddled horse through or for safely using as a ride-through gate.

12 Arenas

To help your horse training go more smoothly, construct safe training facilities. The pens and arenas that you and your horse work in should be constructed of suitable materials, sturdy, and of the appropriate height and size. Footing should be safe, low maintenance, and all-weather when possible.

There are many options when it comes to designing training facilities. Visit horse farms or ranches in your area to see what works and what doesn't work in your particular climate.

I prefer a round pen with a diameter of 66' (20 meters) because it allows me to conduct longeing and long lining lessons as well as ride a horse in balance at all gaits.

Round Pen

One of the most valuable pens you can have is a safely constructed round pen. Besides providing a good place to turn out horses for exercise, a round pen is ideal for conducting training lessons such as sacking out, longeing, ground driving, saddling, ponying, first rides, and rider longe lessons. Also, trained horses can be tuned up and conditioned using work in repetitive circles. What follow are some round-pen specifics that have worked well for me.

SIZE

The size and construction of a round pen depend somewhat on its intended uses. A starting pen that is built solely for initial gentling and first rides might be best at 35 feet in diameter with solid 7-foot walls to ensure maximum control and safety in unpredictable situations. Solid walls tend to make the horse pay exclusive attention to the trainer. Although this can be an advantage with a snorty bronc, there comes a time when the horse must listen to the trainer despite outside distractions. And a small pen like this is not as versatile as a larger pen and thus may be suitable only for a horse farm that has many young horses each year.

A trainer will be less likely to get a foot hung up when riding in a solid-wall breaking pen than in a post-and-rail, plank, or panel pen. Still, being slammed into a solid wall or getting wedged between a horse and a wall can be dangerous. Scaling a 7-foot solid wall is difficult for a horse and next to impossible for a trainer. In addition, the lumber required to construct a solid-wall pen can be cost-prohibitive.

In contrast to a starting pen, a training pen designed for routine longeing, driving, or riding should be 66 feet in diameter to allow a horse sufficient space for balanced movement. A 66-foot-diameter circle corresponds to the 20-meter circle used in

Above left: This portable panel pen has solid walls on the lower portion that keep a horse's legs from getting tangled in the rails. Note the scuff marks on the bottom of the panels inside the arena. *Above right:* A 200' diameter round pen such as this is used as a general-purpose riding pen and for working cattle.

dressage training. Routinely asking a young horse to perform in a circle smaller than this can cause soreness and stress and can lead to unsoundness. A larger pen can make it more difficult for a trainer to control the horse. The walls of a training pen are often shorter and more open than those of a starting pen.

Larger round pens — those more than 200 feet in diameter — are useful for training horses to work cattle and can double as an arena for general riding and conditioning.

On most horse farms, the training pen gets many more hours of use than does the starting pen, so if you have to choose, it's probably better to opt for the larger pen.

CONSTRUCTION

Once the site for the round pen has been selected, the footing needs to be examined. If necessary, the ground should be graded to ensure a level training surface and proper drainage.

Then proceed to dig your postholes. To measure for the postholes, affix one end of a 33-foot rope to the ground at the center of the pen. Walk the rope in a complete circle as a final position check. Decide where the gate will be and mark holes for the gate posts about 5½ feet apart. (If you plan to use a small tractor to work the footing in the pen, make the gate 8 feet or wide enough for the tractor.) Then mark twenty-eight spots at about 7-foot intervals along the circumference of the circle for the rest of the postholes. Posts that are set approximately 7 feet apart, center to center, can use 8-foot boards and railroad ties with the ends trimmed.

The 10-foot posts are set about 3½ feet into the ground, leaving 6½ feet exposed for the wall. The 12-foot gateposts will extend 8½ feet above ground level. If the posts are tilted outward at an angle 5 to 10 degrees from the vertical, the rider's knees and feet will be less likely to take a beating when an inexperienced horse crowds the rail. Depending on the soil type in your area, you may need to use a concrete mix to stabilize the bases of the angled posts.

Set all of the posts except two, so that a truck and/or tractor can fit inside the pen for dumping the footing and leveling it. In many parts of the country, native soil is inappropriate for training surfaces, as it is hard and drains poorly. Four to 6 inches of decomposed granite or sand provides adequate cushion and drainage in most locales and for most training situations. Choose a footing that won't become compacted so it will have good shock absorbency. To end up with 4 inches of footing in a 66-foot-diameter

My 66' diameter round pen has walls that slant out about 8 degrees. Four 2x8 pine boards are nailed on the inside of 8" diameter treated posts. The posts are 10' long and set 3½' into the ground. Sand footing is held in by railroad ties cut to fit between the posts.

round pen, you'll need about 45 cubic yards (67 tons) of sand or decomposed granite. (See Footing, later in this chapter.)

Cut both ends of twenty-nine of the 8-foot railroad ties at about a 70-degree angle so they wedge between the bases of the posts to hold the sand in the pen. Placing the 6-inch side of the ties on the ground allows the 8-inch side to act as the most effective sand barrier. The ties should be wedged in from the inside of the pen and should be flush with the inside surface of the posts. You might want to stack a second railroad tie on top of the ground tie to further prevent footing from being pushed out of the pen at the rail.

Supply list

for a 66' training pen with 5'6"x5' gate

QTY.	ITEM
66-FOOT TRAINING PEN	
28	7–8" diameter 10-foot pressure-treated posts
2	7–8" diameter 12-foot pressure-treated posts
113	2x8 pine boards, 8' long (for four-board-high wall)
30	6"x8"x8' used railroad ties (60 if two layers are used)
1	230' of ½" diameter cable with turnbuckle and 4 clamps (2 pieces 115' each)
650	5" screw shank nails
5'-6" WIDE x 5' TALL GATE	
2	2x8 pine boards, 12' long (for gate horizontals)
1	2x8 pine board, 10' long (for gate verticals)
1	2x8 pine board, 8' long (for gate diagonal brace)
34	carriage bolts (appropriate length for the thickness of boards used for the gate)
2	10" strap hinges with 16 wood screws (#12 x 1½")
1	Gate hook with screw eye
MISCELLANEOUS	
65	tons of sand (44 cubic yards)
30	sacks of dry-mix concrete
2	gallons of wood preservative

The boards for the round pen walls can be rough cut or planed, green or dried, treated or not. The cheapest boards are usually untreated rough-cut wood. Although rough-cut lumber has the advantage of being stronger, it can also be a source of splinters, and it soaks up more paint or preservative than planed wood. A "2-inch" planed board is actually

Above: **When constructing a round pen, leave out two posts until the footing material has been delivered.** *Left:* **As horses work, their hooves push the footing toward the edge of the pen. Railroad ties hold in the footing.** *Bottom left:* **The ends of the railroad ties are cut at a 70-degree angle and are wedged into place between the posts.** *Bottom right:* **The railroad ties are set flush with the inside edge of the posts.**

only about 1½ inches thick. Green lumber can shrink or warp after you have nailed it into place, while dried lumber should retain the straightness it had when purchased. Treated lumber resists rotting, but not sunlight, and adds more cost.

The 8-foot boards are cut to fit and nailed on the inside of the pen using 5-inch galvanized screw shank nails. The spiral-ribbed surface of these nails provides a better grip in the posts than smooth shank nails. You may want to drill pilot holes in the boards to prevent splitting. Nailing the boards on the inside of the pen not only results in more strength, but in the event a horse leans heavily or falls against the rails, the boards are also less likely to pop off. This also keeps the rider's leg from being thumped by every post when trying to control a frightened or bolting animal. Use four or five boards as desired between each set of posts, except for the opening between the two taller gateposts.

To keep the angled walls of the round pen from sagging outward, use a ½-inch steel cable with a turnbuckle for support. One end of each of the 105-foot cables is wrapped around each of the gateposts and affixed with a U-clamp. The cables encircle the outside of the entire round pen about 4 inches from the top, resting in small notches made in each post with a chisel, saw, or ax. They meet across the pen from the gate, where the turnbuckle joins them together. Final tightening is made with the turnbuckle.

The outward pull created on the gateway posts by the cable must be counteracted, or else the gate opening would continue to widen. A railroad tie affixed between the tops of the gateposts provides a stabilizing lintel for the opening. Chiseling a hole in each end of the tie and carving corresponding projections on the tops of the gate posts at least 7½ feet

How many panels do you need to make a round pen?

You can also make a round pen out of panels (see chapter 11, Fencing, for information on panels).

1. First determine the diameter of the round pen you want. This will depend on whether you are going to use the round pen for initial groundwork (40–50 feet), extensive longeing (65 feet), riding (65 feet or larger), or cutting/working cow horses (80–200 feet).
Ex.: 65-foot-diameter round pen for general use

2. Multiply the diameter by 3.14 to get the circumference in feet.
Ex.: 65 × 3.14 = 204.1-foot circumference

3. Divide the circumference by the length of panel you want to use. For round pens 50 feet or smaller, consider using 10-foot panels to get a more circular shape. For pens larger than 50 feet, 12-foot panels should work fine.
Ex.: 204.1 ÷ 12 feet = 17 panels

You can add a 6-foot by 9-foot bow gate to the seventeen panels and end up with a round pen slightly larger than 65 feet in diameter, or you can substitute a 12-foot combination panel/bow gate for one of the seventeen panels and end up with a 65-foot round pen. Either way, you'll have a nice-size training pen for longeing and early rides.

A ½" steel cable runs along the top outside edge of my round pen to stabilize the angled walls.

View of the gate from inside my round pen. The railroad-tie lintel over the gate counteracts the pull of the perimeter cable. The gate opens in; when closed, it is flush with the pen walls. The weight of the gate and the angle of the walls make latching unnecessary except in very windy weather.

high creates mortise-and-tenon joints. An alternative lintel could be made with a piece of ½-inch steel cable and a turnbuckle attached to the tops of the two gateposts. For safety, 2×6 boards should cover the cable from both sides.

The gate is constructed out of the same dimensional lumber as the round pen walls and is designed to be flush with the inside of the round pen. Ten-inch strap hinges are screwed to the inside of the gate, allowing it to swing inward. A heavy latch is located on the outside of the gate; for convenience, a light barrel latch is located on the inside.

If the pen is to be used for turnout, it may be a good idea to treat the rails with an antichew product to discourage chewing.

Gate detail

Made of 2x8 boards, the same as used to build the round pen.

Bolted together with carriage bolts, smooth heads on the inside of the gate.

The two vertical boards are set in 3" from the ends of the horizontal boards so the gate lies flush with the round pen rails (otherwise the verticals would hit the posts).

Sliding barrel latch keeps the gate from blowing open. The horizontal board just above it protects horse and rider from the latch.

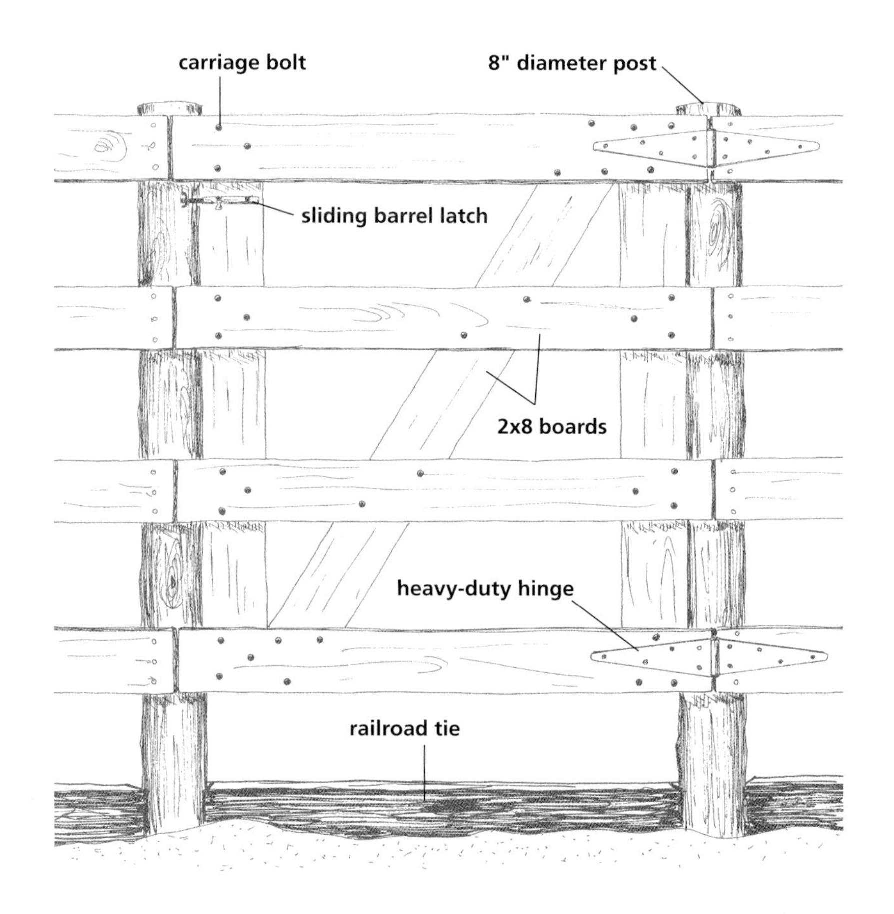

Outdoor Arena

The size and type of riding arena you construct will depend on the type of riding you plan to do. See the chart on page 202 for some suggested dimensions for various activities.

If you plan to build an indoor arena, prepare the site as you would for an outdoor arena. (See chapter 9, Outbuildings, for more information on the arena building.)

FENCING

If your arena fence is at least 6 feet tall, it will discourage horses from putting their heads over the rail as they are turning near the fence. The fencing should be very strong if you plan to ride young horses. Often dressage rings have no official exterior fencing but just an 18-inch-high visual barrier. The perimeters of dressage rings are commonly marked by cones, plastic chains, or ropes or with PVC pipe and cinder blocks.

The shape of the arena will depend on your training goals. Rectangles allow you to ride your horse deep into the corners and teach him to bend or turn. Oval arenas or rectangles with rounded edges are more appropriate for driving and jumping and are easier to disc and harrow. Gates should be flush on the inside of the arena, and the latch should be operable from horseback.

My 100'x200' arena with decomposed granite footing is ideal for working Zipper on his exercises year-round. A full-sized dressage arena (66'x198') can be marked off inside when desired.

Above: **At 3½' tall racetrack railing is acceptable for a riding arena but not for turnout.** ***Below:*** **This large outdoor jumping arena provides an attractive, safe place to work. Note ideal paddocks in background.**

Arena dimensions

USE	SIZE REQUIRED*
Dressage (small size)	66x132' (20x40 meters)
Dressage (large size)	66x198' (20x60 meters)
Calf roping	100x300'
Team roping	150x300'
Pleasure riding	100x200' to 150x250'
Barrel racing	150x260'
Jumping	150x300' *(depending on number and type of jumps and type of course)*
Driving	150x250'

**Size* refers to riding area: plan extra for banks, swales, and fencing or indoor arena walls.

RIDING ARENA DESIGN AND FOOTING

The working and riding surface in an arena and round pen should be safe and provide cushion and traction to encourage a horse to move forward with energy and elasticity. Poor footing can cause a horse to trip, slip, or fall, resulting in injury to the horse and/or rider.

The type of footing you choose will depend on your climate, the native soil, whether the arena is indoors or outdoors, and what type of activity you participate in. Before footing the bill for a new arena surface or trying quick fixes for problem footing, be sure you have considered important design and management factors. Because every farm has different soil, topography, climate, and arena uses, and because materials (sand, clay, stone dust) vary so much by region, confer with an arena professional in your area for specific advice.

You might attain a desirable riding surface by adding the appropriate footing to the prepared arena area. But maintaining good footing is a dynamic process that is affected by the weather, the amount of use the arena gets, and your management. Once you get your personal home arena set up, it will require regular maintenance (weekly or monthly), addition of footing materials every 4 or 5 years, and possible renovation after 15 years or so.

LOCATION

An arena should be located on dry, level ground that drains well. Anything located downwind of the arena will be the recipient of regular dust deposits, so take this into consideration when choosing arena and round pen locations. If the site has marginal drainage, you can improve drainage with a shallow ditch (swale) cut around the perimeter of the arena. Depending on the terrain, the rainwater collected in the ditch can flow directly onto lower ground or can be drained via French drains or an underground tile system.

LEVELNESS

The arena site should be level with a slight (1- to 2-degree) slope or a slight central crown to allow rainwater to pass through the footing and flow off the base. The slight slope discourages puddling. Be aware that a steeper grade could lead to erosion of the footing during downpours.

SUBBASE

The native soil, or "earth," is often a suitable arena subbase. But because you want a fast-draining, easily compactable subbase, in some situations, such as areas with heavy clay, you will likely have to add other materials to the subbase or find another location for the arena. For outdoor arenas, you could consider adding stone dust to improve the subbase. For indoor arenas, a more suitable subbase would be (depending on the specific materials in your area) an equal mixture of sand and clay or slightly more clay.

BASE

The layer of material between the subbase and the footing is called the *base*. The functions of the base include acting as a protective layer between the earth

Arena layers

When talking about arena design, the terms *subbase, base,* and *footing* are used.

The *subbase* is the earth, the native soil. For an arena, the subbase must be very stable. The *base* is the layer of material between the earth and the footing. The base is usually 4 to 12 inches deep and should be very stable and well compacted. The *footing* is the surface material and is usually 2 to 6 inches deep and ideally does not readily compact.

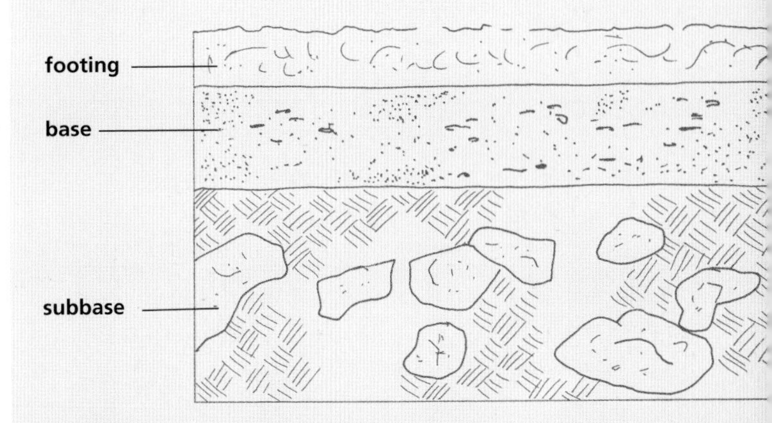

and the footing, giving stability to the arena floor, and carrying rainwater off the arena. The base might be naturally occurring material (such as decomposed granite) or added material such as road base or fine gravel topped with stone dust and clay. The base must contain no large stones, and it needs to be compacted or tamped as hard as concrete. A contractor with a 10- or 20-ton roller (such as used by road crews) can do the necessary packing. If you are adding material to the base, do so in layers, watering and packing in between each layer. Confer with your contractor or arena designer as to whether the materials should be worked in before watering and tamping.

Compaction will be easier to achieve if the material contains particles of various sizes; the smaller particles will fill in the holes among the larger particles. After compaction, let the base settle for a period of 3 months or more before adding the footing so any low or soft spots will be revealed and can be filled and tamped. Particularly in the north half of the United States, a new arena site should be prepared in fall, allowed to settle over winter, tweaked, then finished (footing added) in spring.

Some believe the surface of the base should be left absolutely smooth, while others say that after the base is set, narrow grooves or rills should be cut into the base to help hold the footing in place. Confer with your contractor.

The base layer must be a thick enough to prevent material from the subbase (such as clay or stones) from working up through it into the footing. A 4- to

Above: An arena should be located on dry, level, well-draining soil. _Below:_ When excavated into hilly land, a swale cut on the uphill side can help carry excess rain or snowmelt away from a round pen or arena.

8-inch base is usually sufficient for an arena that is used primarily for flat work. However, the base layer might need to be as deep as 12 inches if an arena will be used primarily for jumping. Some farms have experimented with laying special tough, non-biodegradable cloths between the subbase and the base to keep the layers from mixing. Check with local farms to see what has worked successfully in your climate and soil.

The base must be protected from damage caused by erosion, deep disking, and penetration from hooves. Regular maintenance of the footing protects the base and should eliminate the potential for ruts forming along the rail and associated compaction.

FOOTING

The layer over the base is called the footing, cushion, or surface. Ideal footing, often a mixture of materials, maintains its cushion without compacting. In contrast to base material, for which multiple-sized particles and compacting are desired, footing material should be composed of one particle size so it will not compact as easily.

Depending on the base and use of the arena, the footing could be from 2 to 6 inches deep. Because the function of the footing is to provide a cushioning effect, 3 inches of footing seems to work well for most uses, such as pleasure, dressage, and reining arenas. For driving, use 1½ inches; for cutting, 5 to 6 inches. Jumpers require cushion without excessive depth and, although traction is important, the footing should allow a little slide, so when landing the horse's feet don't "stick." Speed events require a firm footing such as a mixture including stone dust. Reining horses do best on a firm base with a silty, sandy layer with enough give for sliding stops.

Arena footing details

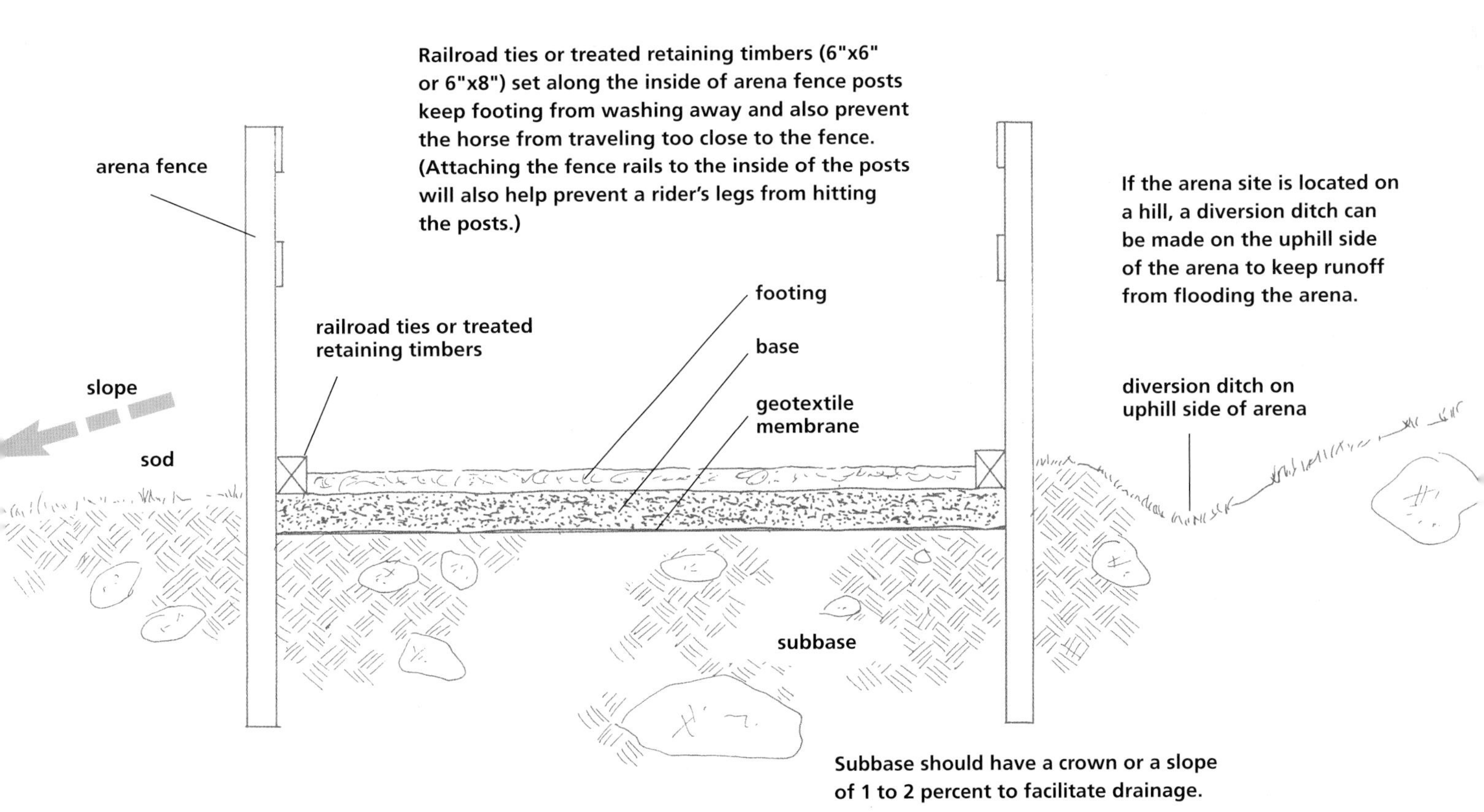

Dressage horses work well on a moderately resilient footing without excessive depth, such as processed wood products or clean sand.

Some footing materials, such as coarse sand and stone dust, are abrasive to the hoof wall and so are unsuitable for working barefoot horses. Wood products and turf are not abrasive and so are suitable for bare hooves. Footing recipes vary widely and include various mixtures of sand, silt, and clay; topsoil and sawdust; simply sand; and various artificial footings. If you are considering an artificial footing, check to be sure that it is environmentally friendly and determine where you can dispose of it if you need to.

Soil Evaluation

Before adding new footing to your arena site or enhancing what is already there, find out what type of soil you are starting with. In order to learn the soil's characteristics, have the soil tested. When you sample, be sure you get a cross-sample from the entire arena. Don't dig down to the base, because the soil character will be different. Have the soil evaluated by a state university soil lab or a private soil lab rather than by the manufacturer of one type of footing that might have a specific interest in sales. Be wary of home-style test kits, because you must know what you are doing or you could get the wrong results, and incorrect decisions could be costly.

A soil sample will show the organic content of the soil. Organic material provides the medium for beneficial soil microbes to grow. Bacterial activities (including secretions) can help the footing absorb moisture, bind the footing particles, and keep the soil resilient and "alive." Also in your soil analysis, you'll learn the mineral makeup of the soil: what percent is sand, silt, and clay and what type of sand the sample contained.

You'll receive a particle-size-distribution, or PSD, report. Using a series of stacked graduated sieves, an analysis is made to determine the percentage of very coarse, coarse, medium, fine, and very fine soil particles in the sample. Fine and very fine soils tend either to pack or be dusty or blow away. Medium- to coarse-size particles are most desirable for the

How much footing do you need?

Footing is ordered in cubic yards of material. To determine how many cubic yards you will need, measure the length and width of your arena in feet; determine how deep in inches you want the footing to be; and convert that to a fraction of a foot. Then use this formula:

cubic yards of material needed =
(length of arena in feet × width of arena in feet × depth of footing in feet) ÷ 27

EXAMPLE:
200' × 100' × 0.33' (4" of footing) =
6600 cubic feet ÷ 27 = 244 cubic yards

Note: 1 dump-truck load = 5–10 cubic yards, so 25–50 truckloads would be needed to deliver 244 cubic yards of sand; also 1 cubic yard of sand = 1.5 tons, so 244 cubic yards of sand = 366 tons.

This particle-size-distribution result, from very coarse (upper left) to very fine (upper right), shows the composition of a typical native soil. It has multiple particle sizes and is therefore prone to compaction; it also has a tendency to finer particles, which are associated with dust.

footing of arenas. While the PSD of your native soil will likely indicate that it is unsuitable as footing by itself, it will guide you in choosing the appropriate soil amendments.

Types of Arena Footing

Footing should drain well, provide traction, cushion, and be relatively dust-free. When adding any footing or additive, don't overdo it. Often, less is more.

Grass or turf. Grass or turf arenas have great aesthetic appeal, but they are not the ideal footing in all instances. Depending on the locale, resilient, tough turf can be difficult to establish and maintain. And with hard use, especially jumping, grass arenas can end up with bare spots, mud holes, and dust tracks. Grass can become very slippery when wet, hard during dry spells, and dangerously hard when frozen. Unlike other footings, grass can't be worked when it is hard. For an arena that would get only light use, a grass ring might be perfect because it offers a great natural footing: good traction and spring. Grass can be a time-consuming choice, though, as it must be regularly weeded, fertilized, watered, mowed, and repaired.

Topsoil. *Topsoil* is a general term for native soil with various components, primarily clay, loam, and sand. Generally, topsoil is not suitable for arena footing, as it usually doesn't drain well, compacts easily, and is dusty. Clay is hydrophilic, which means it has the ability to absorb water and become a very plastic substance. Clay particles tend to be flat ovals that are slippery when wet, so they slide into position with each other and hold their shape when they dry. For outdoor arenas, it is best for clay content to be less than 10 percent. Loam is a rich soil composed of sand, silt, and clay. Sand is discussed below.

One of the most common ways of improving native soil is to disc sand and/or sawdust in with the dirt. This will lighten and loosen the soil and increase its drainage while adding to its cushion.

Sand. The word *sand* is a vague term used to represent a wide variety of materials. Sand particles vary greatly in shape, from ball-bearing round to very angular. Round sand particles interact like a zillion tiny balls, so do not provide the stability required for a horse's landing, loading, and push off. Beach and riverbed sand is often too spherical for arenas. Glacial sand, with its irregular angular surfaces, results in more interlocking particles and therefore better stability and traction.

Sand **is a general term that can refer to beach sand, builder's sand, and everything in between. Choose clean sand of medium-coarse, uniform particle size.**

Unlike clay, sand has more strength when it is wet; in fact, it often becomes too hard. When it is too dry, it loses its binding forces and collapses.

Sand also varies in mineral content. Sand containing quartz is quarried and is often called *hard sand* because it resists breakdown and is long lasting, perhaps 10 years in an arena. In contrast, sand containing feldspar and mica breaks down relatively quickly. Manufactured sand (fine-crushed rock) is not as hard as quartz sand.

Cheap sand has a high silt and clay content, which is referred to as *fines.* Fines are just that, fine particles (those particles below 0.1-millimeter diameter). Clay and silt particles that are 0.001- to 0.005-millimeter in diameter are responsible for dust, so should not make up more than 5 percent of the mixture. Some construction-grade sand contains 50 percent or more fines. Fines create dust and tend to pack, which inhibits drainage. Sand can also contain small gravel or stones. Screened sand has had the larger particles removed so that uniform pieces remain, which makes the sand less prone to compaction. Cleaned sand or washed sand means that the silt and clay have been removed.

Washed concrete sand and *mason's sand* are terms usually associated with medium-coarse, very clean sand (less than 2 percent fines) that usually costs $1 to $3 per ton more than construction-grade sand. Aim for clean, screened, medium-coarse, hard, sharp sand.

Because a horse sinks deep in sand when working, using sand alone, particularly if it is deeper than 6 inches, could stress tendons; start with a few inches and add gradually until you achieve the appropriate depth. It will take 244 cubic yards, or more than 350 tons, of sand to provide a 4-inch cover in a 100-foot by 200-foot arena.

Native decomposed granite from my arena shows various particle sizes. Clean, screened decomposed granite would be better.

Decomposed granite. Native decomposed granite is granite rock that has been ground into small angular granite particles by glacial activity and may contain a small amount of clay silt. Granite is a very hard natural igneous rock formation with a visibly crystalline texture and high quartz content.

Mechanically produced decomposed granite will be similar in structure to the native product, but the amount of silt that it contains can be regulated to meet the use requirements. Decomposed granite can be ideal footing.

Stone dust. Also called *screenings, bluestone, rock dust, limestone screenings,* or white *stone,* stone dust can give better traction than some sands. Minimize dust in the air by choosing a stone dust that has been screened or graded to a single medium to coarse size. This will also mean it won't compact easily. If the existing footing is too deep (for example, sand), some can be hauled away or stone dust can be added to the footing to firm it up. Add a little at a time until you reach the desired consistency, but never use more than a total of 10 percent stone dust. As the name suggests, stone dust can introduce a dust problem unless the product has been carefully selected.

Wood products. National manufacturers use local sources of wood to produce various shredded and shaved wood footings. Wood products add organic materials that usually increase moisture-holding capacity and add cushion to keep footing alive and springy. However, wood products will vary greatly by location. They tend to break down quickly when dry, because they shatter and easily become small wood particles that lead to dust. Wood footings made of smaller particles tend to break down much more quickly than those of larger particles.

If dampened, wood footings last longer. However, wood footing that is constantly wet might decompose via rotting, and wet wood footing can become slick, especially when a large constituent of the wood footing has been pulverized to wood dust. If the base is intact, wood products can improve footing, particularly of indoor arenas. Tan bark, hardwood fiber, and wood chip products tend to freeze later and thaw sooner than the surrounding ground. They don't need to be disced, just lightly harrowed. Besides the high expense of the footing itself, however, processed wood fiber footing requires a well-engineered watering and drainage system in order for it to work at its optimum.

Hardwood products tend to be more durable than pine. However, never use walnut or black cherry products, as they are very toxic to horses.

Rubber products. Ground, crumb, and shredded rubber are used to combat arena hardness. Too much

Wood products vary according to the manufacturer and locale, but they generally increase cushion and moisture-holding capacity.

rubber can result in too much bounce in the horse's step. Rubber products on their own do not create dust, but they will not improve a dust problem caused by another material because they do not absorb water. They will increase soil porosity so water can get into the soil, however. Rubber is lighter than soil: 30 pounds per cubic foot for rubber versus 80 to 110 pounds per cubic foot for soil. This means the footing is lighter and more porous, but it also means that during a downpour, the rubber fraction might float away, and during windstorms, it might blow away. This would not be an issue in an indoor arena.

Rubber footings are usually a recycled material from the tire industry, ideally without the steel-belt portion of the tires. However, metal and other debris have been noticed in some rubber footings, so it is important to choose a product that has been designed for use with horses.

Most rubber footings are black, which reduces the glare of straight sand. They absorb heat from the sun and so can help prevent freezing, but the rubber can be hotter for the horse's feet. Colored rubber products (for example, blue and green) are also available. Determine what coloring agent is used, as some could be toxic. Also check to see where you could dispose of the rubber footing in the event it does not work well for your application.

Coated sand. Polymer-coated sand, a less widely available and more expensive option, offers good cushion, good traction, and minimal dust if organic matter (manure, leaves, bedding) is removed from the footing regularly.

Synthetic fibers. Ground or chopped plastics or other synthetics have recently been marketed as arena footing. They are generally lightweight and could float or blow away outdoors.

Rubber footing products such as these are used to combat hardness.

RUBBER CHUNKS

RECYCLED TIRE PIECES

RUBBER SHREDS

Footing Problems

Footing materials are not the answer to all problems. However, the correct footing or additive might solve a particular problem you are having. Some footings might solve one problem but introduce another. Most footings can't do it all alone, because most footings are not 100 percent dust-free. The exception to this would be clean, premium, decomposed granite. All footings will break down over time, some in a matter of months, some after years.

Certain footing problems tend to occur time and time again. Knowing about potential problems ahead of time can help you design an arena and footing that will minimize problems.

Too hard. Hard footing can lead to joint and ligament problems. Luckily, hard footing is one of the easiest problems to solve. Often just a regular grooming (aeration with disc and harrow) and watering program will improve hard soil. Another solution is to add new footing on top of the hard arena. Also, organic material (such as sawdust or ground bark) or rubber products can be added to the hard soil and mixed in to give life and spring.

How hard is your arena footing? A Clegg Impact Tester, which is a drop hammer that measures the rate of deceleration when an object hits the ground, determines hardness. The drop hammer measures the forces felt by the horse as his limbs contact and settle on the arena surface. Hard surfaces, on which the rate of deceleration occurs quickly, have little "give" and don't absorb energy. Softer surfaces absorb more energy and have a more gradual hoof-impact-to-limb-loading sequence, resulting in a less abrupt deceleration and lower risk of injury to the horse's legs.

The deceleration number will be greater on a harder surface than a softer surface. Values of more than 125 are associated with athletic injuries. Hard-packed or frozen normal soils have values of 175 and higher. A good turf surface has values of approximately 75 to 100. Soil-testing labs can perform the Clegg test on your present arena and then again after renovation.

Too dusty. For both horse and rider, dust is irritating to the eyes, nose, and respiratory tract; it covers buildings, vehicles, and equipment downwind from the arena. Dust can be caused by many factors, including the particle size distribution (PSD), dirty sand, wood dust from decomposed wood footings, percent of organic matter in the footing, type of minerals making up the sand fraction of the soil, watering practices, arena grooming techniques, and overall climate, as well as temporary effects from wind, humidity, precipitation, and temperature.

Gary P. Little, The Sandman

Clean, premium decomposed granite makes an ideal footing for round pens and arenas. Depending on its place of origin, it will vary in color from honey gold to reddish brown.

Dust can be reduced by various methods. Soil particle size can be increased by replacing the footing or amending the footing material so the PSD is more desirable. Natural or synthetic fibers or additives can bind larger particulates to the smaller fractions in the soil. You can increase moisture and reduce evaporation by watering frequently, possibly with the aid of an absorbing agent. Adding organic matter to footing not only increases moisture-holding capacity but also encourages microbes that bind the soil materials to flourish.

Too deep. Footing that is too deep can cause missteps, tendon strain, and other injuries. Most footing that is too deep can be remedied somewhat by watering it regularly. If necessary, some of the existing footing can be hauled away, and/or stone dust or clay can be added to the footing to firm it up. Stone dust may be most appropriate to add to sawdust or wood footings that are too deep, and clay for deep sand. Bonding agents (fibers and polymers) can also be useful additives to help this problem.

Too wet, muddy, or slippery. This problem can be caused by anything from poor planning and design of an arena to an insufficient base or inappropriate footing. Working wet surfaces with a harrow usually speeds up the drying process. Because wet footing often points to poor arena design, there are no patches for this problem. It's usually back to the drawing board.

Freezes easily. In cold climates, footing can freeze into a lumpy mass or a hard surface. Adding salt or a special product to prevent soil freezing and then working it in might be the only way to keep an arena usable in very cold weather. Be aware that salt can be damaging to your horse's hooves and your tack, tractor, and equipment, not to mention the environment.

Inconsistent. Inconsistent footing has varying depth and feel. Some arenas are consistent on the rail but vary greatly in the center. Unevenness is often a symptom of a faulty base. This type of problem will likely require renovation.

Footing Additives

Once the footing is in place, you may need to add products to it from time to time to solve certain problems. Some of the footing amendments that follow are good; others are bad.

Manure, bedding, and compost. Adding fresh manure to arenas usually results in a slick footing that takes a long time to dry. Fresh manure is unsanitary, breeds flies, and harbors parasites. Uncomposted manure and used bedding release ammonia fumes that can lead to respiratory problems in horses. Shavings and straw are slippery and inconsistent, and make the arena hard to maintain; don't add either to footing. Horse manure with or without bedding that has been thoroughly composted into humus has been successfully used as an arena footing additive when combined with native soil, sand, and wood footing products. If used alone as footing, humus (which is largely organic matter) would hold too much moisture. Therefore, through trial and error, add humus in small increments to find the ideal mix.

Commercial compost may sometimes contain heavy metals that can be a health hazard to horses and humans.

Waste oil. Applying used motor oil to arena footing is banned by the EPA, since it is harmful to the environment. Besides potential harm to the soils and groundwater, spreading old tractor or car oil on an arena is very messy for horse, tack, and rider and can cause respiratory problems.

Oil products. Oil products that are vegetable based (palm, coconut, and soybean) and environmentally safe have recently been developed specifically for dust control, as they can bind, coat, and add weight to small particles. How well these products work over the long term remains to be seen.

Salt (sodium chloride). Salt is used to prevent footing from freezing because it lowers the freezing temperature of water. Salt is drying to horses' hooves and can create problems with tack (especially leg boots) and corrosion of farm equipment. And if you use a salt on an outdoor arena, the runoff could kill the vegetation near your arena.

Calcium chloride. Calcium chloride is applied to pull moisture from the air into the soil to combat dustiness. However, since it leads to hoof problems (dryness) and is corrosive to metal (indoor arena structures and equipment) and leather, there are better options to consider that do not have these drawbacks.

Water. Additives that are designed to slow down evaporation by encouraging microbial populations to flourish are applied every time the arena is watered. The microbes produce binding fluids that hold footing particles together and prevent them from drying out.

Fibers. The addition of natural or plastic fibers to existing soil can both aerate the soil and bind it.

Polymers. Polymers designed as a moisture retention aid for use with turf have had varied success in arenas. They are usually made of starch or synthetic acrylic. These water-absorbing products are crystals when dry but soften and swell when moist and

Regular arena watering is much easier when using equipment designed for the purpose.

Smith Irrigation

release their moisture gradually. Depending on the product, about 2 teaspoons of crystals can absorb a quart of water. If improperly applied, polymers can be slick when they are fully engorged with water, and when the arena soil freezes, so will the polymers.

Experts advise to begin application conservatively and increase until desired results are obtained. Usually, polymers are added to the soil at a starting rate of 5 to 10 pounds per 1000 square feet. In some situations, 12 pounds per 1000 square feet might do nothing, but 17 to 18 pounds might do the trick. If a soil is water-repellent (either naturally or because of prior addition of oil or other substance), it might interfere with a polymer's effectiveness. Polymers are also rendered useless by most types of hard water and all salts and so should not be used in arenas where calcium chloride has previously been applied.

Footing Care

To keep your arena footing in optimum condition, groom and water it regularly.

Moisture. You'll need to develop and practice watering techniques so the footing maintains a moisture level of 8 to 14 percent: less for clay, more for sand. A moisture meter like the kind you use to test hay bales can give you an idea of how close you are to maintaining this range.

How much water? To determine the amount of water needed to control dust and provide uniform texture, multiply the square footage of the arena by 0.07. The resulting number indicates the approximate number of gallons of water needed. To estimate how long this will take to distribute on the arena, divide the number of gallons by 5 to find the number of minutes of pumping required. Five gallons per minute assumes an average water pressure of 40 pounds using a ¾-inch hose. If you have lower water pressure, decrease the number to 4. If you have higher water pressure, use the number 6.

Example:
60'×120' arena = 7200 sq. ft.
7200 sq. ft. × 0.07 = 504 gal. water
504 gal. ÷ 5 = 100.8 min., or
1 hr. 40 min. pumping time

As you plan your watering schedule, you need to take into consideration water pressure, the output of the sprinklers, and the need of your particular footing. A common watering problem is too much moisture in the arena center and not enough on the rail. One way you can see how much water each part of your arena is receiving is to place empty tuna cans at various locations and compare their contents after 10 minutes of sprinkling.

You also must learn the water-holding capacity of your soil. Some of this can be determined via soil testing; some must be learned using trial and error. For example, if you water for 10 minutes, you may find it accomplishes absolutely nothing. Twenty minutes might just settle the dust — temporarily. Thirty minutes might be perfect; 40 minutes may be too much and result in slippery spots. Too much watering also results in soaking the base, which results in loss of stability.

One of the best ways to determine how well your watering program is working is to simply use your fingers. Water should penetrate through all of the footing to the base before the arena is used.

Grooming. Discing and harrowing are necessary to maintain the resiliency of most arenas. However, discing too deep can destroy the base and result in undulations in the footing. These furrows are more stressful to a horse's legs than footing that is even but slightly hard. Also, discing can cause footing (such as wood and rubber products) to be worked in too deep, where they aren't useful. With some footings, discing requirements are minimal or not recommended. Harrowing might be all that is necessary.

Daily arena maintenance should include picking up any rocks that have surfaced, removing weeds, and hand-raking trouble spots. Often, it is necessary to hand-rake the footing back onto the track around the perimeter of the arena and at the takeoff and landing spots of jumps.

The "track" on which you ride might eventually become either rock hard or grooved as in a slot-car game, hampering your horse's movement. It may be necessary to frequently disc, harrow, and water the arena footing to keep the conditions ideal for training. (See chapter 10, Equipment, for more information on machinery and equipment.)

Tie Areas

In addition to the tie areas that you have in and around your barn, you might want to construct an additional hitch rail near your round pen and arena. This will come in handy when you want a horse to stand saddled for a while before you ride or if you have several horses you will be riding in succession. The structure the horse will be tied to should be solid and unbreakable, such as an 8-inch post set 3 to 4 feet into the ground. Never tie to rails or boards that could pull off if the horse is spooked. Consider laying rubber mats on the ground in the tie area to prevent erosion from pawing or normal movement.

Below: Harrowing an arena may be all that is necessary to maintain footing. _Right:_ A hitch rail near your arena or round pen makes it convenient to hitch a horse when working another.

Management

PART THREE

13 Land

To make informed land management decisions, consider general aspects of the climate such as the temperature, growing season, and grazing season as well as the expected time and amount of average precipitation.

When planning pasture improvements, renovations, and routine management, contact your local agricultural Extension agent for advice on soil testing, plant varieties, fertilizer and irrigation needs, weed control, and a timetable for planting and harvesting that is suited specifically to your locale.

Soil Testing

One of the first places to start is to have the soil tested to determine if it needs mechanical or nutritional improvement. Ideally, your soil should be fine textured and fairly moist, well aerated, not heavily compacted, and composed of a balanced mixture of sand, silt, clay, and organic matter. Such a soil, often referred to as *loam,* is usually friable; that is, if you grab a handful of the soil, squeeze it tightly into a ball, and then try to break the ball, it should crumble.

If you are lucky, your land will have 3 to 5 feet of topsoil. You will want to protect your topsoil by avoiding intensive agriculture and minimizing erosion by wind and water. Avoid overgrazing, seed and mulch bare spots, and plant trees and shrubs to minimize wind erosion.

To see how well drained your soil is, you can perform a simple percolation test. Dig a hole 1 foot deep and 1 foot wide. Fill the hole with water. Record how long it takes for the water to drain completely. The ideal time is 10 to 30 minutes. Drought-prone soils drain in 10 minutes or less and would be suitable only for dry-land plants. If it takes 3 hours or more for the water to drain, your soil has either a large percentage of clay or an impermeable layer that blocks water flow.

To determine if your soil needs nutritional supplementation or chemical balancing, submit a soil

Ribbon test

With about 2 tablespoons of moderately moist soil, try squeezing out a ribbon between your thumb and forefinger. If you can't, your soil contains at least 50 percent sand and only a little clay. If the ribbon breaks before it is 2 inches long, it contains 25 percent clay or less. If you can squeeze out a 2- to 3½-inch ribbon, your soil contains 40 percent clay or more.

Soil sampling

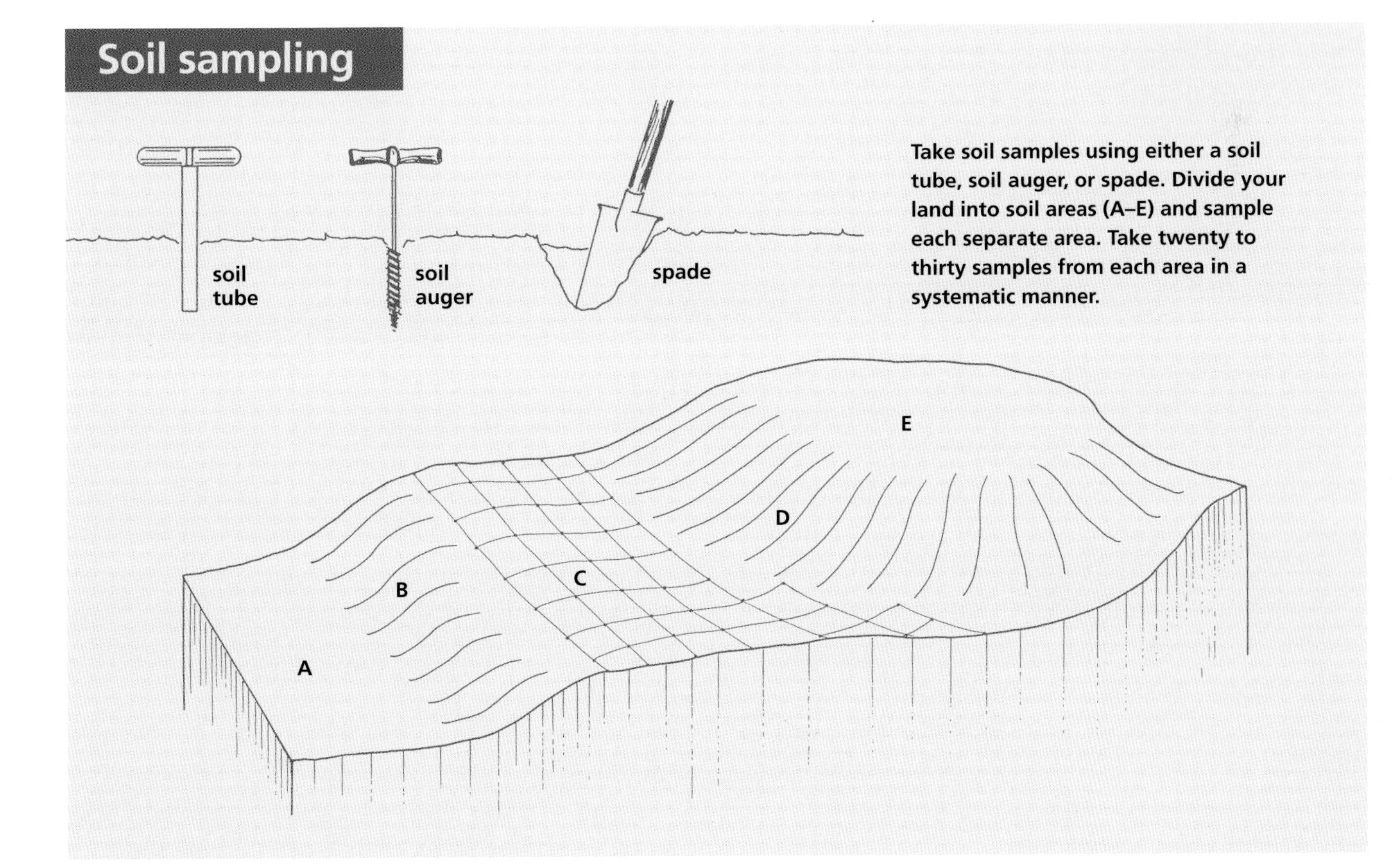

Take soil samples using either a soil tube, soil auger, or spade. Divide your land into soil areas (A–E) and sample each separate area. Take twenty to thirty samples from each area in a systematic manner.

sample to a soil-testing lab. Soil-sampling procedures are fairly standard. First, you will need to determine the various distinct areas of soil on your property by noting the following factors: difference in texture (sand, silt, clay), color, slope, degree of erosion, drainage, and past management (fertilization, cultivating, grazing). Make a map of your land and draw in these areas and number them.

Next, take samples to cultivation or plant-rooting depth (6 to 12 inches, per lab instructions) from each area. For sampling, use a stainless-steel or plastic soil tube or auger or a clean spade; rusty tools can contaminate samples with iron. Systematically take twenty to thirty samples from every uniform area. Avoid very unusual spots, such as waterways, hedgerows, and piles of decomposing organic matter, or sample them separately.

Mix all samples from an area thoroughly in a clean plastic bucket; galvanized steel or brass containers will contaminate the sample with zinc, whereas rusty steel may contaminate it with iron. After mixing, take out enough soil to fill the sample container or a plastic sandwich bag. Ask the soil lab if you should ship the sample as is or if you should air-dry the soil before shipping. Label the sample container with your name, address, and the sample number (corresponding to the area on your map). Fill out the soil information form completely. Pack the sample container.

The results of the soil test will yield some valuable information. Routine tests are usually made for pH, organic matter, nitrate-nitrogen, phosphorus, potassium, calcium, magnesium, lime, micronutrients, soluble salts, and texture. This generally gives an adequate soil profile, unless a special problem is suspected, in which case further testing may be necessary. Ideal soil pH for horse pastures is 6.0.

Soil test results will indicate whether nitrogen or phosphorus should be added to your soil to improve hay and pasture yield. A routine soil test will also indicate the salt and sodium content of your soil. A high level of soluble salts greatly limits crop yield, and correcting saline soils is time-consuming, expensive, and temporary at best.

Making Plans

Whether you plan to use your pasture for year-round grazing or have decided that it would be more profitable to take one cutting of hay before allowing your horses to graze, the principles of raising hay will give you a good idea of how to improve a pasture.

You may be interested in developing temporary pastures, permanent pastures, or both. A temporary pasture is intended for 1 to 2 years of vigorous year-round grazing. Because the plants are never allowed to develop deep roots, they can be more susceptible to damage from hooves. Cool-season plants such as rye, wheat, alfalfa, and clover work well for temporary pastures. Of course, the investment of labor to initiate such pastures is high in relation to the number of years of use.

Permanent pastures are those you plan to use for 10 years or more before major renovation. Along with a well-established base of native plants, they may have an introduced population of perennial grasses. Such a pasture has a well-developed root structure that tolerates temperature and moisture stress and resists mechanical damage and erosion.

If you want to harvest hay from some of your land, whether you do it yourself or hire a custom baler to do the work for you will depend largely on three factors: the size of your property, if you have the necessary equipment, and if you enjoy field work.

Figure that you'll need a tractor, mower, baler, and pull-type stacker, equipment that will cost you a minimum of $20,000. Such an investment wouldn't be warranted unless you had 10 acres or more of good hay ground.

Custom farming terms can vary greatly, but generally, if you contract with someone to harvest your hay, you'd pay all expenses associated with the care and maintenance of the land and the harvest and you'd give one-third to one-half of the hay to the contractor.

If you want to do part of the work yourself or pay cash for the work, you can often hire a custom hay farmer at a certain amount per acre to cut and rake the hay plus the cost of baling and stacking the hay.

Above: After harvest, grass hay fields can provide late-summer grazing when managed wisely. *Right:* If you own hay fields, you can hire a custom hay producer to cut, rake, bale, and stack the hay for you.

Left: Baling hay is not for the faint of heart, as balers can be idiosyncratic. *Above:* This self-propelled stack wagon picks up individual bales in the field and builds a stack of approximately 160 bales, which can be delivered by the wagon or a retriever truck.

Improving Your Pasture or Managing Your Hay Field

The story of the hay in your horse's feeder may have started 2 to 3 years ago when a field was prepared and seeded. Often, hay is used in rotation with other crops such as corn and barley. While such grains deplete the soil of its nitrogen, legume hays such as alfalfa restore this necessary element to the soil. Because of this rejuvenating effect, along with the fact that premium hay is a good cash crop, alfalfa is a very popular hay to raise. Furthermore, a large, well-established, and properly managed grass or grass/alfalfa pasture can produce a good crop for first-cut hay and then be used again for grazing.

Regularly evaluate your fields and pastures to encourage vigorous, dense, diverse, and desirable plants and to discourage overgrazing, spotty grazing, weeds, bare spots, and soil erosion.

To establish a hay field or a pasture, you follow a similar procedure.

PREPARING THE SOIL

Fields can be worked using conventional or alternative tilling. With conventional tilling, the field is cultivated deep with a plow, disc, chisel, or rotary tiller and can be further worked with a mulcher to mix and aerate compacted soil. Following this, the field is floated (leveled) with a disc and/or harrow to smooth out rough spots and to prepare it for seeding. Hay and pasture plants do best in a fine, firm, clod-free seedbed, not an overly soft or clay-pan field.

Due to the high cost of fuel, labor, and equipment, alternative methods of tillage have been developed. Minimal tillage might consist of a light disking and harrowing, resulting in less compaction due to less traffic. If you do not till deeply, erosion caused by wind or water is greatly reduced, but seeds must then be drilled. You may find it best to hire a custom farmer with a renovator or a cultimulcher to work your field. With these implements, he can cultivate, chop, aerate, seed, and fertilize all in one pass.

ADJUSTING pH

Soil pH indicates acidity. Neutral soils have a pH value of 7.0. Sour (acid) soils have lower values and sweet (alkaline) soils have higher values. In some areas you may need to add limestone to reduce soil acidity and thereby increase yield. An acid soil may interfere with a plant's ability to absorb nutrients. A soil pH test indicates if you need to add lime but does not tell you how much to add. The amount of lime needed will be determined by a pH buffer test.

Lime, which is a base, will raise the pH (decrease the acidity). It may need to be added every 2 to 20 years, depending on leaching. Adding lime is especially important where there is 25 inches or more of rain per year, which can leach lime and minerals out of soil and leave it acid. Lime can be added on top (even over snow), but tilling it in is much better. For this reason, address soil pH issues prior to planting. In contrast to lime, nitrogen fertilizers increase soil acidity.

FERTILIZING

If a soil test indicates the need for additional nutrients, fertilizer containing specific ratios of nitrogen, potassium, and/or phosphorus can be added. Fertilizing can get seedlings off to a vigorous start, ensure consistently higher yields, and help plants withstand stresses from insects and winterkill. Nitrogen increases plant growth. A productive, established pasture might require 100 to 180 pounds per acre of nitrogen per year. Remember that legumes, such as alfalfa, fix their own nitrogen in their root nodules, so they require little or no nitrogen application compared to grasses.

This grass-alfalfa field is ideally suited for horse hay and is prime for cutting.

Understand the labeling

If fertilizer label reads 32-10-10, it means 32 percent nitrogen, 10 percent phosphorus, and 10 percent potassium; the rest is filler. So if the fertilization recommendation is 120 pounds of nitrogen per acre, you would spread 375 pounds of the fertilizer per acre (120 pounds ÷ 0.32 = 375 pounds).

Phosphorus is essential for good root development, and phosphates sweeten grass, so horses will tend to eat less palatable grasses. Potassium increases a plant's resistance to drought and disease.

Applying the right fertilizer at the right rate and at the right time increases plant yield, increases a plant's water-use efficiency, and decreases weed problems by making the desired plants so vigorous that the weeds cannot get established. Fertilizer can be broadcast and left to dissolve, or it can be tilled into the soil. Usually, fertilizer is applied when preparing the soil to give seedlings a good start and then once more during the growing year. (Follow this same principle when using manure or humus as fertilizer.)

SEEDING

When selecting hayseed, choose the type of hay you wish to grow as well as the variety of hay that does best in your local area. Extension agents have results of field trials that test varieties and mixes for factors such as yield, resistance to drought, and the ability to withstand root rot, wilt, and winterkill. The quality of the seed is dictated by purity (very low weed and other seed content) and germination rates.

Hayseeds are drilled into the field to a depth of about ¼ to ½ inch in ideal soils and no more than

Straight alfalfa is too rich for most horses, and at full bloom these plants are past ideal cutting time.

¾ to 1 inch in sandy soils. Broadcasting seed on top of the soil requires 50 percent more seed than drilling.

Although an alfalfa hay field usually remains productive for about 5 to 6 years, and grass fields often longer, not all of those years are equally productive. The key to getting a new field established is taking advantage of winter snows and spring rains by seeding in the fall.

Sometimes, in order to get a stand of hay or pasture growing, a nurse crop (also called a *cover crop* or *companion crop*) is planted along with the hayseed. A nurse crop, such as oats, will emerge ahead of the more vulnerable hay seedlings. While the oat plants tower over the developing hay seedlings and protect them from the rays of the sun, the root structure of the oats adds to soil stability. The nurse crop is harvested the summer following seeding, at which time the young hay is ready for the sun. Often, however, the pasture or hay doesn't grow vigorously enough for grazing or a hay crop until the second year.

Hay Varieties

Grasses such as brome, bluegrass, timothy, rye, tall fescue, wheatgrass, and orchard grass are hardy perennials that usually decrease in digestible nutrients as they mature. Tall fescue is widely adapted; has good tolerance to wet, dry, or alkaline soil conditions; and can withstand a lot of traffic.

Orchard grass produces excellent-quality forage but will not tolerate drought, wet, or alkaline soil. If irrigation water is not available for the entire season, smooth brome and wheatgrass can be used. Bermuda grass (in the South) and tall fescue (in the North and South) resist trampling, so are good varieties for "exercise" pastures.

Legumes such as alfalfa, clover, and bird's-foot trefoil are more nutritious and don't require nitrogen fertilizer, but they are less hardy and impractical when used alone for pasture. With pasture mixes, plants mature at different times and exhibit a range of abilities to withstand stresses such as drought, flood, heat, and cold. If a blend of grasses and alfalfa (or other legume) is preferred, the legume should not exceed 25 percent of the mix for horses.

IRRIGATION

Heavy irrigation at longer intervals develops better root systems and hardier pasture than does more frequent light watering. Sandy soils have lower water-holding capacities and therefore require more frequent irrigation, while clay soils are not porous and hold water longer, but much of the water is unavailable to plants because it is held so tightly by the soil. Loam soils generally have the highest plant-available water-holding capacities.

Most pasture and hay fields need somewhere around 24 to 36 inches of water per year, so you may need to supplement your natural water supply to get the maximum yield from your pastures. Irrigation equipment can be expensive and is labor-intensive, but the yield per acre of hay grown with irrigation is often twice that of hay grown without it.

To consider using irrigation, you must have access to a large water source and in some areas "water rights" from a water-use agency. You can irrigate by

Hay varieties and characteristics

HAY VARIETIES	POSITIVE ATTRIBUTES	WHEN TO CUT	POTENTIAL PROBLEMS
ALFALFA	High-quality protein, especially for growth; a good source of calcium; highly palatable	First flower	Needs well-drained soil; shatters when dry; can contain too much protein for some horses; possibility of blister beetles; excess calcium to phosphorus for young horses
BIRD'S-FOOT TREFOIL	Does well in poorly drained soils; similar nutritionally to alfalfa	Early bloom	Low yield; may have lower palatability
RED CLOVER	Does well in poorly drained soils; high-quality protein	Early to midbloom	Difficult to put up well; notoriously dusty and possible toxicity from mold
ORCHARD GRASS	Early start, high yield, safe feed	Boot stage	Can get tough and unpalatable after early bloom
TIMOTHY	Does well in poorly drained soils; safe for idle adult horse	Boot stage	Not drought-resistant; when only hay fed, not enough energy for working horse and marginal in crude protein, calcium, and phosphorus for working horse
BROME	Drought- and trample-resistant; safe feed	Early to midbloom	May be unpalatable if too mature and fed alone; can be low in protein, calcium, and phosphorus

ALFALFA

ORCHARD GRASS / BROME

TIMOTHY

Irrigation water and equipment are expensive, and irrigating is labor-intensive, but it greatly increases yield.

surface flooding, furrows, or sprinklers. Surface flooding is not good for rolling pastures; the land must be leveled and graded with a slope of 0.1 to 0.4 feet of grade per 100 feet of length. Furrows should be designed and regulated so they can be irrigated and drained efficiently, because standing water (for more than 24 hours) encourages mosquito and horsefly breeding. Sprinklers are difficult to manage in areas of high wind and lose a lot of water to evaporation, so they are best when used in areas of high humidity. Sprinkler systems are the most expensive irrigation method initially but require less labor to operate, and they allow for the most even distribution of water.

The moisture that is required to grow good hay can also contribute to its demise. Although hay grows quickly in rainy country, it may be difficult to find a gap in the weather pattern big enough to allow harvesting and baling. Dry, sunny regions with adequate irrigation water are great hay-growing areas.

WEED CONTROL

A weed is sometimes defined as a plant out of place. Not all weeds are bad — some are fine additions to a horse's diet, being palatable and containing higher crude protein than grasses and higher trace elements than legumes. Such safe and good weeds include dandelion, buckhorn plantain, chicory, kochia, and lamb's-quarters.

Some plants, while not bad, compete vigorously with desirable grasses and are either unpalatable or of poor nutritional quality. Examples are downy bromegrass (cheatgrass), ironweed, horseweed, and willows.

Other plants, while not poisonous, are considered bad because they have a harmful mechanical effect on horses and they displace good grasses. This category would include plants such as foxtail, thistle, and others that cause lacerations and sores in the mouth. Burdock burrs mat in a horse's mane and tail, resulting in a time-consuming and unpleasant grooming session.

Noxious weeds, those that are officially being controlled by local, state, or federal legislation, include leafy spurge, cheatgrass, yellow star thistle, salt cedar (tamarisk), Russian knapweed, and yellow toadflax.

Still other plants contain toxic components that make them poisonous to horses, and these plants should be eradicated from pastures and hay fields. The list of poisonous plants will vary according to locale, so check with your agricultural Extension agent. Some examples are locoweed, senecios (ragworts and groundsels), buttercup, poison and spotted hemlock, water hemlock, deadly nightshade, curly dock, horse nettle, stagger grass, atamasco lily, mustard, hairy vetch, Mexican poppy, dogbane, hound's tongue, milkweed, perilla mint, chokecherry, red maple, and acorns. It is also important to

BURDOCK

Bindweed (background), foxtail (foreground), thistle, and burdock are bad but not poisonous. Wooly loco is poisonous to horses and should be eradicated from pastures and hay fields. *Bottom right:* When given a choice, Aria avoids the wooly loco and grazes the lush grass-mix pasture. When forage is scarce, however, horses will eat poisonous plants.

avoid planting highly toxic plants such as yew, rhododendron, and oleander in or around horse enclosures.

The best defense against weeds is to establish a good, vigorous stand of grasses and legumes that can compete with weeds and to avoid overgrazing pastures: overgrazing invites weeds. Mowing can impair weed growth by removing the larger-leafed, more vigorous weeds that shade the developing desirable plants. If the plants are cut before the seedpods are mature, it can also prevent the start of a new cycle of weeds. Most fields tend to clean themselves of weeds with strategically timed pasture mowing or after the first cutting of hay during the second year. Try to buy weed-free hay, or you may inadvertently introduce new weeds to your property each time you feed.

Burning eliminates mature vegetation and seeds and kills some parasites, but it can be a safety hazard and an air polluter. It is usually done in spring. Check to see if your local ordinances require a burning permit.

Points to remember when using herbicides

- Check to see if you need an applicator's license.
- Read directions carefully and mix according to manufacturer's directions.
- Don't think that if a little is good, a lot is better; you could do permanent damage to your land.
- Remove all animals from the area and leave them off as long as directions indicate.
- Be aware of wind drift as you apply; you could inadvertently kill your newly planted fruit trees or destroy your garden or your neighbor's.
- Beware of contaminating streams and other water sources.
- Use only approved herbicides for riparian areas.
- Wear a respirator.
- Wear rubber boots, not leather shoes or boots. Wear rubber gloves.
- Keep your dogs and cats confined for as long as the directions indicate.

Using Herbicides

It is difficult to control weeds chemically in a legume/grass hay field or pasture, as many of the herbicides would kill the desirable plants as well as the weeds. Herbicides can be used very effectively, however, in grass pastures. You will need to choose a selective herbicide specifically for your situation. Try to control weeds by other means, but if you must use herbicides, carefully monitor every step or you could ruin your pasture or harm the environment.

DISEASE AND INSECT CONTROL

Bacterial, viral, and fungal diseases and some insects can affect plant yield and may kill the entire field. The selection of the appropriate variety of hay or grass for the local growing conditions is often the best preventative. Genetic research has provided us with varieties that are resistant to specific diseases and pests.

Particularly in the Southwest, second-cut or later alfalfa hay that has gone to bloom may contain blister beetles, which can be lethal to horses. The toxin in these insects, cantharidin, is so deadly that just a few beetles, dead or alive, can kill a horse. If you live in an area where blister beetle poisoning has been reported, confer with your county Extension agent for assistance in identification of the beetles and purchasing certified blister-beetle-free hay.

Predator insects such as wasps, praying mantises, and ladybugs have been used successfully to kill aphids and grubs. In most cases, if your climate will support these predator insects and their prey in the field, the predators will show up on their own. Introduction of predator insects is usually not needed.

RODENT AND SNAKE CONTROL

Rodents, including mice, gophers, rabbits, ground squirrels, and groundhogs, can present problems in a pasture or hay field. First, their holes can pose a threat to your horses as they gallop across the pasture. Second, the mounded dirt from their burrows can end up in your baled hay. In addition, rodents can carry dangerous diseases, and a large population can put a real dent in the forage production of your field. Depending on your situation, it might be impossible or undesirable to try to eradicate rodents completely, but you can keep them under control by keeping a couple of cats and dogs on your acreage and allowing them access to the problem fields. Natural predators of rodents are hawks, owls, coyotes, and snakes.

Be sure you know your snakes. Frequently, people kill harmless and helpful snakes, thinking they are rattlers or other poisonous snakes. Some snakes, such as bull snakes, are natural predators of both rodents and rattlers. Seek out a herpetologist in your area if you need help in identification. (See chapter 15, Sanitation, for more information on rodent control.)

MOWING

In between grazing periods, it may be necessary to mow the ungrazed portions of the pasture to discourage weeds, encourage regrowth of desirable

Left: With the mowers set at 4", I uniformly mow the pastures two or three times during summer to prevent the undesirable plants from taking over or going to seed. *Below:* Richard attacks a small trouble spot with a hand mower — often a more expedient approach than getting out the tractor.

plants, and discourage the plants from getting too mature and going to seed. This may need to be done only one or two times per year, but it results in a higher-quality pasture. Usually, clipping the grasses to 4 inches is sufficient. In areas with cold, snowy winters, do not mow in late fall. Instead, leave the tall stems to catch snow.

HARROWING

The principle behind harrowing a pasture after a grazing period is that it evens out rough spots and distributes manure clumps. If you are in a dry, sunny climate, harrowing dries the manure and might help to kill the parasite eggs. In humid climates, however, harrowing the manure in pastures spreads the parasite eggs over a larger area while still allowing them to be viable, and so in effect increases a horse's chances of reinfestation. In such a situation, collecting manure and composting it is best. The other alternatives are to let manure accumulate away from grazing areas and to harrow and leave the pasture vacant.

AERATING

If your land is compacted, you can provide the plant roots with more oxygen by aerating the soil. An *aerator* is a heavy spiked drum rolled along behind a tractor.

HARVESTING HAY

The first growth of the second year may be quite weedy and not the best feed for horses, either as hay or as pasture. Removing the first cut, baling it, and using it for cattle hay is a good choice. The second cut from the second year will probably be the first crop suitable for horses. The third year marks the beginning of the prime years for an alfalfa or alfalfa-mix field. After 5 years, due to the death of some of the alfalfa plants, the field will have an uneven growth pattern and decreased yield. With a mixed field, the grasses gradually take over, and after 4 to 5 years, grass will dominate the field. You may be able to slow the transition by reducing your nitrogen fertilizer applications.

Cutting Guidelines

When to cut hay is critical. Usually it is determined by plant maturity, but other methods involve evaluating crown regrowth after first cut and simply using predetermined calendar dates. No matter which method is used, always keep an eye on the weather. Hay makers hope for dry but not overly hot days when the hay is in the windrows (that is, the cut hay lies in long strips throughout the field). Extreme heat or wind can result in dry, brittle hay. Rain or damp weather prevents hay from drying thoroughly and usually results in bleached or moldy hay.

Using plant maturity as the guide, there exists a trade-off between maximum yield and maximum quality. The premium hay grower chooses the optimum time when the plants are at their nutritive peak. Leaves contain the most protein. Young, immature grass plants have a high leaf-to-stem ratio, so are generally high in protein and low in fiber, resulting in excellent hay but fewer bales per acre. Mature plants, with a low leaf-to-stem ratio, have a lower protein content and higher fiber content. Although this results in a greater number of bales per acre, the bales are of lesser quality.

Legumes, such as alfalfa, should be cut when the first flower appears in the field: that is, the first flower on a representative plant in the field, not an odd plant along a ditch or field edge. Another way to gauge cutting time is before 1/10 bloom, which is when one out of ten buds have bloomed on the plants. On very large operations, cutting is started at mid- to late bud stage so that cutting will be complete by midbloom at the very latest.

Because most grass fields are cut only once, the farmer often waits until the plants are very tall and seed heads are mature. This results in a high yield and safe roughage, but one with low nutritive value. Ideally, grasses should be cut at the boot stage, when the seed heads are just emerging from the stem. The emerging head will be short, compact, and resilient, not 3 inches long, dry, fuzzy, and shedding seeds.

Mixed hays, such as grass/alfalfa, are cut using the maturity of the alfalfa plants as a guide. Each day a

This grass-alfalfa field has matured past the optimum stage for cutting. The alfalfa is in full bloom and the grass seed heads are full blown and releasing seeds. If put up properly, it would make a safe but less nutritious hay.

plant stands after first flowering or past the boot stage, crude fiber increases and crude protein decreases by about 0.5 percent.

Using bloom as the sole indicator of plant maturity can be misleading in some situations, as moisture, clouds, temperature, and the stage of the plants at the previous cutting affect bloom. If the first cutting was mowed at the bud stage, for example, and adequate moisture was available for regrowth, then the field could be cut every 35 to 40 days after the first cutting. Using such guidelines helps to ensure that there will be three cuttings. Hay cut early is usually of high quality and is followed by a fast regrowth and decent second- and third-cut yields. If a field is cut three times, approximately 45 percent of the year's yield will be in the first cut, 30 percent in the second cut, and 25 percent in the third cut.

Horsemen are opinionated on which cutting is best to buy. Although there are some differences in the cuttings, the quality of the hay is much more important than the cutting. From a nutritional standpoint, all cuttings can result in prime horse hay. With alfalfa, there will be some variation in protein content between cuttings. Although first-cut alfalfa hay is reputed to have large, tough stems, this is true only if the hay was too mature when cut. If first-cut hay is mowed at the prebloom stage, the stems will not be coarse and the nutritive value will be high. More weeds do tend to appear, however, in first-cut hay.

Second-cut alfalfa hay is usually the fastest growing because it develops during the hottest part of the season, and it usually has more stem in relation to leaf. Of all cuttings, second cut tends to be the lowest in crude protein, but its 16 percent average is adequate for all classes of horses.

Third-cut (and later) alfalfa develops a higher leaf-to-stem ratio because of slower growth during the cool part of the season. Therefore, third-cut hay will usually have the highest nutritive value. Horses that are not accustomed to a good, leafy alfalfa hay may experience flatulent (gaseous) colic or loose stools.

Mixed hays from all cuttings will have similar nutritional values except that with a grass/alfalfa mix, the first cutting will contain a larger proportion of grasses than will the other cuttings.

Curing Hay

Most hay today is mowed, conditioned (stems crimped so they will dry faster), and put in a windrow all in one operation. This results in less manipulation of the hay and less leaf breakage and loss. The hay dries in the windrow until the moisture is out of the stem. The level of dryness can be determined by giving a handful of the hay three twists. If the stems pop as they break, the moisture content is about right for baling. Scraping the green covering off a stem will also reveal if the stem is still wet.

Raking or turning the windrow rolls the hay on the bottom of the pile to the top. This may be necessary in humid climates, if hay has been rained on, or if the stand was unusually dense and the windrows

To determine if hay is ready for baling, grab a handful and give it three twists. If the stems pop, it is probably dry enough to bale.

are heavy. Raking will facilitate further drying but may contribute to leaf loss. It is essential that raking be done when the hay has adequate moisture, such as with an early dew, which will prevent leaf shatter and loss.

BALING

Once the hay in the windrow is determined to be at the appropriate moisture level, the hay should be baled. In semiarid climates, the hay might need morning dew to help hold the leaves on the stems, which may require the hay grower to get up at 3 a.m. and bale for the few hours when baling is optimum. Baling throughout the heat of the day works well in more humid locations, as long as the hay is thoroughly dry.

Bale size is dictated, for the most part, by the automated bale wagon that will be used to pick up and stack the hay. The currently popular wagon requires a 40-inch-long bale weighing approximately 65 to 70 pounds. The tightness of the bale can be adjusted. Tight bales handle well, stack well, and shed weather better. A too-dry bale must be baled tight in order to retain its leaves, but too-wet hay that is baled tight will result in heating and molding.

Bales are generally left in the field for a few days to cure or sweat, particularly if adequate dew was on the hay during baling. Often, you have to gather the bales because rain is in the forecast or because you need to irrigate the next cutting. Today stacking is generally done with automated bale wagons, resulting in tight, stable stacks with staggered joints. A tall stack results in fewer top and bottom bales, the ones commonly lost to weathering and ground moisture. Side bales generally do not get drenched during a rain, so they dry out adequately. The middle bales are protected.

If the bales contain too much moisture, they can ferment and create heat. The heat can be great enough to result in spontaneous combustion, causing an entire stack to catch fire. You can check the internal temperature of a bale by simply cutting the strings and passing your hand between some flakes. It should feel cool. If it is only slightly warm, the problem is minimal and may not result in any spoilage or fire.

Richard checks the hay in a windrow to see if the bottommost hay is dry enough for baling or if it needs to be turned by raking.

To take the temperature inside a stack, push a pipe 2 feet down into the stack and then drop a thermometer on a string into the pipe and let it hang there for about 10 minutes. This allows the pipe and the air in it to attain the temperature of the stack. If the temperature registers over 160°F, the hay should not be loaded in a barn and will likely no longer be suitable for horse hay, as the heat makes undesirable changes in the carbohydrates in the hay. The heated hay could be used for mulch or cattle hay.

Improving the Land

A good pasture manager respects, appreciates, and cares for the land. To apply John F. Kennedy's stirring words to pasture management, "Ask not what your land can do for you; ask what you can do for your land." Do not eke out the last iota of nutrition that the land can possibly offer. Leave enough reserve so the pasture can rejuvenate. Protect the land from the damaging effect of overgrazing by horses.

DAMAGE BY HORSES

Horses are wasteful and gluttonous, and their hooves can be very damaging to the land. A horse will eat, trample, or damage at least 1000 pounds of air-dry forage per month. What does this mean in terms of carrying capacity? Two acres of productive irrigated pasture may hold one horse per month during the growing season, while it might take 30 to 60 acres of dry rangeland to support a single horse.

Horses go for the young plant growth and succulent roots, letting weeds go to seed and mature plants go to waste. They defecate in certain areas and then will never consider eating the plants growing there unless forced to by starvation.

In addition, during wet periods (natural and irrigation) especially, the hooves of horses ruin root structure and can turn a field into a sea of mud and then a plain of dirt when it dries. Horses left on a pasture too long paw to reach tender roots, thereby destroying a plant's ability to rejuvenate. Horses should therefore be put on pasture or hay fields when the growth is optimum — about 4 to 8 inches high, depending on the grass species — and then the

Right: **Horses can turn a pasture into a barren wasteland if not properly managed.** ***Below:*** **Plant selection, cross fencing, and quick rotational grazing all contribute to a healthy pasture and healthy horses.**

Grazing management guidelines

- Cross fence to create several smaller pastures and rotate among pastures to prevent overgrazing.
- Graze pastures when grasses are 6 to 8 inches tall.
- For horse safety, graze after pasture grasses are taller, more mature, and higher in fiber.
- Think "early summer" rather than "spring." Young grass or fresh regrowth is higher in carbohydrates and so would be more likely to lead to colic and laminitis.
- Horses will eat over 6 pounds of pasture per hour and graze 16 out of 24 hours every day. They do not require that much grass, so to preserve your pastures and maintain your horse's optimum weight, limit the number of hours your horse is on pasture.
- When 50 percent of vegetation is gone and 3 inches of grass remains, remove horses. After removing horses, mow all pasture to a uniform 4 to 5 inches.
- Scatter grass seeds on bare spots.
- When grass regrowth reaches 6 to 8 inches (2 to 6 weeks), return horses. Remove horses again when grass has been grazed to 3 inches.
- During rest periods (usually winter), keep horses in sacrifice pens.

plant's growth in the field should be closely monitored. Remove the horses when 50 percent of the forage has been ingested or damaged.

Rapid pasture growth occurs early in the season, then starts slowing until about 12 weeks into the season, when most plants have matured and gone to seed. Plants need to be regularly monitored so horses can be removed before the majority of the grass has been grazed down. In the meantime, weeds should be mowed before they go to seed to encourage desirable plants to take over. Bare spots should be reseeded and protected from traffic and grazing by removing horses or installing temporary electric fencing. Overgrazing causes pasture damage that in the long run costs much more to repair than the feed costs it saves.

If the horses' grazing rate is greater than the field's ability to regrow, they should be put on another pasture so the grazed field can rest until it returns to the 4- to 8-inch height. After the first cut of hay is taken off a grass field, if the season isn't such that a second cut can be expected, grazing the field for several months in late summer and early fall is often cost-effective and won't harm the field if it is not overstocked.

Rotational grazing means that land must be divided up into more pieces, requiring more fences,

Left: **Graze pastures when grasses are 4" to 8" tall.**
Above: **When pastures wane in fall, or when 50 percent of the grass is gone or 3" remains, remove horses.**

Well-managed pastures help maintain property value and foster goodwill among neighbors.

waterers, and labor. It does, however, result in a higher stocking rate per acre of land. For example, 1 acre per horse could be sufficient if horses were rotated among three pastures or between two pastures and a set of holding pens. Controlled, intensive grazing results in the highest yield, as horses are only allowed to graze for 3 to 5 days before being moved. This system would be best implemented using temporary electric fences to subdivide pastures.

Continuous grazing is typically practiced on farms or ranches with very large pastures. This minimizes the amount of fencing and waterers, but it can result in seasonal forage shortages and areas that are permanently contaminated with feces and permanently overgrazed. Moving the location of salt and water in such pastures will minimize spots of overgrazing. Continuous grazing yields a low stocking rate: one horse requires two to three times as many acres, or more, for continuous grazing as he would for rotational grazing.

Overgrazing stops root growth

GRASS PLANT GRAZED (%)	ROOT GROWTH STOPPED (%)
10	0
20	0
30	0
40	0
50	2–4
60	50
70	78
80	100
90	100

Horses can share pastures with other livestock without the risk of parasite contamination between species, and it is possible that cattle or goats might eat some of the mature vegetation left by the horses.

Other Livestock

Horses can be mixed or rotated with other livestock to maximize the use of the pasture. If you run cattle and horses together, aggressive horses might chase calves, or horned cattle might go after meek horses, but usually there is no problem at all. If you let cattle rotate with horses in a pasture, they may clean up some of the mature grasses left behind by the horses. Because horses and cattle have different parasites, the life cycles of horse parasites will be broken during the time the cattle are on the pasture. Sheep, by contrast, which tend to eat the center of a plant and leave the tall, tough outer leaves, don't really contribute to the health of a horse pasture.

Parasite Contamination

Horses shed parasite eggs in their feces, then the larvae hatch on the pasture and are ingested by grazing horses. Parasite larvae are very tough, and it takes extreme dryness, heat, or cold to kill them. The more crowded and overgrazed a pasture, the worse the parasite recontamination problem. One horse on 5 acres won't have as large a parasite load as will five horses on 1 acre.

To reduce recontamination, depending on the pasture size, manure can be removed, left where it falls, or harrowed. Since a horse produces 50 pounds of manure per day, on small pastures and paddocks it is best if manure is picked up daily. It can be hauled away or composted and then spread on horse pastures. On very large pastures, horses tend to allocate certain areas for grazing, defecating, and lounging, thereby instinctively avoiding the parasites in their manure. In such pastures it may be best to leave the manure piles where they are rather than harrow and spread the larvae throughout the pasture. In midsize pastures in temperate climates, once the horses are removed in fall or early winter, the manure in the pasture can be harrowed to expose parasite eggs to freezing weather and to spread the manure more evenly so it decomposes over the winter. Otherwise, larvae that survive the winter immediately begin reinfecting horses in spring, especially young stock.

In southern pastures, the timetable is reversed. The summer is the best time to keep horses off pasture and to harrow the pastures to expose the larvae to the hot, dry weather. When the horses are returned to the rested and harrowed pastures in November, the recontamination rate should be decreased.

No matter how good a pasture is, it needs to lie dormant for some months each year, especially during the muddy months. Mud plus hooves equals lost shoes, pasture damage, and weed invasion. Mud is also an excellent breeding site for insects and a harbor for bacteria and fungi that lead to thrush, rain rot, and scratches. During the wet months, horses can be moved to pens or stalls.

Water

Good-quality water is of vital importance to the health and well-being of you and your horses, as well as for the proper functioning of your acreage. The drinking water must be not only safe for horses to drink but also palatable, so horses will want to drink it. In addition, suitable water is needed for washing horses, feeders, and blankets; hard water or water of poor quality makes these chores more difficult.

Water Sources

Become familiar with the source of your water. Does it come from a city or municipal source or is it from your own well? If the latter, where is the watershed that feeds your well and other surface water on your land? Are there waste dumps or landfills nearby? Are there industrial pollutants in the air, water, or soil? Private landowners are sometimes eligible for a government cost-sharing, rural, clean-water program.

Well water. If you have your own well, it should be located uphill and away from livestock areas. Most wells are drilled holes 6 to 8 inches in diameter and from 50 to 400 feet or more into the ground. The hole is lined with steel casing to a depth dictated by your health department, but usually about 100 feet. The well casing extends about 18 inches above the ground and has a watertight cap. The soil around the exposed steel casing should be graded to carry surface water away from the well.

Cistern. You might want to consider installing a concrete or plastic cistern, 1000 gallons or larger, uphill from the barn to provide water to hydrants by gravity flow during a power failure. If you have a well that produces a very low flow, you can rig your well pump with a timer so water is pumped into the cistern for short intervals throughout the day. A float valve in the cistern can be used to turn off the well timer when the cistern is full. That way, when you need a lot of water to fill troughs or give horses baths, you won't pump your well dry, because you'll be drawing water from the cistern. The cistern will refill at the pace you set the timer at for your well's capacity. A cistern can also supply firefighters with extra water should they need it.

The bottom of the cistern must be located higher than the faucets in the barn in order for water to flow by gravity alone during a power outage. Be sure to install a filtered air vent in the top of the cistern to keep it from collapsing when water is drawn out or from straining the pump when water is being pumped into the cistern.

You will need to test your well water and cistern at least annually for bacterial contamination and quality. Your county health department laboratory usually performs bacterial contamination testing. Your state agricultural university or a private laboratory often provides water quality testing.

Community-supplied water. If your acreage's water comes from a community source, it should arrive at your property in an acceptable form. Community-supplied water must adhere to EPA guidelines and be analyzed at regular intervals throughout the year. The results are usually printed and made available to the public. It would still be a good idea to test the water for purity, as contamination can occur anywhere along the way or on your property. Also, baseline mineral and pH values provide impor-

As a pond dries up, as shown here, the stagnant water becomes a concentrated source of disease-causing organisms.

tant reference in the event your young or pregnant horses experience problems that may be attributed to water chemistry imbalances.

Natural water sources. Creeks, springs, and streams can provide a fresh supply of drinking water, but they also may be a source of contaminants from upstream. Ponds can serve as watering spots but may become stagnant. Because ponds don't have free-running water refreshing them, and because birds, rodents, and other animals frequent them, ponds can become reservoirs of disease-causing organisms such as those associated with equine protozoal myelitis (EPM) and West Nile virus.

In addition, ponds can collect agricultural runoff and have quite high nitrate levels (see page 241). Ponds with high bacteria, nitrate, and phosphorus levels often develop an overgrowth of algae. Certain algae are toxic and can cause sudden death. To prevent algae buildup, decrease runoff contamination with diversion ditches or uphill retention ponds, maintain a buffer zone, clean out the pond bottom periodically, and, if appropriate, consider stocking the pond with algae-eating fish. When in doubt about the suitability of natural waters on your property, contact your county health department to have the water tested.

Also, be sure the approach to and footing around watering holes are safe. During the winter, for horses turned out on pasture, you may need to break the ice on a pond or stream several times a day.

SAFEGUARDING NATURAL WATERS

When waste enters water, bacteria proliferate, the oxygen balance is disrupted, and native residents of the water (including fish, aquatic plants, invertebrates, and amphibians) can die. Whether you have a lake, pond, river, stream or creek, or ditches on your property, manage them responsibly because all waters are connected.

Wetlands (subirrigated "swampy" lowlands) filter pollutants and help prevent flooding and erosion, so are a valuable resource to be protected. Keep horses off wetlands when they are wet.

Riparian areas (streamside vegetation and soils) can be ruined by horses. Manure, urine, overgrazing, destruction of trees, and muddy banks lead to less protective vegetation, warmer water temperatures, and less fish and wildlife habitat. Limit the horses' access to streams. Maintain a buffer zone of grasses, brush, and trees around creek and pond edges to help filter nutrients from excess runoff before it enters the water.

IRRIGATION WATER

If you plan to irrigate your pastures or fields, you should evaluate the quality of your irrigation water for its soluble salt content, sodium content, bicarbonate concentration, and toxic elements. Excess salt

Due to agricultural runoff, irrigation water is not necessarily suitable drinking water for horses. In addition, some canals pose safety hazards for horses.

makes moisture less available so that even though a field appears to have plenty of moisture, the plant roots are unable to absorb it, and they experience a physiological drought. Water with a high sodium content affects the physical properties of the soil, eventually making the soil hard and compact and increasingly impervious to water penetration. Water that is high in bicarbonate tends to result in a higher sodium hazard.

Two toxic substances that sometimes occur in water are boron and chlorine. Toxic elements can affect the pasture or hay fields immediately or after they have accumulated over a number of years.

Horses should have easy access to free-choice, pure, palatable water with minimal mineral content.

Water Quality

No matter what the source, the water for your horses must be pure and palatable and contain minimal mineral matter. There should be no harmful organisms or decomposing organic matter in your horses' water. This means that the well (and cistern if you have one) must be free from contamination and that the water vessels you use for the horses are kept scrupulously clean of old feed, algae, dead animals, and dirt.

To test your well for bacterial contamination, obtain a special sterilized container provided by your county or local health department and take a sample according to the instructions provided with it. The sample must be representative and should not be from a new or inactive well. A well should be thoroughly pumped before sampling.

Far left: **Domestic and stock wells should be disinfected regularly as stipulated in local health department specifications. To accomplish this, bleach is usually diluted in water.** *Left:* **The solution is then poured into the well.**

The laboratory will test it primarily for coliform bacteria, a large category of bacteria associated with intestinal discharge of humans and domestic and wild animals. If your test results show the presence of coliform bacteria, your well has been contaminated since its last sanitization. Some common sources of contamination are a crack in the well house roof, walls, or floor; an improper well cap that lets runoff into the well; and a bird, mouse, or feces that fell into the well when the well cap was left open.

If your sample shows contamination, the health department will probably tell you whether the water is safe for humans or animals and send you chlorinating procedures. Because you will be without water for a whole day during chlorination, you should draw emergency water or arrange for the delivery of other water.

Typical instructions are as follows: Mix 1 gallon of 5 percent unscented household chlorine bleach with 10 gallons of water and add that to the well. Water should then be run from the well through every pipe to all water outlets, hot and cold, including all household outlets (don't forget toilets), hydrants, and barn fixtures and spigots. Let the water run until you smell chlorine at each outlet, and then turn off the water. When the chlorine water is in all the pipes, turn off your pump. Let the chlorinated water stand in the pipes, receptacles, pressure tank, and pump for 8 to 24 hours as instructed. Do not use any water during this time. Then open all receptacles and let the water flow until the odor and taste of chlorine are gone. After a specified period of time (days or weeks), retest the well. Plan for a routine well check once a year.

If you need to chlorinate a large volume of water in a cistern or other storage tank in order to safely use it, determine the volume in the pressure tank, the pump, the pipes to the cistern, and the cistern itself. Check the chlorine dosage rate recommended by your health department. If you add too much chlorine to your water storage, it can kill plants, be unpalatable to animals, and may even harm your horses. If you don't add enough, it won't disinfect. Chlorine is very volatile and dissipates rapidly.

Fresh water from clean streams or springs or water that has just been drawn from a good well is bright, has a pleasant taste, and is naturally aerated. Air promotes digestion in horses by helping digestive juices permeate feed. Foul-smelling water indicates the presence of sewage gases from decomposing organic matter. Horses have a keen sense of smell and detect tainted water easily. A bucket of water in an untidy stable or a stale trough often absorbs impurities, such as carbonic acid. Horses usually refuse to drink such bad water.

Rainwater from roofs often contains soot, dirt, and other impurities. A green or yellow color to the water often indicates fermenting or decomposing plant matter. Iridescence is often due to the presence of petroleum products. Muddiness, from clay, for example, poses no big problem, but it is always best to try and provide clean and clear water.

MINERAL MATTER

The mineral content, as well as pH and levels of salts, sulfates, nitrates, and metals, will be revealed by a chemical water analysis that can be performed at university or private laboratories. Hard water, usually from deep wells and some deep rock springs, has a large concentration of calcium and magnesium relative to sodium, whereas soft water, such as distilled water, rainwater, or surface water springs, streams, and rivers, has a large concentration of sodium relative to calcium and magnesium. As you look at a chemical evaluation of your water, keep in mind that what might constitute good drinking water for you and your horse may not be the best water for washing. (See under Hard Water.)

Although it is acceptable to give horses drinking water that contains moderate amounts of calcium and magnesium carbonates, water with excessive

Effects of excess salt in water

SOLUBLE SALTS (MG/L)	EFFECTS ON HORSES AND CROPS
<1000	Excellent water
1000–3000	Very satisfactory for animals; may cause temporary or mild diarrhea in animals not accustomed to it; low palatability; may have adverse effects on many crops
3000–5000	Satisfactory, but may cause temporary or mild diarrhea in animals not accustomed to it; often refused. Must be used with very careful management practices on salt-tolerant plants
5000–7000	Reasonably safe, except for pregnant or lactating animals; not suitable for crops
7000–10,000	High risk with pregnant, lactating, and young horses
>10,000	Unsuitable for animals or crops

From P. N. Soltanpour and W. L. Raley. Evaluation of drinking water quality for livestock. Service in Action. Colorado State University Extension Service Quick Facts No. 4.908, 1982.

As a natural water source dries up, the mineral content becomes concentrated. In addition, runoff can increase nitrate levels.

levels of these compounds can have an astringent, laxative, or dehydrating effect. Some animal owners and some veterinarians have suggested that hard water can cause urinary calculi ("stones" in the horse's urinary tract), but currently there is no data to support the theory.

Water that contains high levels of soluble salts such as sodium chloride, calcium chloride, magnesium chloride, and some of the sulfates can also have deleterious effects.

OTHER IMPURITIES

Other undesirable substances in water for horses are sulfates, nitrates, and toxic substances.

Water with a high sulfate level might indicate contamination by waste or septic intrusion. It is associated with *E. coli,* which can be toxic to plants, and its bitter taste makes the plants unpalatable for horses. High concentrations can have a laxative effect.

High nitrate concentrations in water are often caused by agricultural runoff, yet because it is colorless, odorless, and tasteless, you won't detect it. Be sure you are preventing nitrogen loss in your pastures and fields by using proper manure management and fertilization practices.

Nitrate (converted to nitrite) interferes with the oxygen-carrying capacity of the blood causing methemoglobinemia (blue baby syndrome). Pregnant women, children, and animals can be very susceptible to high nitrates and so should not drink water that has more than 10 mg/L of nitrate. Mature animals should not drink water with more than 100 mg/L of nitrate.

Arsenic, selenium, barium, cadmium, and mercury above recommended limits must be removed. Check with your county Extension agent and your veterinarian for specific limits and suspected problems in your area.

pH

You should also have the pH of your water assessed. Absolutely pure water has a neutral pH of 7.0, and wells usually range from 6.5 to 8.0. Some metals, such as lead and zinc, are more soluble in acid water.

If water with a pH of less than 5.0 (acidic) runs through lead pipes, it may result in corrosion and subsequent ingestion by your horse of excess lead, which can be toxic. Water that has been softened by an ion-exchange water-softening unit and runs through lead pipes can pick up lead and result in lead poisoning. Today plastic water pipes are used in most residential and agricultural applications, so lead leaching is not a problem.

The calcium and magnesium in hard water with a pH greater than 8.5 (alkaline or basic) tend to precipitate out, causing a white, crusty residue or film. The residue interferes with washing and rinsing and can leave a scale on horses, blankets, and buckets.

HARD WATER

If hard water is used to bathe horses or wash blankets, it makes thorough cleaning difficult and leaves behind a white residue from the precipitated calcium and magnesium. You can overcome hardness in wash water by using water-softening equipment or adding a softening compound.

A popular water-softening device is the common household salt-filled water softener, which exchanges sodium ions for the calcium and magnesium ions that cause water hardness. Although the water is made suitable for cleaning purposes, it becomes very high in sodium (salt) and is often not suitable for drinking water or for watering plants. Therefore, water from such a softener and from some softened community water would be appropriate for washing

Water hardness[a]

RELATIVE HARDNESS	GRAINS[b]	mg/L OR PPM
Soft	0–4.5	0–75
Moderately hard	4.5–7.0	75–120
Hard	7.0–10.5	120–180
Very hard	>10.5	>180

[a] Calculated based on calcium carbonate and magnesium carbonate.
[b] 1 grain per gallon (gpg) = 17.1 mg/L or ppm.

horses and blankets but not for drinking unless its sodium content does not exceed the EPA recommendation of 20 ppm for people on sodium-restricted diets. You can soften batches of water that are specifically going to be used for washing horses, blankets, and equipment by adding, as needed, sodium hexametaphosphate, a water softener sold under the trade name of Calgon. (Do not confuse Calgon with Calgonite; the latter is a cleaner made specifically for automatic dishwashers and contains several harsh detergents.) Sodium hexametaphosphate adds sodium to the water to make the sodium concentration higher than the calcium and magnesium; this makes it a better solvent. Less expensive softeners such as washing soda and trisodium phosphate are very alkaline, which is undesirable for use on hair and skin. In addition, they precipitate minerals from the water, resulting in a sludge or film in the wash or rinse water. Sodium hexametaphosphate, by contrast, holds the minerals in suspension in the water so they cannot form scum. It is tasteless and nontoxic and has a neutral pH.

This water softener is useful in two ways: to increase the effectiveness of soap and to act as a thorough rinse. With cold, warm, or hot water, sodium hexametaphosphate helps soap or shampoo do a better job of cleaning, as it prevents the dingy, insoluble scum from forming. As a rinse for a horse's hair or a blanket, a sodium hexametaphosphate solution removes the graying dullness left by previously deposited soap residues. It has a superior ability to combine with and sequester oily and greasy substances, which prevents them from reacting with the horse's skin or becoming trapped in the fibers of a blanket.

Here are some especially handy uses for a sodium hexametaphosphate solution: to sponge away the outline of bridle and saddle from a horse that has just finished working; to rinse the mane or tail without shampooing; to dampen a stable rubber for use as a dust magnet in the final stages of grooming.

How much sodium hexametaphosphate to use depends on the hardness of the water. One teaspoon per gallon of water would be adequate for naturally soft water with a hardness of 4.5 grains or less per gallon. Two tablespoons per gallon would be more appropriate for very hard water with a hardness score of 10.5 grains or more per gallon. At those rates, the 4-pound box available in most grocery stores goes a long way. Store the box of sodium hexametaphosphate granules in a cool, dry place and mix as needed.

Watering Devices

If you are putting in hydrants in a temperate climate, be sure they are the freeze-proof, self-draining type and located where they are not accessible to horses. Drain and roll up hoses after each use. Troughs are fine if you have a large number of horses, but if just one or two horses are drinking from a trough, the water could become stagnant. Clean troughs regularly. If troughs are to be used in the winter in tem-

Clockwise from top left: Bracket-mounted stall water bucket; insulated bucket holder; trough.

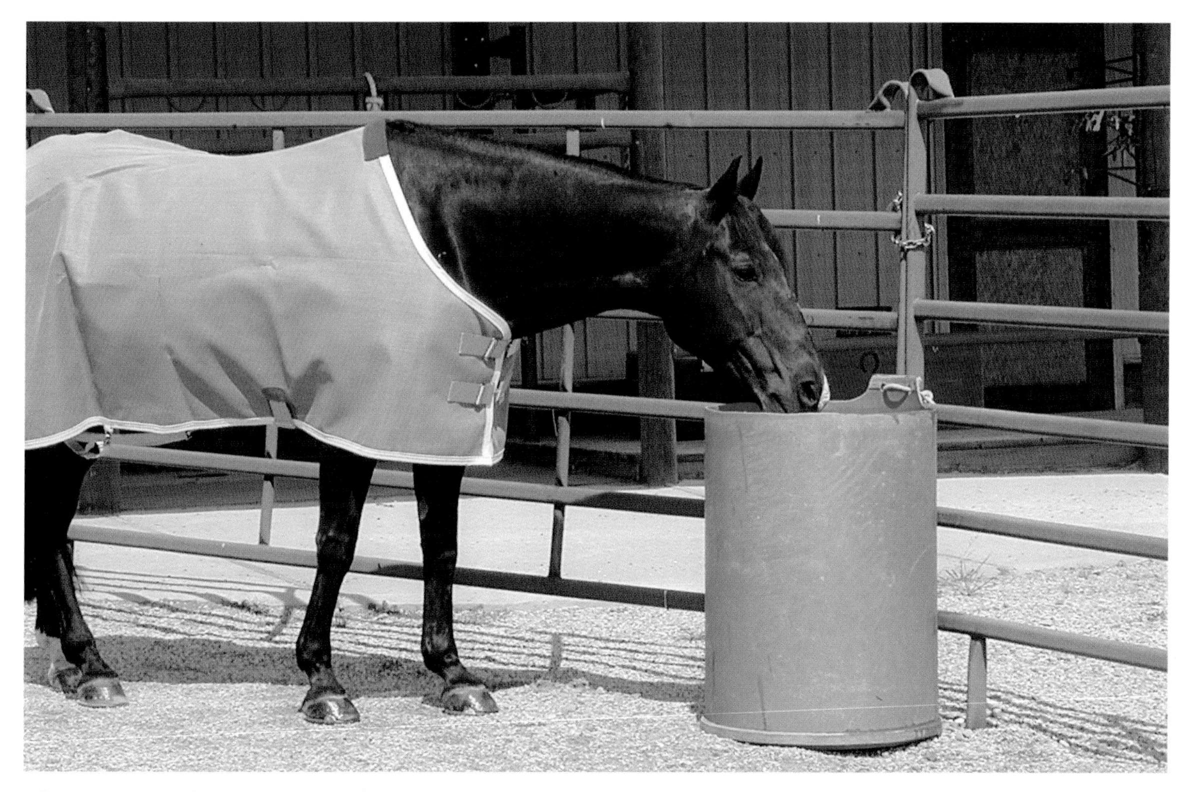

Above: **Zipper takes a long draw from the water barrel in his pen. The top was cut off this 50-gallon vanilla barrel, so it can be attached to the panels with a tab.** ***Right:*** **I chop a hole in the ice on the creek once or twice a day during winter to ensure that horses on pasture can get a drink.**

perate climates, they can be insulated and kept from freezing with a caged tank heater. Otherwise, you'll need to check the troughs at least twice a day for freezing and break the surface ice if necessary. Waterers made from plastic barrels that held a non-toxic substance (such as vanilla or vinegar) hold about 35 to 40 gallons, which is suitable for a single horse pen.

Buckets work best in stalls if they are mounted with a bucket bracket, which prevents the bucket from being tipped over by the horse. Freeze-proof buckets can also be used. Automatic (heated, if necessary) waterers can be shared between two stalls, two paddocks, or two pastures. Provide plenty of waterers so that all horses have the opportunity to drink an unlimited amount of pure, fresh water. (See Hill, *Stablekeeping* [Storey, 2000], for more detailed information on waterers.)

15 Sanitation

Good sanitation practices are important for your horse's health, your family's health, and relations with your neighbors, and help you meet your legal obligations. Sanitation involves management of manure, flies and other pests, mud and moisture, and hazardous wastes. The more conscientious your sanitation and land management plans, the more productive your land will be, the healthier your creeks and ponds will be, and the more you will be able to enjoy natural wildlife.

Good sanitation practices are essential for suburban horse-keeping. A well-run horse acreage adds interest and a special ambience to the neighborhood.

No matter whether your horses are pastured or stabled, they will produce generous amounts of manure and urine daily. In an enclosed barn, the added waste products of respiration of the skin and lungs help to make the environment an ideal breeding ground for bacteria.

Urine contains urea and hippuric acid, both of which release ammonia, a volatile gas, into the air. The pungent vapor can be injurious to the eyes and lungs of horses and humans. To minimize ammonia, feed adequate but not excessive protein (minimal alfalfa hay), clean stalls regularly, and provide adequate ventilation.

Waste products can also be destructive to flooring and walls, tack, and horses' hooves. The combination of dung and urine is a perfect medium for the proliferation of bacteria that can begin destructive processes on leather. Dung and urine can also break down the integrity of hoof horn. When certain fecal bacteria ferment, their secretions can dissolve the intertubular "hoof cement." In addition, moist manure softens, loosens, and encourages the breakdown of hoof horn cells, even more aggressively than water or mud does.

Wherever there is manure, there are parasite larvae. Horse parasites leave the horse host via the manure, then reinfest a new host; the life cycle is then repeated. When a horse eats from manure-contaminated ground, he continually reinfects himself. Parasite larvae can do great internal damage to a horse as they migrate through the tissues. Deworming horses every 2 months decreases the number and viability of parasite eggs shed, but daily removal and proper management of manure are the best ways to break the parasite life cycle.

Bedding

Bedding should be clean, dust-free, nontoxic, absorbent, not slippery when wet, soft enough to encourage a horse to lie down without developing fetlock and hock sores, easy for you to handle, available, economical, and something that a horse won't eat. Thinking ahead of time about what type of bedding you will use will probably affect your choice of stall flooring and vice versa. Availability greatly affects price. In the bedding chart, price estimates are general. You may find wood shavings more economical in timber areas and straw bedding less expensive in farming areas.

The bedding with the highest water-absorbing capacity is not necessarily the best bet. Extremely absorbent bedding sops up too much urine, and the horse stands or lies in the soggy mess, and bedding

Traditional bedding, in order of absorbency

BEDDING TYPE	LBS. WATER ABSORBED[a]	COST[b]	COMFORT	CLEANLINESS
PEAT MOSS	10.0	High	Thick, soft bed, usually dust-free	Sodden, heavy to shovel, difficult to find, expensive
PINE CHIPS	3.0	Low	Rough	May contain foreign objects
OAT STRAW	2.8	Low to medium	Good if not crushed	Good
PINE SAWDUST	2.5	Low	Warm, soft bed	May contain foreign objects
WHEAT STRAW	2.2	High	Good if not crushed	Rarely dusty
BARLEY STRAW	2.1	High	Short stems, not elastic	Often damp or dusty
PINE SHAVINGS	2.0	Low	Fluffy bed	May contain foreign objects
HARDWOOD CHIPS, SAWDUST, OR SHAVINGS	1.5	Low	Can be rough	May contain foreign objects and toxic hardwoods
SAND	0.2	Low	Soft but abrasive	Dusty

[a] Per pound of bedding; [b] depends on availability in your area.

with very little absorbency allows too much moisture to pass through to the flooring. In most cases the ideal bedding has an absorbency of between 2.0 and 3.0 and is free from dust, mold, and injurious substances.

Softwood products, such as pine sawdust, shavings, and chips, are commonly used for bedding. They are absorbent and give the barn a fresh smell; however, wood bedding takes longer to compost than straw. *Sawdust* made up of large particles is produced when logs are sawed into lumber. Sawdust from smaller saws, such as those in cabinet shops, may be too fine and dusty for bedding.

Shavings are thin, small slices of wood produced by the planing or surfacing of lumber. Shavings are available at sawmills and cabinet shops. Shavings are difficult to remove from a tail or mane.

Chips are small, coarse pieces of wood produced by the drilling, shaping, turning, or molding of lumber; they are not as comfortable or as absorbent as sawdust or shavings. Hardwood products are generally undesirable because of their poor absorbency and in some cases, such as with black walnut, a dangerous toxicity. Horses merely coming in contact with such shavings have experienced founder and death.

Straw bedding is traditional, comfortable, and readily available but often dusty, and many horses eat their straw bedding. Wheat straw, because of its high glaze, is not as absorbent as oat straw and so does not become as slimy and sloppy when wet and is less palatable to horses; it therefore may be safer to use with a horse that overeats. Oat straw is bright but it becomes slippery when wet and it is too palat-

CONVENIENCE FACTORS	PALATABILITY	OTHER
When dry, light; when wet, very heavy	Low	Difficult to see manure; gets soggy; associated with thrush; good compost value; fast composting
Usually hauled bulk, then must be shoveled	Low	Can be drying to hooves
Light bales but slimy when wet	High	Horse often gets "straw belly" from eating bedding; fast composting
Bags OK; bulk must be shoveled	Low	Can pack and be drying to hooves; can ferment when wet and heat hooves
Light bales; heavy when wet	Low; but OK if eaten; awns irritate eyes and mouth; can cause colic	Good compost value but makes large manure pile
Light bales; heavy when wet	Low	Can cause colic
Very light, but large volume required	Low	Can be drying to hooves; hang in mane and tail; slow composting
Hauled bulk, then must be shoveled	Inadvertently ingested with feed; black walnut toxin can cause founder	Toxic effect if black walnut is present; shavings difficult to remove from tail
Heavy	Low, but can be ingested with feed	May result in colic or hoof damage; sand particles can work their way into the bottom of the hoof wall

able. Barley straw should be avoided because of the sharp, barbed awns that can become lodged in a horse's gums. Straw is the preferred bedding for foaling stalls but is very slippery on wood floors. All beddings, even peat moss, have the potential to be dusty. To prevent respiratory problems in your horse, be selective and don't purchase dusty bedding material. Bedding can be purchased in bulk or in bags and needs to be stored in a dry, weatherproof shed.

Compressed wood pellets are a recent innovation in bedding. Some pellets absorb up to four times their weight in liquid. Use of this product differs from traditional bedding, and recommendations vary according to manufacturer. To begin, you would spread approximately five 40-pound bags of pellets in a 12-foot by 12-foot stall and sprinkle the pellets lightly with water, which causes them to expand. Each day when you clean the stall, you will remove the solid waste and leave the wet bedding. After 1 to 2 weeks, you will remove the wet areas and add more product as needed, approximately 2 to 5 bags per month.

Newspaper pellets are made of processed and compressed shredded newspaper. They have similar properties to wood pellets but are not as widely available.

Kilned clay granules, a manufactured sand bedding, resembles cat litter. The granules have a large surface area, which soaks up and evaporates urine. This recently introduced bedding is reported to be nonflammable and dust-free, but it requires a specialized barn cleaning system. It has been suggested that if a horse ingests the bedding, sand colic could result.

Cleaning a Stall

It is far easier to clean a stall when the horse is out for exercise than with the horse in the stall. Remove the dung piles using a steel or plastic fork with tines spaced to pick up the manure yet let the bedding fall through.

Moisture is the main cause of odors in a stall. Search for spots of wet bedding and remove them. Expose the stall floor to dry by banking the remaining good bedding against the stall walls. Use a stiff broom to sweep some of the oldest bedding back and forth over wet areas to absorb the moisture, and then scoop up that bedding with a shovel. Use a stall freshener product to absorb any remaining moisture and odors.

Hydrated lime (calcium hydroxide) used to be the old standby for deodorizing and drying out stall floors. It lowers the acidity of the urine and causes dirt particles to clump, thereby allowing air to get to them and dry them out. It is cheap and readily available, but not especially safe or effective. In fact, hydrated lime is highly alkaline and can irritate your skin or your horse's, especially when damp. Horses that eat from heavily limed floors can suffer mouth, throat, and lung damage.

Products containing zeolites or blends of diatomaceous earth and granular clay are safer, far more absorbent than lime, and better at reducing ammonia odors. They are nontoxic to people and animals, nonflammable, and environmentally friendly.

After the application of a stall freshener, let the stall floor dry all day, with barn doors and windows open if possible. In the evening rake the dry "old," but still usable, bedding back in the area of usual defecation and urination. Add fresh bedding, if needed, to the area where the horse lies or stands. If you have a spare stall, move your horse there and let his regular stall dry thoroughly. Before you return your horse to his stall, pick out his hooves and give him a good brushing and check his blanket, if he wears one.

Every week or two, depending on how heavily the stall is used, remove all the bedding and start fresh.

Zeolites

Zeolites are a group of naturally occurring minerals, hydrous silicates, that were deposited as a result of volcanic activity millions of years ago. Zeolites have a honeycomb-like structure that gives them a large surface area. This enables them to absorb tremendous amounts of odors. Foul-smelling gases latch on to dust particles that have a positive molecular charge. Zeolite molecules have a negative molecular charge, so they act like magnets to attract dust particles, thus helping clear the air of odors as well as dust. In a similar manner, zeolites trap positive ammonium ions directly from urine, which makes them especially effective at reducing ammonia odors.

The honeycomb structure of zeolites, with millions of tiny pores, enables them to absorb up to 60 percent of their weight in water. Also, zeolites do not become dangerously slippery when wet, which is important if your stall flooring consists of wood or solid rubber mats. There's no need to let the stall dry out completely when using zeolites. Just remove wet bedding, cover damp floor areas with 1⁄16 to 1⁄8 inch of a zeolite product followed by a layer of dry bedding, and return the horse to the stall. The zeolites will absorb any remaining moisture and odors and hold them until the next stall cleaning.

If odors are mounting but you don't have time to clean stalls, you can sprinkle a zeolite product on the wet spots or place an open-top container near the stalls to absorb odors.

Studies have demonstrated that zeolites are essentially nontoxic to people and animals, whether ingested, on the skin, breathed, or in the eyes. Zeolites are nonflammable and environmentally friendly and can be safely handled around horses' water and feed with bare hands, which is especially nice for those suffering from chemical sensitivity. Zeolites have a neutral pH of 6.0 to 8.0 and, as an added benefit, used zeolites are loaded with ammonium ions, which makes them an excellent slow-release fertilizer for gardens, plant beds, yards, fields, and potting soil.

Managing the Manure Pile

Manure production on even the smallest horse farm requires constant attention. Understandably, poor sanitation practices can be a source of neighbor dissatisfaction. As land- and animal owners, we have the responsibility to be good stewards of our animals and of the environment. Depending on the government regulations in your area, your horse property might be classified as an animal feeding operation (AFO), which would require you to follow a specified management plan to protect the land and groundwater. Some of the criteria used in determining if you fall under government regulations relate to how many horses you have in what amount of space, the size of the confinement area, and whether the confinement area has forage in it. But even if your farm is not classified as an AFO, you need to develop an appropriate and conscientious sanitation plan.

A 1000-pound horse produces approximately 50 pounds of manure per day or a little more than 9 tons per year. In addition to this, a horse produces from 6 to 10 gallons of urine per day, which, when soaked up by bedding, can constitute another 50 pounds daily. Thus, if you have five average-sized horses living in pens, they will produce more than 45 tons of manure per year. If you keep them in bedded stalls, you will have 90 tons of manure and used bedding to manage each year.

About one-fifth of the nutrients a horse eats is passed out in the manure and urine. Manure is composed of undigested food, digestive juices, and microorganisms. The bacteria make up as much as 30 percent of the mass! Because urine (usually as soaked bedding) is a liquid, it contains more dissolved nutrients that are readily available than do feces.

If the manure is properly handled, about half of those excreted nutrients can be used by pasture or crop plants in one growing season, with the balance being used in subsequent years. Horse manure is considered valuable manure because it is "hot," or capable of breakdown by composting. A ton of fresh horse manure (without bedding) will supply the equivalent of a 100-pound sack of fertilizer (approximately 14-4-14; exact nitrogen-phosphorus-potash ratio will vary), as well as providing valuable organic matter and trace elements.

Even if manure is not to be used as a fertilizer, it must be properly managed in order to control odor,

Above: **With consolidation, added moisture, and mixing, this sprawling manure pile would be more likely to compost properly.** ***Right:*** **This tidy pile is composting; the humus will be ready to spread in a few months.**

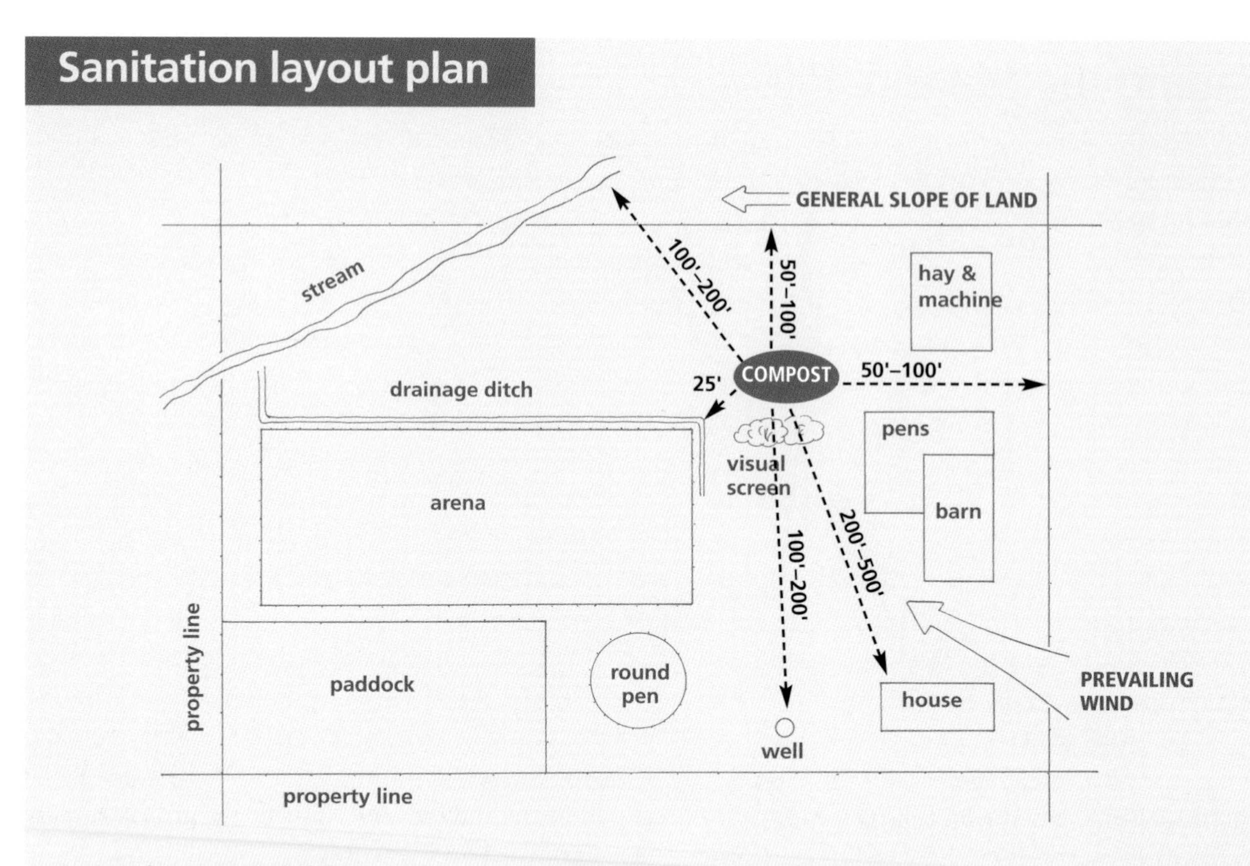

remove insect breeding areas, kill parasite eggs and larvae, and prevent nitrogen and phosphorus in the manure from contaminating water. Nitrates entering the groundwater can affect streams and wells. Nitrates in drinking water above the EPA maximum contaminant level can cause health problems, particularly in infants. Although horses tolerate elevated nitrate levels somewhat better than humans do, they are more susceptible to nitrate poisoning than most other monogastric (single-stomach) animals.

If excess phosphorus runs off the land during rainfall or snowmelt and settles in your pond or lake, it can lead to eutrophication. This is a process whereby excess nutrients in water cause excessive and unhealthy aquatic plant growth, particularly of algae.

HANDLING MANURE

There are basically five ways to manage manure: give it away, let it lie, haul it away, spread it fresh, and compost it.

Give it away. If you are in an area where gardening is popular and you make access to the manure pile easy, you might be able to give away all the manure your horses produce.

Let it lie. This second option relates to manure deposited on pastures and is viable only if you have very large pastures. If you had four horses on 100 acres, they would tend to create areas where they defecate, areas where they lie and rest, and preferred grazing areas. Generally, given enough room, horses do not contaminate their eating areas. But few of us are fortunate enough to have so much pasture. On

Left: **A covered can on wheels makes it convenient to collect manure regularly.** *Middle:* **If you can't compost on your farm, accumulate manure and bedding in a large receptacle for sale, transport to a compost site, or disposal.** *Right:* **Although not ideal environmentally, manure is picked up with the trash in some areas.**

medium-sized horse operations, larger groups of horses tend to be turned out on smaller pastures. When nine head are turned out on 2 acres, even for a few hours each day, the manure accumulates quickly. Yet it is impractical to manually pick up manure on a daily basis from a 2-acre pasture. The best option is to let it lie where it fell. If you can afford to leave a pasture vacant for a year, harrow the manure to spread it more evenly over the pasture. But you wouldn't want to do this and then immediately use the pasture for grazing, because you would have just helped parasites continue their life cycles by distributing the parasite eggs and larvae over the entire pasture.

The other three manure management options are more suitable to small acreage and begin with daily collection. Daily collection or twice-daily collection is best for most small horse operations.

Agronomy is the science of land management. Once manure is collected, it can be hauled away, spread immediately on a hay field or cropland at agronomic rates, or composted for later distribution as humus. Agronomic rates of fertilizer application will vary according to local conditions; your county Extension agent will be able to help you determine what is optimum for the land, the environment, and your operation.

Many small horse operations produce more manure than their gardens and pastures require for fertilization. If you have four horses and a 1-acre pasture, your pasture does not need and should not get all of the manure the four horses produce. Here are your three options.

Haul it away. Some refuse-collection services are specially designed to handle manure or are willing to haul it along with other trash. Putting manure in a specially designated Dumpster that is emptied weekly works in some areas, or you could haul manure to the landfill yourself and pay, for example, $5 per cubic yard to dump it. Manure that is hauled off to a landfill soon becomes buried, and without access to oxygen it won't decompose. Instead, it generates methane gas, which is not beneficial for the environment, so this is not a good option.

Spread it fresh. Spreading fresh manure daily is the least environmentally responsible choice, due to the probability of agricultural runoff leading to high nitrates and water contamination. It is far better to compost. However, if manure must be spread daily, it should be distributed on land that will not be grazed by horses for at least a year. Even on cropland or hay fields, it should be spread thinly and/or harrowed to encourage rapid drying, thus eliminating favorable conditions for fly larvae and decreasing odor. Be sure you have determined an agronomic application rate for your land so that you know when

Left: **Loading fresh manure onto a spreader for immediate application to land is suitable only for farms with large tracts of cropland or pastures not grazed by horses.** ***Right:*** **Horse manure loses 50 percent of its bulk by composting.**

your daily spreads must stop. Try to apply manure and compost at times of greatest plant growth so the plants can use the nutrients. Avoid spreading manure and compost on frozen soil because it could be blown or washed away.

Although few problems are encountered when applying fresh horse manure to established grass pastures, fresh horse manure should never be applied to a newly seeded hay field or pasture or to a garden or newly planted trees, since it can burn plant tissues. If it will be used in these or similar situations, horse manure should be composted from 6 to 8 weeks before being applied. For best results with new plants, work composted manure into the soil at least 4 weeks before seeding or transplanting. Using raw manure on gardens is not recommended due to the risk of *E. coli* bacterial infection.

Compost it. The best and most environmentally friendly method of dealing with horse manure on a small acreage is daily collection and composting. Composting is efficient, convenient, and environmentally responsible. The manure does not have to be hauled off the property every day or week, and composting reduces bulk by up to 50 percent while concentrating nutrients.

Composting releases nitrogen and other nutrients slowly, so little nitrogen and phosphorus leaches or runs off, minimizing environmental pollution. In addition, compost does not have an unpleasant odor to most people and is pleasant to handle. A properly composted manure pile will kill parasite eggs and many weed seeds, prevent flies from breeding, and result in a good-quality soil enhancer and fertilizer.

The maintenance of a compost pile requires air, moisture, and temperature control. It would be ideal to have three compost piles: one to which fresh manure is being added daily, one that is in the process of decomposing, and one that is fully composted and ready to spread. Assume that 300 to 600 cubic feet of space (depending on the type of bedding you use) will be needed to store a year's worth of manure from one horse.

Before starting a pile, check your local zoning ordinances. Be sure the pile is out of sight and smell of residences and downwind from the stable and the house. It should be located at least 150 feet away from waterways, including streams, irrigation ditches, and wells. Locate the pile so that it is convenient for daily dumping and periodic hauling and so that you can reach it with a hose for watering.

If possible, the piles should be located on a sloped (concrete) floor with 4-foot walls. The fresh pile is usually left open for convenient daily addition. Because an open pile is subject to drying by the sun and leaching of nutrients by rain and melting snow, the other piles could be covered.

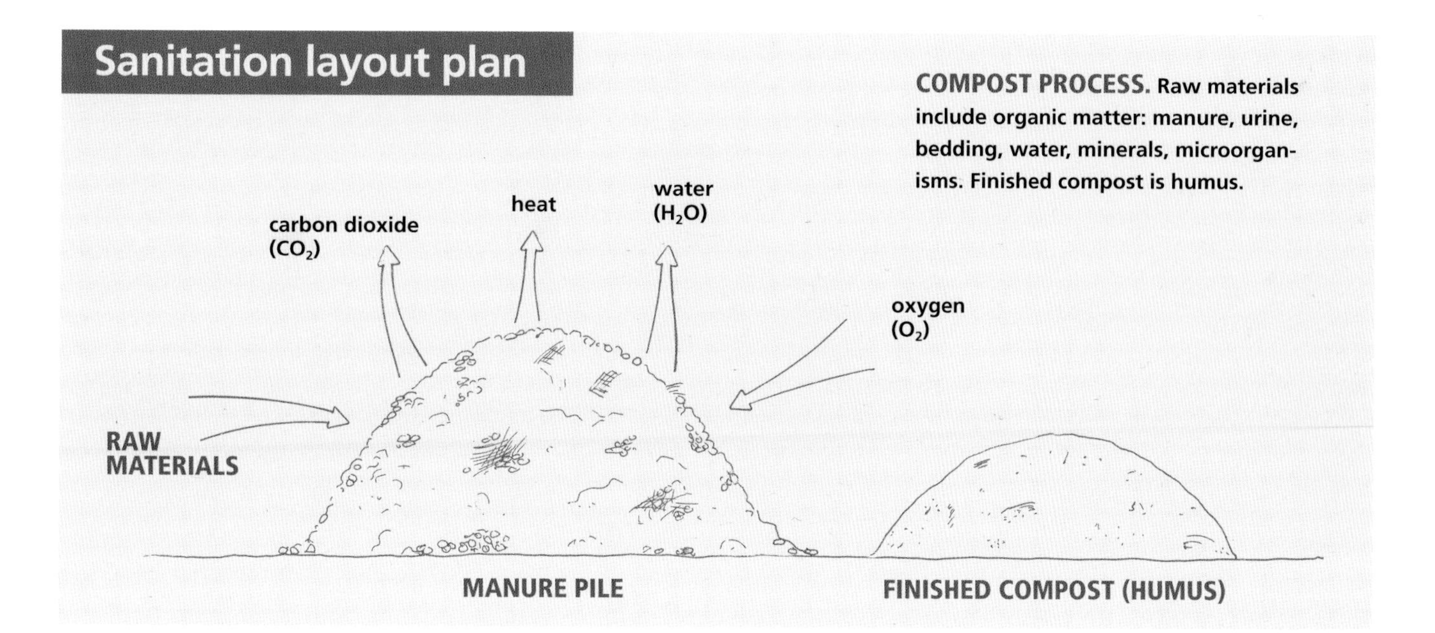

The composting and composted piles should be kept uniformly moist — about 50 percent, like a wrung-out sponge. A dry pile dehydrates the microorganisms; a soggy heap smothers them. Whether you cover and water your compost piles will depend on your local precipitation. In arid or semiarid climates, you'll want the pile uncovered so it can benefit from precipitation and you can add water to the pile as needed. In very wet regions, the compost pile could become too soggy and allow undesirable leaching into surface waters, and so might do better under a roof or cover. Plastic covers retard oxygen exchange and smother the bacteria, so a tarp might be a better choice. Earth covers are too heavy, reducing the necessary pore space. Open piles or those with geotextile fabric covers seem to fare best. An open pile should be about 4 feet high and 4 to 6 feet wide or sized so that you can reach it with a fork to

Our manure fort is built into a hill below the horse barn, which makes for convenient dumping.

A section of fence helps contain the pile, and a sprinkler hose laid over the top allows us to add moisture when necessary.

turn the compost. Once the pile has been formed, add to its length, not to its height, to form a windrow. When the compost process is completed at one end, that end is ready for use.

Decomposition of manure begins with the formation of ammonia as urinary nitrogen decomposes. The effectiveness of composting depends on the size and shape of the pile, the degree of compaction, moisture content, and aeration of the manure pile. A large pile with a flat or concave top is ideal because it retains its own heat and the top captures moisture from precipitation. A small pile with a dome would not hold its own heat as well, and its rounded top would tend to shed water. A fluffy, occasionally turned pile (for aeration) makes the best environment for the aerobic composting microorganisms. Aeration can be accomplished by hand with a fork or with the loader on a tractor. You might turn the pile three or four times in the first 2 months, then once a month until it's done.

Passive aeration can be attempted by inserting PVC drain tiles into the pile either horizontally or vertically. These 4-inch-diameter pipes have rows of ½-inch holes drilled in them. Use 10-foot sections and leave the ends exposed.

A carbon-to-nitrogen (C:N) ratio between 25:1 and 30:1 is optimum for composting. The lower the C:N ratio, the hotter the compost. Horse manure has

Below: The only way to know whether a manure pile is chilly or cooking is to take its temperature. *Right:* We spread humus twice a year on vacant pastures that won't be grazed for 6 months.

Compost pile troubleshooting

WHAT	WHY	DO THIS
Compost smells bad.	Needs more oxygen.	Turn pile with a fork or tractor or insert PVC drain tiles.
Compost is soggy.	Too much rain or snow.	Cover or add leaves or bedding and mix.
Compost is too dry.	Not enough rain.	Put on a sprinkler or soaker hose, then mix until uniformly moist.
Compost is hotter than 150°F.	Too much nitrogen, pile too large.	Add leaves or bedding or make smaller piles.
Compost is cooler than 110°F.	Not wet enough, pile too small, not enough nitrogen, too much bedding.	Add more water, make pile larger, add more raw manure or fresh grass clippings.

a C:N ratio of 30:1 to 50:1. If you add sawdust (400:1) or straw (80:1) as bedding, the C:N ratio will be higher (cooler). To keep the C:N ratio in the ideal range, select an amendment with a lower carbon content or a higher nitrogen content. Grass clippings, at 17:1, make a good addition to lower the C:N ratio.

If you use hydrated lime to dry stall floors, the negligible amount in composting bedding will not affect a manure pile significantly one way or another.

Microorganisms essential for composting thrive at temperatures of 100°F to 150°F. At 130°F and higher, most bacteria, viruses, fungi, protozoa, weed seeds, and fly larvae are destroyed.

To kill most parasite eggs, the heat of the manure pile should be maintained at 145°F for at least 2 weeks or at lower temperatures for longer periods of

Compost pile calendar

MONTH	MANURE PILE A	MANURE PILE B	MANURE PILE C
JANUARY	Sell, store, or spread	Turn once; water as needed	Turn twice; water as needed
FEBRUARY	Sell, store, or spread	Turn once; water as needed	Turn once; water as needed
MARCH	Start pile	Sell, store, or spread	Quit adding; turn once
APRIL	Turn twice; water as needed	Sell, store, or spread	Turn once; water as needed
MAY	Turn twice; water as needed	Sell, store, or spread	Turn once; water as needed
JUNE	Turn once; water as needed	Sell, store, or spread	Turn once; water as needed
JULY	Quit adding; turn once	Start pile	Sell, store, or spread
AUGUST	Turn once; water as needed	Turn twice; water as needed	Sell, store, or spread
SEPTEMBER	Turn once; water as needed	Turn twice; water as needed	Sell, store, or spread
OCTOBER	Turn once; water as needed	Turn once; water as needed	Sell, store, or spread
NOVEMBER	Sell, store, or spread	Quit adding; turn once	Start pile
DECEMBER	Sell, store, or spread	Turn once; water as needed	Turn twice; water as needed

time. The only way you will know if a manure pile is at the proper temperature is to test it with a compost thermometer. If the pile is too cool, it may be too dry. If a pure manure pile is too hot, you can moderate its heat so the composting cycle lasts longer by adding leaves or bedding.

The final phase of composting is a curing stage of about a month wherein the microorganisms finish degrading the more complex organic compounds and useful soil-nitrifying bacteria repopulate the compost. As the bacteria die and decompose, they release their stored nitrogen. As the fiber breaks down, carbon dioxide and water are released, decreasing the bulk of the manure by up to half.

The end product of composting is humus, the dark, uniform, finely textured, odorless product of the decomposition of organic matter that is so valuable as a soil conditioner and additive.

The process of decomposition of a manure pile can take anywhere from 2 months in summer to 5 months in winter, and the quality of the resulting humus will vary. Composting horse manure could become a lucrative cash by-product of your horse operation, because landscapers, mushroom growers, and worm farmers seek good-quality humus. You could also use it to enhance the soil on your property; humus improves plant growth, makes the soil softer and more aerated, and increases the soil's ability to hold water.

It is best to spread manure or compost four times during the pasture-growing season — spring to early fall — at the rate of ¼ inch per application, or approximately 1 inch total per year.

Space for manure/compost

NO. OF HORSES	SPACE NEEDED
1	8'x8'x8'
Up to 5	Two or three 12'x12'x8' compost bins (turned by hand)
5+	One 30'x30'x8' bin for three compost piles (turned by tractor)

Flies and Other Pests

Stable flies, horseflies, deerflies, horn flies, face flies, mosquitoes, and ticks are all bloodsuckers and create problems for horses and their owners. Mosquitoes, ticks, and other insects are carriers of diseases such as West Nile virus, Lyme disease, and pigeon fever (dryland distemper) and can become a management nightmare for horse owners.

STABLE FLIES

Stable flies are the same size as common houseflies. Both males and females draw blood, commonly feeding on the lower legs, flanks, belly, under the jaw, and at the junction of the neck and the chest. When they have finished feeding, they seek shelter to rest and digest. Their bite is painful; some horses have such a low fly-tolerance threshold that they can be driven into a frenzy or can panic and run. Even rather tough horses may spend the entire day stomping alternate legs, causing damaging concussion to legs, joints, and hooves and resulting in loose shoes and loss of weight and condition.

Stable flies breed in decaying organic matter, and moist manure is a perfect medium. A female often lays twenty batches of eggs during her 30-day life span, each batch containing between forty and eighty eggs. The eggs hatch in 21 to 25 days. When the eggs hatch, the adult flies emerge ready to breed. (If you have seen small flies and thought they were immature stable flies, you were probably looking at a different type of fly.) The number of flies produced by one pair of adults and their offspring in 30 days is a staggering figure, in the millions. That is why prevention is the best way to keep the fly population under control. It is based on removing breeding grounds, controlling moisture, and using insecticides and other insect-control measures.

Prevent breeding. Manure management and moisture control are two key ways to remove breeding grounds. Remove manure and wasted feed daily from stalls and pens, and either spread it thinly to dry or compost it. Keep moist areas to an absolute minimum. Be sure there is proper drainage in all

Left: Minimize mud and standing water to dramatically increase sanitation and hoof health and to decrease insect populations. *Below:* A fly predator on the edge of a dime.

Spalding Laboratories

facilities. Repair leaking faucets and waterers. Eliminate wet spots in stalls and pens by clearing away bedding, adding stall freshener, and providing adequate air circulation via natural airflow or fans to dry stall floors.

Kill larvae. Fly predators can break the life cycle of flies. This method of biological control is safe and nontoxic and, if properly implemented, requires much less labor for a greater degree of control than many insecticide-based methods.

Fly predators are tiny, nocturnal, stingless wasps that lay eggs in the pupae of the common housefly, the biting stable fly, the horn fly, the lesser housefly, the garbage fly, and the blowfly. The wasp eggs use the contents of the pupae as food, thereby killing the fly before it can develop. The wasps stay within 200 feet of where they hatched and work while you sleep. They are harmless to animals and people. Methods of control involving insecticides must be carefully implemented or they will wipe out the predator population along with the flies.

Chemical larvicides can be applied to manure piles or fed to horses. Those fed to horses pass through the digestive system undigested and begin their work on the larvae in manure. The distribution is unequal in the manure, however, and the larvicides have an active life of only about a day.

Trap adults. Baits, including sticky paper, sweet fluids, and sex attractants, can be used in areas of heavy accumulation.

Five lines of defense in the war on flies

PREVENT flies from breeding.

↓

For flies that manage to breed,
PREVENT larvae from hatching.

↓

If some larvae succeed in hatching,
CAPTURE adult flies immediately.

To deal with flies that avoided the traps,
KILL those that remain.

↓

For flies that survive all previous methods,
PROTECT your horse from them.

Top, left to right: **Reusable bottle fly trap; disposable bag fly trap; flies lined up on sticky tape; sticky flypaper.** ***Right:*** **A horsefly trap (by Horse Pal Fly Trap) keeps horse flies from bugging Sassy.**

Bag and jar traps use attractants and can capture thousands of flies. Some use *muscalure,* a sex attractant (pheromone) to draw the flies. Others require the addition of fish or meat. Rotten-bait traps, commonly used with a 1- or 2½-gallon jar, can be smelly and must be emptied, then restocked.

Disposable bag traps are more convenient. Just scent them with the accompanying tube of sex attractant, add water, and watch the trap catch up to 10,000 flies.

Flypaper is available in strips, coils, or on reels. Some papers contain sex attractants; others are just sticky. Flypaper is an inexpensive, disposable way of mechanically catching flies.

Horseflies and deerflies can be trapped using a specialized visual attractant.

Kill adults. Much less will have to be done with insecticides and other chemical control measures if manure and moisture are handled properly. The indiscriminate use of any form of *insecticide* (a chemical that kills flies) or *repellent* (a chemical that keeps flies away) can result in the development of resistant strains of flies and can harm horses, humans, and the environment.

Fly bait kills flies that eat it. Flies are attracted to the poison bait because of an enticing sugar base and/or a sex attractant. Fly bait can be used in hanging bait stations or as scatter bait on lawns and around buildings. But with this method, there is great risk of children or other animals (birds, puppies) eating the bait.

There are many forms of insecticides and repellents. Long-term residual insecticides last up to 6 weeks and are applied on fly resting sites, such as rafters and bushes. Fogs and mists are to be used daily and are either expelled into the barn air using an automatic timer or applied to the horse's body with a handheld mister. Impregnated strips are useful for enclosed areas such as tack rooms, feed rooms, and offices.

Fly sheets and masks come in many styles and provide protection from flies, ultraviolet rays, and dirt. Dickens *(left)* sports a model with ear and neck protection, while Zinger, the matriarch, is stylish in her tropical blue, pink, and green ensemble.

Protect your horse. You can further protect your horse from flies and other biting insects with fly gear and repellents. Repellents are available as sprays, lotions, wipe-ons, gels, dusting powders, ointments, roll-ons, shampoos, and towelettes. Repellents contain a substance irritating to flies, such as oil of citronella, and most contain some amount of insecticide (mostly pyrethrins and permethrins) as well.

Repellents can be water, oil, or alcohol based. Oil-based repellents remain on the hair shaft longer but the oil attracts dirt. Water-based repellents don't last as long but attract less dirt. To increase the lasting effect, some water-based repellents are made with silicone, which coats the hair shaft and holds the repellent in place longer. Alcohol-based repellents dry quickly and so are good for a fast touch-up, but the alcohol can have a drying effect on both hair and skin.

Repellents can also contain sunscreen, coat conditioners (lanolin, aloe vera), and other products, which increase lasting power. How long a repellent lasts depends on the weather, management, exercise level of the horse (how much he sweats), and grooming (brushing, blanketing), and whether the horse rolls.

In addition to spray-on, wipe-on, and stick repellents, impregnated strips and tags can be attached to halters. These are especially helpful in controlling face flies, which have sponging mouthparts and feed on mucus around the eyes and nostrils. Some degree of relief can also be afforded the horse by using fly shakers attached to the crown piece of a halter or the brow band of a bridle. These strips mechanically jiggle the flies off a horse's face when he shakes his head. Mesh fly masks prevent face flies from landing around the eyes and ears. Cool, open-weave fly sheets keep flies from pestering the horse on his body. Mesh leg boots can eliminate stomping.

MOSQUITOES

Although birds are the reservoir for West Nile virus, mosquitoes can transmit the disease to horses and humans. Birds and rodents are the reservoir for sleeping sickness — western equine encephalomyelitis (WEE), eastern equine encephalomyelitis (EEE), and Venezuelan equine encephalomyelitis (VEE) — and mosquitoes can transmit these diseases to horses. Mosquitoes and other biting insects spread equine infectious anemia (EIA), a disease of the nervous system.

Luckily, there are vaccinations for West Nile and sleeping sickness; there are no vaccinations for EIA, however, so sanitation is still key. Try to keep your horses off pasture at dawn and dusk during prime feeding times for mosquitoes and other bloodsucking insects.

Many of the good sanitation practices that reduce fly populations will also discourage mosquito breeding. But other specific practices can significantly decrease mosquito breeding grounds. Mosquito larvae can hatch in 4 days in standing water, so be vigilant about eliminating puddles and rain collectors, and frequently change the water in troughs.

- Throw out old tires that tend to collect rainwater and make ideal mosquito breeding conditions.
- Discard cans, buckets, drums, bottles, and any other vessel that can hold water.
- Fill in or drain any low spots that turn into puddles.
- Regularly check drains, ditches, and culverts to ensure they are clean of weeds and trash so water will drain properly.
- Monitor your irrigation equipment and ditches regularly.
- Repair leaky pipes and outdoor faucets.
- If you have a swimming pool or wading pool, manage it responsibly.
- Change water in birdbaths and empty plant pot drip trays at least once a week.
- Keep grass mowed and shrubbery trimmed around barn and house so adult mosquitoes will not find refuge there.
- Keep lights off as much as possible.
- Keep gutters clean and free of debris and leaves so they drain completely.
- Flush fresh water through any barn drains at least weekly.
- Clean out and refill watering troughs at least once a week.
- Contact your local Extension agent for other suggestions suitable to your particular location, such as biological controls. For example, nontoxic bacterial larvicide granules and donuts can be added to ponds or wetlands.

TICKS

Rodents are the reservoir of Lyme disease, and black-legged ticks are the vehicles that transmit the disease to horses and humans. Currently, there is no approved vaccine against equine Lyme disease. The chances of you or your horse becoming infected with Lyme disease in the Northeast and certain parts of the Midwest can be twenty-fives times that in other parts of the United States. Keep horses clear of tick-infested areas as much as possible and carefully examine each horse at least once a day for presence of ticks, especially during tick season in your area.

Remove nymphs (immature ticks, usually seen in spring) and adult ticks (usually seen during summer and fall) immediately. Ticks can attach anywhere on a horse, but common areas are on the chest, near the rectum, and along the base of the mane. Ticks should be removed by grasping the tick as close to the horse's skin as possible using a tick remover or fine tweezers and then pulling straight up with a slow, steady force. Do not pinch the tick with a thumb and forefinger and squeeze it as you pull it out, as this could squirt the diseased material from the tick into the horse. Once you have removed the tick, apply an antiseptic (alcohol or antibiotic ointment) to the bite site.

RODENTS AND OTHER WILD ANIMALS

Rodents that can create problems in and around a horse barn or pasture include mice, rats, gophers, prairie dogs, marmots, pack rats, moles, and shrews. Not only are they hosts for disease-causing parasites, but they can also cause damage and health problems themselves. Rodents need to gnaw to keep their incisors worn down, so they chew nonedible things such as electrical wires and tack. They can carry bubonic plague, typhus, and rabies and are reservoirs for

Exponential increase

Imagine this: 1 male + 1 female + 1 year = 1000 mice.

salmonella, Lyme disease, and sleeping sickness. They can damage tack and make a mess in a feed room. Ideally, your feed room should be rodent-proof and all feed stored in rodent-proof containers.

Although poisons and baits can be used to control rodents, good sanitation and cleanliness, plus a few cats, work best! Be sure to vaccinate your barn cats regularly, especially against rabies. Terriers are good rodent deterrents but are more prone to digging than cats. If a mouse or two sneaks into a tightly enclosed area (such as a tack room) where cats are not allowed free access, simple traps baited with a dab of peanut butter will eliminate the population before breeding begins.

Gophers, prairie dogs, and marmots all create burrows in pastures and subsist on the pasture vegetation; this destroys plants and results in holes dangerous for horses and people. A few of these rodents are a natural part of the ecosystem, but when a population grows so large that it renders a pasture unusable, then it is past time to do something about it. Natural predators of rodents include hawks, owls, and snakes. Providing desirable habitats on your property for predators may encourage them to take up residence.

Rodent-control measures for pastures range from trapping live for adoption, to sonic deterrents, to poison and kill traps. Choose the method that you are most comfortable with.

Other wild animals can also be a nuisance or a hazard to your horse's health. Skunks, which are part of the weasel family, can carry rabies, yet are an important part of rodent control since mice, moles, and shrews are part of their diet. Opossums carry and shed the protozoan that causes EPM, so contact between horses and opossums should be eliminated. Raccoons are nocturnal rummagers that can make a mess in your barn and can carry rabies.

The best way to keep rodents, skunks, opossums, and raccoons from being attracted to your horse farm is to keep things tidy: garbage containers should be secured, pet food and water bowls should be put away, and barn doors should be closed during cold weather and at night.

BIRDS

If you leave you barn door open, it will be an invitation for birds to nest. A few nesting swallows helping with fly control is a nice thing, but when your barn is invaded by sparrows or pigeons, the health hazard and noise will be extreme. Bird droppings carry salmonella, and birds often carry mites, fleas, and lice. Screen openings and keep doors and windows closed during nesting season; your barn cat can also help control birds.

Mud Management

Mud is home to bacteria, fungi, and other organisms that cause diseases such as thrush, rain rot, scratches, and abscesses, and flies and mosquitoes breed in mud and puddles. Mud is damaging to the environment, so avoid overgrazing, use gravel in high traffic areas and pens, divert rainwater from traffic areas, and pick up manure daily.

An old wives' tale suggests that if you have a horse with poor-quality or dry hooves, you should let the water trough run over to force the horse to stand in the mud. While the basic notion might sound logical, in fact mud can be harmful to hooves. Excess water absorbed by the hoof weakens the layers of hoof horn and results in soft, punky hoof walls that peel and separate. A weak hoof can spread out like a pancake. Too much moisture also makes a horse's soles soft and susceptible to sole bruises and abscesses.

The effects of repeated wet/dry conditions are even more damaging. Research has found that the condition of hooves worsens during hot, humid weather, especially where horses are turned out at night. Typically, horses walk around in dew-laden pastures all night and then either are left out where the sun will dry the hooves or are put in a stall where the bedding dries the hooves. Horses that receive daily baths or rinses or those that repeatedly walk through mud and then stand in the sun experience a similar decline in hoof quality. In both cases the hoof is undergoing a stressful moisture expansion/contraction that results in cracks, splits, and peeling.

If you want a firsthand example of how drying such situations can be, stick your fingers in some fresh mud past your fingernails and let them dry. Your "hooves" (fingernails) and "coronary bands" (cuticles) will probably show signs of drying and cracking after just one episode! Mud has the effect of drawing out moisture and oils and tightening pores — as in a poultice or a mud facial.

Another common problem with horses that stand in dirty, wet footing is the condition called thrush. Thrush is a decomposition of hoof tissues that usually starts in the clefts of the frog. Anaerobic bacteria, those that survive and flourish without air and are present almost everywhere on the farm, are sealed in the clefts of the frog by mud and manure and are kept moist by urine and muddy conditions. The bacteria destroy the hoof tissue and produce a foul-smelling black residue that you will never forget once you have smelled it. Erosion in the clefts of the frog can become so extreme as to reach sensitive tissue and cause lameness.

One of the best ways to preserve your horse's hooves is to keep them clean and dry.

When buying new land, inspect it carefully for old dumps that might be hidden by vegetation. This is one of many trailer loads of junk we hauled off our new pastureland.

Hazardous Substances

Guarding your horse against chemical poisoning is essential, because horses like to investigate unknown things by nibbling and tasting.

All toxic substances must be stored in tight, well-labeled containers. This includes rodent poison, insecticides, herbicides, and antifreeze. Although lead-based paints are no longer sold, you might find them on older horse buildings or fences, and horses should be prevented from contact with them. Be sure to read all product labels thoroughly and follow directions carefully. Any unlabeled substance should be discarded in a safe manner.

Chemical poisoning can often occur unknowingly. Don't feed treated grain or seeds that were meant for planting, or you may be giving your horse a dose of mercury! Although treated grains often have a pink or reddish hue, sometimes they look just like feed grain.

Don't give a horse feed that was meant for cattle, sheep, or goats. Often these ruminant feeds contain urea, a source of nonprotein nitrogen designed for ruminants, which should not be fed to horses. And some cattle feeds may contain growth stimulants that can be permanently damaging to the nervous system of the horse.

Don't let horses near junk or vehicles. Using lips and teeth to inspect things, horses may ingest toxic paints, antifreeze, or battery fluids. Ethylene glycol, the toxic substance in antifreeze, can cause kidney failure and death. Protect horses from all fumes: vehicles, paints, and solvents. Don't apply insecticides or herbicides near feed or water areas, and be aware of wind drift when you are spraying.

Septic System

Protect the septic drain fields and leach fields for your house and barn by eliminating vehicle, tractor, and horse traffic from the area. The weight of a tractor, truck, or trailer would compact the soil, making the field less effective. In addition, the force delivered by galloping hooves could dig up and damage drain tiles.

Disease Prevention

Disease and infection are usually spread either by physical contact, contaminated feed or water, or airborne antigens. If you have disease or infection on your acreage, you must work closely with a veterinarian to bring it under control. A combination of treatment, disinfection, and quarantine can eliminate the spread of disease or infection and eradicate the disease-causing organisms.

Sunlight, especially with dry, hot air, is a powerful disinfectant. Specific chemicals are effective against certain organisms. Your veterinarian will advise you what to use.

Besides using healthy management practices every day, you can minimize problems by routine immunization and quarantine. All horses should receive yearly immunizations according to the recommendations of your veterinarian. All new animals should be quarantined on arrival and observed for at least a week before they are mixed in with resident horses. Any horses that leave the farm temporarily, and especially if they have been exposed to a large number of other horses, should be quarantined on return.

Carcass Disposal

It is a fact of life that someday your horse will die. If your horse dies of a disease, your veterinarian can tell you what the procedures are for disposal of the diseased carcass. The options for disposing of a 1200-pound carcass are limited. Cremation (about $1000) is expensive and not widely available for horses because of their size. Rendering plants that handle dead horses are few and far between, and it might be difficult to get timely pickup service. If you are lucky enough to have a large-animal disposal service in your area, get a phone number and keep it current and available. Your veterinarian, farrier, or Extension agent should be able to provide you with contact information. If you live in a suburban area, your best option might be to have your veterinarian arrange for euthanasia and disposal of your horse's carcass. Some landfills have a special section for animal carcasses (approximately $150 for a horse).

If you have sufficient room, local ordinances allow it, and you have access to a backhoe, you might consider burying your horses on your property.

Environmental responsibility reminders

- ☐ Divert roof runoff away from horse living areas to minimize waste materials washing into ditches, streams, and ponds.
- ☐ Minimize overgrazing and its effects (erosion and weed proliferation) by creating sacrifice pens that you can use for turnout when pastures need rest.
- ☐ Cross fence pastures and rotate.
- ☐ Limit access to creeks and streams to preserve riparian areas and to decrease amount of waste deposited in or near the water.
- ☐ Create a windbreak wildlife habitat from downed trees and branches.
- ☐ Compost manure.
- ☐ Control weeds.
- ☐ Fertilize at the optimum rate; don't overfertilize.
- ☐ Reseed stressed areas and bare spots.
- ☐ Protect trees.
- ☐ Minimize mud.

Burial at an animal cemetery is also an option but an expensive one (about $1000) that is not available in all areas. If you have a large enough tract of land, the best solution might be to prepare a final resting place for your horse on your own property. Local zoning, health, and environmental protection regulations vary widely, so check before you dig.

Burial of dead animals should not result in contamination of groundwater, and the grave needs to be deep enough so as not to encourage or permit access by vermin, scavengers, and other potential vectors of disease.

If it is legal to bury a horse on your property, you will likely need to hire a backhoe operator to dig an 8-foot-deep hole that is approximately 6 by 10 feet. This will ensure that there will be 3 to 4 feet of earth on top of the carcass. Locate the hole at least 100 feet from any water source and on ground high enough so that the bottom of the hole does not contact groundwater.

Disasters

Depending on where you live, certain natural and man-made disasters may impact your horsekeeping. Whether wildfire or flood, drought or blizzard, earthquake or hurricane, your farm could lose power or accessibility or you may need to evacuate. Mental and physical preparedness is critical. Prepare your facilities, know what to do, and remain calm.

Fire

A fire can be one of the most devastating of life's experiences — physically, financially, and emotionally. Knowledge, preparation, and forethought, however, can prevent most barn fires. Fires need a source of heat, oxygen, and fuel. While the heat that starts a barn fire can be simply sunlight or friction, more commonly it is stacking hay too close to lights that are left on or electrical failure from rodent-damaged wires or from improperly managed lights or portable appliances. Fires can also be started by lightning, the open flame of a match or cigarette, the spontaneous combustion of hay or bedding, and not properly attending the burning of ditches or garbage.

Most horse barns are well aerated, providing a good source of oxygen to spur on a fire. Fuels for a fire can be gases, liquids, or solids. Propane, gasoline, alcohol, liniments, paints, hoof dressings, hay, bedding, grain, tack, and combustible building components can all add fuel to a fire.

Hay makes ideal fuel for a fire, so it's best to store only a few days' supply in the horse barn.

FIRE PREVENTION

Fire prevention begins with sound building design and proper maintenance. Strict adherence to safety practices is essential.

- No smoking in any buildings. Put up signs and enforce the rule with no exceptions. Provide sand-filled containers well away from the barn for guests to dispose of their smoking materials.
- Keep hay and bedding separate from the stable and tack room, with a 100-foot buffer zone. Be sure all hay is well cured.
- All appliances should be disconnected when not in use and be routinely inspected. This includes radios, clippers, water heaters, pipe-heating tape, treadmills, and bug zappers. Stall lights should be in cages or heavy glass covers. The improper use of heating units — electric, kerosene, or propane — and infrared lights is among the most common causes of barn fires today.
- Be sure wiring complies with the National Electrical Code; run wire in conduits so rodents (or horses) cannot chew it. Protect outlets and switches from dust and moisture with spring-loaded weatherproof covers; replace broken faceplates; keep panel boxes covered, dry, and dust-free; don't overload plugs or circuits.

Spontaneous combustion

For the first 2 or 3 weeks after hay is cut, bacterial action creates heat, especially in alfalfa and clover. If the temperature of a haystack is more than 150°F, it should be checked frequently for an increase. If it reaches 175°F, it will likely begin charring. At 185°F, it should be moved out of the barn, with firefighters present. Often, a stack will smolder until it reaches the oxygen at the outer edges of the stack, and then it will explode or burst into flames. Damp bedding or grain can similarly combust spontaneously.

Create a defensible space

A defensible space is one in which vegetation has been managed to reduce fire threat to homes and buildings and to provide access for firefighters to defend the buildings. Defensible space is especially important in areas where wildfires are common, but is helpful on any farm during time of fire.

To calculate defensible space using the chart below, first determine the slope of your land and the type of vegetation you have. A *transit* (telescoping sighting instrument on a tripod) is the most accurate instrument for measuring slope.

$$\text{rise} \div \text{run} = \text{slope } \%$$

or

$$\text{change in elevation} \div \text{measured distance} = \text{slope } \%$$

To estimate slope, lay a 50-inch stick (which is the *run,* or measured distance) down the slope you want to measure. Put a carpenter's level on the stick and raise the end of the stick until it's level. Measure the distance from the end of the raised stick to the ground (this is the *rise,* or change in elevation). For example, if the rise is 4 inches and the run is 50 inches, the slope is $4 \div 50$, or 8 percent.

A pitch of 0 to 20 percent (0–9 degrees) is considered flat to gently sloping; a pitch of 21 to 40 percent (8–18 degrees) is moderately steep; a pitch over 41 percent (over 18 degrees) is very steep.

Recommended defensible space around buildings

VEGETATION	SLOPE		
	flat	moderate	very steep
Grass	30'	100'	100'
Shrubs	100'	200'	200'
Trees	30'	100'	200'

So, for example, if your acreage is moderately sloping grass, you'll want to keep a 100-foot border mowed around all buildings, and dense shrubs at least 200 feet away from buildings. Also ensure that there is no continuous stand of trees within 100 feet of buildings.

Remove all dead plant material from the defensible space. Also, consider removing some trees and shrubs to decrease the amount of flammable vegetation on your property. For more information on fire protection and defensible space, visit our resource guide at www.horsekeeping.com.

Maintain a fire strip around all buildings.

- Maintain a conscientious rodent-control program. Rodents need to continuously wear down their incisors, so they chew constantly. Electrical wires are a favorite target and become a fire hazard if frayed or disturbed.
- Use fireproof materials wherever possible. Although they may be more costly, they can often result in lower insurance rates. Install roofs with Class C fire resistance or better. Firewalls or sliding fireproof doors between sections of the barn will minimize the spread of fire. They can be located between storage and stable areas and can be used to divide the barn itself into smaller sections. Consider using concrete or block walls or fire-retardant wood, paints, sealers, and coatings.
- Use steel and masonry as much as possible in the construction of the barn. If using wood, choose a fire-retardant type that is pressure-impregnated with mineral salts to reduce the flammable elements of wood. The roof should be of unburnable metal or a treated wood. Fiberglass panels, in contrast to glass windows, have a higher resistance to heat and will not explode like glass.
- Locate fire extinguishers at each door and so that you won't have to travel more than 50 feet from anywhere in the barn to reach one. Mount the fire extinguishers in plain sight and do not use them as a coat or blanket rack.
- Check extinguishers regularly to be sure they are fully charged.
- Keep a garden hose by each hydrant to use on Class A fires. Note that if you try to use water to extinguish a Class B fire, you could cause the fire to spread, and with a Class C fire, the water stream may conduct electricity, and you may receive a shock.
- Buckets of sand are useful to smother all types of fires.

Choose an appropriate extinguisher

The National Fire Protection Association classifies fires three ways. Class A fires involve ordinary combustible materials such as wood, paper, cloth, rubber, and many plastics. Class B fires are fires of flammable liquids, oils, greases, tars, oil-based paints, lacquers, and flammable gases. Class C fires involve energized electrical equipment where the electrical nonconductivity of the extinguishing media is of importance. All three types of fires can occur in barns. Choose a fire extinguisher that is rated ABC, such as a 3A:40B:C, which weighs from 8 to 10 pounds and is light enough for most people to handle easily. A unit of that size will project a stream of chemical about 15 feet for 8 to 15 seconds. You should have a fire extinguisher of this size or larger for every 3000 square feet of barn space.

Practice!

Before recharging a fire extinguisher, practice with it so you'll be comfortable using it. You can practice aiming at a cardboard box, or if your local codes allow, set a small fire in a safe area such as in the middle of a gravel driveway. Pull the safety pin, stand 7 to 10 feet from the fire, aim at the base of the fire, squeeze the trigger, and spray with a back and forth motion across the base of the fire until the fire is extinguished. When you are done practicing, have the fire extinguisher recharged so it is ready to use.

Left: The area around this light switch is covered in cobwebs and dirt, which could prompt a spark. *Right:* An overhead sprinkler system is a valuable fire-fighting system.

- Keep your barn clear of dust, cobwebs, and chaff, especially on light fixtures. Keep debris from accumulating in and around the barn. This includes loose hay, manure, lumber, oily rags, and twine. Manure should be removed daily to a site away from the barn.
- Grass around the barn should be mowed regularly so there is at least a 40-foot fire strip along each exterior wall.
- Locate any gas pumps or storage tanks at least 50 feet from the barn.
- Park machinery and vehicles at least 12 feet from the hay and stable.
- Store combustible fluids, such as insecticides, clipper wash, pesticides, and veterinary supplies, in tight containers and in small quantities. Store large quantities away from the barn.
- Lightning rods, antennas, and wire fences that are attached to the barn must be grounded. A lightning rod should have an Underwriters Laboratory (UL) label ensuring that the rods and the grounding element have been made to safe specifications.
- Consider installing a fire detection system. Household smoke detectors don't work well in barns, as dust can make them inoperable. Choose one that is specially designed for barns and is easy to recognize and hear. You can have it sound in the barn, the yard, the house, and/or an alarm company.
- Consider installing a fire-fighting system such as sprinklers if your water supply system can support one. If you are on city water, a sprinkler system would probably work, but a domestic well would likely not provide the necessary volume and pressure. Be sure the sprinkler system is freeze-protected, because if it is not, once it freezes, breaks, and thaws, your barn will be flooded. Automatic chemical-spraying systems are self-contained and work much as the dry-chemical handheld extinguishers do.

EMERGENCY PREPARATIONS

Prepare for the worst, and practice. It may take several hours to run through your plan the first time, because you will undoubtedly come up with more questions that need answers. And certain parts of the plan may need changes. After a few run-throughs, things will go smoothly and you will gain an added measure of confidence that will help you in the event of a real emergency.

- Be sure your address is clearly visible from the road where your driveway joins the street at the entrance to your property.
- Design an emergency fire plan. Have a fire drill regularly so that all family members understand the priorities and dangers. In general, the goals are to protect humans, horses, and equipment and buildings, in that order.

In case of fire

Immediately notify your local fire department at: ____________________

Say: "I have a stable fire at" ____________________

If necessary, give brief directions to your location: ____________________

Critical Steps

- Evacuate humans and horses from the area as quickly as possible and be sure children are kept away from the fire. If manpower is available, fight the fire.
- If possible, promptly put out small fires with extinguishers or by smothering them with blankets or sand.
- Be sure the front gate is unlocked so fire equipment can enter your property.

Then your actions will depend on the number of people you have available.

- Have halters and lead ropes by each stall and have extra halters in your house and horse trailer. Evacuate horses in a planned manner, never risking a human life for a horse. Lead horses out. Don't spend a lot of time trying to blindfold a horse, as it often does not make a difference. Never try to chase a horse out of a barn, as he will likely run back in. Be calm and your horse will pick up confidence from you. But if a horse refuses to leave, walk away from the horse — do not jeopardize your life.
- Put horses in an enclosed area or tie them to a safe, strong post so they cannot return to the barn or race around, endangering firefighters.
- Keep all entrances clear of vehicles and equipment, as well as the areas surrounding the fire.
- Once all horses and humans have been removed, reduce the oxygen to the fire by shutting all the windows and doors. Move all equipment and machinery. Protect nearby structures by hosing them down and soaking the base area of the buildings as well.
- During all of these procedures, be sure all children are kept away. Without fully understanding the dangers of a fire, a child may reenter a barn to save a pony or horse (which may have already been removed from the barn) and be overcome by fumes or smoke. Fumes and smoke, not fire, cause most fire fatalities. Resist the temptation of going back into any part of the barn to get your favorite trophy or saddle.

Fire Hazards

- Electrical appliances, including heat lamps, space heaters, and fans
- Flammable liquids such as kerosene, gasoline, paint, pesticides, cleaning agents, and fertilizers
- Excessively long grass surrounding the barn
- Trash and manure stacked against the barn
- Dirty, dusty light fixtures, fuse boxes, and switches
- Hay stored in the barn, especially too close to light fixtures
- Poor ventilation in the hayloft
- Tractors and other machinery stored in the barn
- Any clutter that would make your property inaccessible to fire trucks

Safeguards

- ☐ Test fire detectors monthly.
- ☐ Check fire extinguishers every 6 months.
- ☐ Recharge fire extinguishers as needed or, at a minimum, every 6 years.
- ☐ Stage regular fire drills.
- ☐ Diagram your barn so that in case of an emergency, you or any other person in the vicinity will be able to act as quickly and efficiently as possible. On the diagram, plot your barn's floor plan. Include the location of exits, the fire alarm, fire extinguisher(s), utility shut-offs, first-aid equipment, and equine inhabitants. Also keep an up-to-date count of the number of horses in your care along with a list of their owners' names, addresses, and telephone numbers.

- Know where and how to shut off electricity, gas, and water and do so if necessary.
- Post phone numbers for the fire department and police next to the phone. It also helps to have a prepared statement nearby that includes your address and directions to your property and barn that you can read if your thoughts are scattered.

Natural Disasters

Other emergencies such as floods, earthquakes, hurricanes, winter storms, wildfires, and winds can affect your horsekeeping by cutting you off from electricity, water, or roads. When disaster strikes, you need to make an assessment and a plan that will ensure the safety and well-being of people and animals, secure your facilities, and reestablish the functioning of normal routines as soon as possible.

Design a specific plan for the type of emergency that would most likely occur in your area and then rehearse it with everyone who lives on your property or is involved with your barn. If you live in a forested area and wildfire is your biggest threat, your disaster plan may involve vacating to less wooded area on your land or evacuating altogether. Here is where trailering practice will prove its worth. (For more information on trailering, see Hill, *Trailering Your Horse* [Storey, 2000]).

Firefighters encourage horse owners to evacuate their animals early because the number of horse trailers on the road in a wildfire evacuation zone often outnumbers fire vehicles, clogging highways and hampering efforts. If immediate evacuation is necessary, it is often best to use a grease marker or spray paint to write your name and phone number or address on the side of your horse and turn him loose. Horses are often better able to fend for themselves loose than when left in an enclosure.

If winter storms are common, your plan should include alternate sources of power and water. If flooding is a threat, know where the nearest high ground is on which you can relocate your horses.

Stay calm and focused. Other people and your horses will gain confidence from you.

Know first aid for people and horses. You may be required to take care of medical emergencies.

Know how to turn off electricity, water, and gas.

Obtain emergency procedure information from your local authorities that will inform you what supplies and equipment you should have on hand, how to report downed or broken utilities, what to do in the meantime to avoid further catastrophe, how to treat contaminated water, and how to safely stay warm. (For detailed information specific to horse facilities, see Hill, *Stablekeeping* [Storey, 2000].)

Alternate power source

A gasoline-powered generator can supply the power necessary to run a water pump and lights. One gallon of fuel will run a generator for 2 to 8 hours. Because of exhaust fumes, a generator should be operated outside the barn, so you will need 12-gauge or heavier extension cords with the correct plug configurations for connecting the generator to the water pump, heaters, and lights. Also, your pump and lights will have to be configured by an electrician to accept auxiliary power.

17 Security

Safe, secure facilities minimize risk and liability for both you and your horses. An estimated forty thousand horses are stolen each year. It is essential that you be able to prove identity and ownership of each of your horses in the event of theft or escape. You can further protect yourself from loss by having insurance appropriate for your situation.

Follow safe practices, maintain safe fences, and keep necessary paperwork in order.

• Loose horses that leave your property can cause damage and traffic accidents that could prove to be your liability. Install a secure perimeter fence around the portion of the property where horses are contained. That way, if a horse gets out of a pen or the barn, there will be a fence to prevent him from leaving the property. Make sure all gates in your perimeter fence are closed at all times to prevent horses from leaving your property. Consider using locks on all perimeter gates.

• Put tools away; keep barn aisles and passageways clear.

• Constantly keep an eye out for broken boards, protruding nails, broken latches, and discarded twine on the ground.

• Perform daily fence checks on electric fence components and make a thorough weekly inspection of all fences.

• Keep your dogs under control and train them to respect horses and to stay away from the barn and pens. Do not allow loose dogs to roam your property — it is just too risky.

• Post NO TRESPASSING signs on all sides of your property.

• Train your dog to alert you to the presence of intruders. Yard lights that come on at dusk or are triggered by motion sensors can be a great deterrent. Consider installing a professional security system or placing surveillance cameras in any areas that you feel might be more apt to be burgled.

• For their safety, horses should never be turned out wearing halters; in addition, a haltered horse is usually easier to steal.

• Protect your horse farm from theft by installing an entrance gate that can be securely locked and cannot be lifted off its hinges.

• Install a lighting system around the barn and paddocks.

• Keep the tack room secure. A tack room with no windows and secure, lockable doors helps to minimize theft. If your tack room has windows, cover them with sturdy steel grills. Mark all tack with a brand or identification number, keep it out of sight when you're not using it, and lock it up when you're gone. Check with your insurance company to see if certain types of locks are required to validate theft coverage. If you keep your barn clean and organized, you'll be more likely to notice when things are missing or out of place.

• Record the identification of horses, tack, equipment, and other valuable items by photo, video, and serial and registration numbers.

• Secure your horse trailer with a hitch lock, wheel clamps, or electronic tracking device.

Top: **This two-way slam latch can be padlocked shut.** ***Bottom:*** **If your trailer is left unattended, a hitch lock may help safeguard it from theft.**

- Keep several copies of current photos and written descriptions of all horses for immediate distribution to police, local veterinarians, farriers, and others.
- Form a neighborhood association so you can share information and help watch for suspicious activities.

Horse Identification

Horse ID has two purposes: one is to prove that a particular horse belongs to a particular person and the other is to prove that a particular horse is a particular horse.

Registration papers for a purebred horse provide the physical description of the horse, including color, markings, and photo or drawing identification as well as the breeder or owner's name and address.

An ID certificate or brand card might be available or required in your state as a permanent record of your horse's identifying characteristics such as age, breed, sex, color, markings, scars, brands, and possibly blemishes or hair whorls too. The horse is shown either by photo or drawing from each side and from the front and back. The markings are not only indicated visually but also described in writing. There may be a U.S. Animal Identification Plan requirement for horses in the future.

Branding identifies a horse as the property of a person or a farm/ranch.

Your personal brand must be registered; first check with the state brand board as to the availability of a brand, rules of registration, and placement of a brand. There is hot branding and freeze branding.

Hot-iron branding is used on western ranch horses as well as by some breed registries. The brand is registered with a brand board in the horse owner's name, so it serves as proof of ownership. An iron is heated by fire, gas, or electricity and held against the hide (hip, shoulder, neck) of the animal to burn a permanent scar into the skin.

Horse identification can prove a certain horse belongs to a certain person. On U.S. ranches, hot branding has been used for hundreds of years.

Freeze branding uses cold (from liquid nitrogen) to make an indelible mark or number on a horse's hide; the hair under the brand either fails to grow or grows in white. The brand is registered in a database, which would prove that the horse belongs to the owner of the brand. Depending on the database and how it must be accessed, finding freeze-brand records can be less handy than finding hot-brand records.

Lip tattoos are another option. The Jockey Club Thoroughbred Breed Registry requires all Thoroughbred racehorses to have a tattoo applied to the inside of the upper lip. The tattoos last for 4 to 5 years, fading over time.

An electronic microchip can be implanted in the nuchal ligament of a horse's neck to be read with a radio frequency scanner. Chips are not noticeable, which is good for the sake of appearance, but because a thief cannot see them, they don't act as a deterrent like a visible brand does. Microchips indicate ownership but can identify a particular animal only if there is an appropriate scanner available.

To prove definitively that a particular horse is a particular horse, you can have blood typing or DNA typing done on each of your horses. Although this is not a widespread option right now, some breed registries have blood-typing records and offer DNA testing.

The best course of action is to keep a folder on each of your horses and regularly update it with current photos from both sides, the front, and the rear and close-ups of any unusual scars, markings, or brands. A horse's hair coat changes between seasons and as he ages. Include a detailed written description of the horse along with a copy of his registration papers. Ideally, have all of this information on one sheet of paper and have several copies so that you can distribute them to authorities if necessary.

Horse Insurance

The purpose of insurance is to protect you against financial loss, not emotional loss. When it comes to insurance for your horse operation, there are basically four types to consider: horse mortality, horse medical, liability, and property. To determine what type of insurance and coverage is appropriate for you, work with an agent who has a thorough background in equine insurance and writes with companies that have an A or A+ rating. Keep in touch with your agent and notify him or her immediately when you have a claim.

HORSE MORTALITY INSURANCE

Horse mortality insurance is a form of term life insurance for your horse. It pays the value of the policy if the horse dies of natural or accidental causes (with certain exclusions). It might also pay if the insurance company agrees that a severely sick or injured horse should be euthanized.

Full mortality insurance covers all causes of death, including illness and injury, and may include proven theft. A thorough physical examination is usually required before a full mortality policy is issued, and even if the horse passes the examination, the insurance company may have standard exclusions for certain causes of death.

The policy amount is the value of the horse. Mortality insurance rates are based on the type of policy and the value of the horse. Rates will vary according to the breed, age, and use of the animal.

The value is usually the purchase price of the horse or, if homebred, twice the stud fee. Sentiment and replacement costs are not part of the value. A horse's policy value can be increased by factors such as winnings and sale of offspring, and these things will have to be proved and submitted to the insurance company. Generally, insurance underwriters take care in establishing the true value of the horse at the time the policy is written so that there aren't problems at the time of loss. If an insured horse dies or is seriously ill or injured, notify the insurance company immediately and according to the specific instructions outlined in the policy in order to avoid difficulties with your claim.

A mortality insurance policy might be a good idea if you can't tolerate the financial risk of losing your investment in the event of a horse's death, serious injury, or theft.

Equine insurance general information

TYPE	COVERAGE	REQUIREMENTS
Full mortality	Death by sickness or accident; may cover proven theft (90–100 percent); humane destruction clause allows vet to destroy horse without approval by insurance company; postmortem may be required; death by poisoning may or may not be covered.	Veterinary certificate less than 30 days old. Owner application. May be available for all ages or specified range, such as 2–14.
Surgical only	Pays veterinarian's surgery bill for nonelective and noncosmetic surgery; also some (e.g., 25–30 percent) associated costs such as anesthesia, medication, bandages, and hospitalization.	Must carry full mortality.
Major medical	Pays surgical and nonsurgical costs such as lab and diagnostic tests, medical treatment, surgery, and postsurgical treatment.	Must carry full mortality.
Personal liability	Covers if horse gets loose and is hit by car, damage to car or injury to passengers; covers if horse damages vet's equipment, etc.	
Commercial liability	Lessons, boarding, etc.	
Care, custody, and control	Injury or death of animal in your care.	Must own farm where horses not owned by you are boarded; must operate with approved standard of care.
Professional errors and omissions	Trainer, instructor, farrier, judge, etc., covered if commit error in providing professional services.	

Note: This information is based on research with many equine insurance companies.

MEDICAL INSURANCE

Medical insurance usually covers the cost of surgery and medical treatment due to accident, illness, or injury. Medical insurance is generally available only for horses already covered by a full mortality policy. Although these policies have a large deductible and high premiums, most major medical policies pay the entire cost of the treatment (over the deductible) up to the policy amount. Medical insurance requires an application and a completed and signed veterinary exam form.

Surgical-only insurance is less expensive than major medical because customarily it pays only for the veterinarian's surgery bill in nonelective surgery, such as colic. In addition, some associated costs such as portions of X-rays and drugs might be paid. But anesthesia and board probably won't be covered, resulting in a net coverage of the total expenses of only about 50 percent. Often, elective surgeries such as castration are not covered. Policy costs are a flat fee per horse per year. There is usually a small deductible and a maximum claim amount.

LIABILITY INSURANCE

Your homeowner's policy will usually cover public liability and property damage if the horse use is "recreational." But be sure you understand what your particular policy means by the difference between "recreational" horse use and a horse business. A business is indicated when you file a Schedule

COMMENTS	PREMIUMS (ESTIMATE)
Minimum policy usually $100–$150; binding when check and satisfactory application received; might have a guaranteed renewal clause for 1 year or longer.	Based on age, breed, and use. Usually 3–7% of the value of the horse. Be sure to read exclusions to coverage and requirements to keep policy in effect.
Maximum claim might be $8000; may have guaranteed renewal; often does not cover castration, caslick repairs, or postmortem.	About half the premium of major medical.
Usual benefits $3000–$5000 with some policies having guaranteed renewal; no preexisting conditions.	$150–$250 per horse per year with a deductible of $250–$500 with or without copay.
Most homeowner's policies cover; if not a homeowner, can get horse owner policy, or if professional, can get commercial policy.	Homeowner's policy variable according to coverage desired; horse owner policy for renters (e.g., $350/year for three horses).
May be cost-prohibitive for small operations.	Variable; $800/year and up.
Limits up to $200,000 per animal and $500,000 loss per policy per year.	Depends on size of operation.
Overall coverage up to $1,000,000 worldwide; up to $100,000 per horse.	$600 and up per year according to the number of students and/or horses.

F (Farm) or a Schedule C (Business) with the IRS. Often, however, even if you consider yourself a hobby farmer, the insurance company may treat you as if you were running a horse business. If one of your horses is used for showing, parades, or rental, or if your facilities are used for boarding, training, instructing, or breeding, you may have to obtain additional liability coverage with a supplemental business policy.

It is important that all horsekeepers be covered by a personal or professional liability policy as appropriate for the situation. Homeowner's policies in some states might cover the owner if someone was injured on the property or if a horse got out on the road and damaged a vehicle or injured a person. If you run a small lesson or training business at your acreage, you should be sure you have adequate liability coverage. To decide whether additional liability insurance is appropriate for your horse farm or ranch, contact various insurance companies and review their policies. Read the terms carefully and ask the agents how various scenarios would play out. Use examples of situations you would most likely encounter at your farm and ask whether you would be covered. Before you invest any money, be sure your insurance agent is experienced, reliable, and qualified.

The cost of liability insurance will depend on the type of policy, the amount of coverage, and the terms and exclusions. Various terms are used in naming

Insurance application

The owner application asks questions such as these:

What type of fencing do you have?

How often do you deworm and what type of dewormer do you use?

Has there been infection or contagious disease on the premises in the last 12 months?

How many of your horses have died or been destroyed in the last 3 years?

Are there any encumbrances on this horse?

Is there any indebtedness due because of the change of ownership on this horse?

Have you ever been canceled or refused insurance?

Is this horse insured now? Has he ever been insured? With whom?

Who will be providing care for this horse?

Veterinary examination

The veterinary exam for insurance will vary depending on whether the policy will contain riders for surgical or major medical. Some of the things the veterinarian will check include the following:

- Temperature, pulse, and respiration
- Condition of eyes
- History of colic
- Evidence of nerving, blistering, firing
- Evidence of laminitis
- Past lameness
- Past surgeries
- Fecal exam results
- Scars
- Male: castrated or both testicles evident
- Female: Mare in foal; past breeding or foaling problems; reproductive exam
- Any vices or bad habits
- Deworming program, parasite problems, colic problems
- Any congenital abnormality or deformity
- Results of Coggins test
- Current medication
- Any other medical facts that should be brought to the attention of the company

and describing types of coverage. Some of them overlap or are very similar, and others use the same words and mean different things. The following is a general guideline to types of policies.

Personal liability, public liability, general liability, premises liability. These terms describe liability coverage in homeowner's policies. Such policies may have limitations, because they are not written expressly for the needs of a horse owner. They are designed to cover human bodily injury and property damage claims on premises, yours or someone else's. Such policies should cover a visitor to your home who gets kicked or the passengers and vehicle in an accident caused by your loose horse. They should cover the cost of the claim as well as your legal expenses. The choice of coverage might be something like $300,000, $500,000, and $1,000,000. Most general liability policies do not cover things that happen to the horse. They usually specifically exclude coverage for damage to any animal (injury or death) that you have in your care, custody, or control. If you rent or lease the farm you live on, you can purchase horse-owner policies in lieu of homeowner's policies.

Commercial liability. Professionals, those who receive money for an equine activity, are not covered by a personal liability policy. Some states have adopted laws that release professionals from liability for equine-related accidents. Before deciding whether you need additional liability insurance, be sure to thoroughly understand the verbiage of your state's laws and find out if the law related to equine professionals has been tested in court. Whether you give lessons at your home or run a full-service training facility, in order to be fully covered you'll probably need a commercial liability policy. Ask your agent whether the name of the insured on your policy should be yours or that of your business.

Care, custody, and control. This type of insurance covers injury or death of an animal that is in your care when the mishap takes place. For example, it should cover a claim that your negligent feeding practices were responsible for a horse that died of colic. The limits of coverage for this type of insurance can range from $5000 to $200,000 per animal.

Professional errors and omissions. With this type of policy, coverage is for liability arising out of a negligent act, error, or omission in rendering or failure to render professional services. If you do something wrong while you are working with or caring for a horse or client, whether or not you are aware of it and whether or not the result is apparent immediately or later, this type of policy should cover you.

The best insurance against losses from accident and illness is conscientious care and employment of safe procedures. If additional protection is required, carefully consider the various forms of equine insurance available today.

Safe practices in and around the horse farm decrease your risk of liability.

PROPERTY INSURANCE

Your horse property insurance should cover your horse facilities, equipment, and tack. If you are a hobbyist, this should be covered on your homeowner's policy, but be sure to check. If you use your facility or horses for a business, then you will need to purchase a business property insurance policy. Ask your agent if the policy should carry your name or the name of your business as the insured.

For either your homeowner's or business property insurance policy, you will need to have an itemized inventory of your tack and equipment. This will help you determine how much coverage you need and will also help you identify items that are stolen or destroyed. List the major items with a brief description and serial or registration numbers if applicable. List the value and include receipts if possible. Photographs or videotapes of your tack room inventory can also assist with identification and description.

Keep one copy of the inventory and visual supports with your insurance policy at home and one in your safety deposit box.

18 Routines

Horses are creatures of habit and are very healthy and content when good management practices are implemented on a regular basis. Establishing daily, weekly, monthly, and yearly routines will increase your horse's well-being and minimize your veterinary bills. These routines can also bring a sense of contentment and order to your own life. Use calendars, lists, record books, file card boxes, or whatever it takes to keep you organized and on track. Your horse will thank you for it!

Daily routine around the barn

MORNING

- ☐ Visual exam
- ☐ Feed hay
- ☐ Feed grain
- ☐ Clean stall or pen
- ☐ Check water
- ☐ Sweep barn aisle
- ☐ Hands-on check
- ☐ Exercise

EVENING

- ☐ Feed hay
- ☐ Feed grain
- ☐ Clean stall/pen
- ☐ Check water
- ☐ Sweep barn aisle
- ☐ Visual exam
- ☐ Late-night check

The visual exam

A visual exam is a routine that allows you to assess the overall health and well-being of your horse. If you see something abnormal, palpate or take vital signs. (For more information, see Hill, *Horse Health Care* [Storey, 1997].)

OVERALL STANCE AND ATTITUDE: Head up and eyes bright, eager for feed or lethargic, inattentive, or anxious?

LEGS: Wounds, swelling, or puffiness?

APPETITE: Finished all of his feed from the previous feeding?

WATER: Has he taken in a sufficient amount of water?

MANURE: Is it well formed yet will break, or is it hard and dry or loose and sloppy or covered with mucus or parasites or filled with whole grains?

STALL: Are there signs of pawing, rubbing, rolling, or thrashing? Or wood chewing?

Plan to store a maximum of a week's supply of hay in your horse barn. Richard restocks on Saturday mornings.

Weekly tasks

- ☐ Restock hay supply in horse barn.
- ☐ Check grain supply.
- ☐ Check bedding supply.
- ☐ Dump and scrub all waterers, troughs, tanks, tubs, buckets.
- ☐ Scrub feed dishes.
- ☐ Mow grass around buildings.
- ☐ Strip stalls.
- ☐ Check veterinary supplies.
- ☐ Check upcoming farrier and veterinarian appointments and get ready for them.

Seasonal tasks

Keep track of all periodic appointments and seasonal tasks on a barn calendar and keep the calendar updated; check it regularly.

SPRING

- ☐ Spread manure.
- ☐ Keep horses off pasture.
- ☐ Spray early weeds.
- ☐ Check for ticks.
- ☐ Get out the shedding equipment.
- ☐ Get fly gear ready.
- ☐ Wash winter blankets, repair, store.
- ☐ Tune up mowing equipment.
- ☐ Make routine veterinary appointment — dental/immunization.

SUMMER

- ☐ Monitor pasture until it is 4 to 6 inches tall.
- ☐ Set out salt and mineral blocks on pasture.
- ☐ Set out water troughs on pasture or check creek or pond.
- ☐ Assign fly sheets and masks.
- ☐ Introduce horses to pasture gradually.
- ☐ Mow weeds in pasture.
- ☐ Keep fire strip mowed around all buildings.
- ☐ Repair facilities, paint fences.
- ☐ Purchase year's supply of hay.
- ☐ Ride!
- ☐ Protect your horse from the sun and insects.
- ☐ Monitor weight of pastured horses.
- ☐ Monitor pasture to prevent overgrazing and damage to trees.
- ☐ Make sure fall and winter blankets are ready to go.

Purchase a year's supply of hay during the summer and have it delivered in stacker loads to save costs and ensure consistency in your horse's diet.

FALL

- ☐ Remove horses from pasture while there is still vegetation in the field.
- ☐ Harrow manure in pastures.
- ☐ Remove bots eggs.
- ☐ Allow horses to gain about 5 percent body weight in temperate climates to prepare for winter.
- ☐ Wash fly sheets and store; assign blankets.
- ☐ Winterize tractor and truck.
- ☐ Get snow-removal equipment ready.
- ☐ Winterize barn pipes.
- ☐ Oil hinges and latches to prevent freezing.
- ☐ Give manure spreader a checkup.
- ☐ Install snow fence as needed.

WINTER

- ☐ Make winter shoeing plans.
- ☐ Monitor winter water intake.
- ☐ Increase roughage when below freezing.
- ☐ Keep winter feeding areas clean of old feed.
- ☐ Clean tack and get ready for spring!

Sample management calendar

MONTH	FEED	VET/FARRIER	EXERCISE	GROOMING	OTHER
JANUARY	Increase the hay 10% for every 10 degrees below freezing	Farrier appointment	Out during day	Clean hooves daily all year	Turn compost as needed all year
FEBRUARY	Same	Deworm	Out during day		
MARCH	Same	Farrier appointment; beware of excess moisture on hooves	Out during day	Shedding	
APRIL	Gradually increase grain as work increases	Same as above; deworm; flu, WEE/EEE, tetanus shots; float teeth	Out during day	Check for ticks	Spread manure
MAY	Increase grain; decrease hay	Farrier appointment; West Nile per veterinary recommendation	Out during day	Clean sheath or udder	
JUNE	Begin grazing; rotate grazing of pasture as needed	Deworm including bots	Out at night		Buy year's supply of hay
JULY	Monitor grazing	Farrier appointment; remove bot eggs	Out at night		
AUGUST	Monitor grazing	Deworm/bot eggs	Out at night		
SEPTEMBER	Organize winter water plan, if necessary	Farrier appointment; remove bot eggs; flu and rhino booster	Out during day		
OCTOBER	Decrease grain as work decreases	Deworm including bots	Out during day	Winter tail care; body or trace-clip	
NOVEMBER	Increase hay 10% for every 10 degrees below freezing	Farrier appointment	Out during day	Blanket if working through winter	Spread manure
DECEMBER	Same	Deworm	Out during day		

Summer Task: Choosing Good Hay

Good-quality horse hay should be leafy, fine stemmed, and adequately but not overly dry. Because two-thirds of the plant nutrients are in the leaves, the leaf-to-stem ratio should be high. The hay should not be brittle but instead soft to the touch, with little shattering of the leaves, since lost leaves mean lost nutrition. There should be no excessive moisture that could cause overheating and spoilage.

Good-quality hay should be free of mold, dust, and weeds and have a bright green color and a fresh smell. In some instances, however, hay color may be misleading. Although the bright green color indicates a high vitamin A (beta-carotene) content, some hays might be somewhat pale green to almost tan due to bleaching yet still of good quality. The interaction of dew or other moisture, the rays of the sun, and high ambient temperatures cause bleaching. Brown hay, however, indicates a loss of nutrients due to excess water or heat damage and should be avoided.

Dusty, moldy, or musty-smelling hay is not suitable for horses. Not only is it unpalatable, but it also can contribute to respiratory diseases. Moldy hay can also be toxic to horses and may cause colic or abortion. Bales should not contain undesirable objects or noxious weeds. Check for sticks, wire, blister beetles, poisonous plants, thistle, and plants with barbed awns such as foxtail and cheat grass.

Top to bottom: **A nice grass-mix hay; when shopping for hay, cut open several bales so you can see, smell, and feel any undesirable heating; the hay on the left might cost twice as much as the hay on the right, but it is probably ten times better for your horse. Although price is not always an indicator of quality, it does not pay to skimp on hay quality. The hay in the bottom picture has been baled with a lot of thistle, which some horses will not eat: try to buy weed-free hay whenever possible.**

Hay varieties and characteristics

HAY NAME	DRY MATTER %	DIGESTIBLE ENERGY (KG DRY MATTER)	CRUDE PROTEIN[a]	CRUDE FIBER[a]	CALCIUM[a]	PHOS-PHORUS[a]
ALFALFA early bloom	90	2.5	20	23	1.4	0.3
CLOVER (all varieties)	89	1.9–2.2	14–22	21–31	1–2	0.23–0.33
GRASS HAYS[b] (all early bloom)	86–95	1.8–2.2	8–12	31–34	0.3–0.6	0.2–0.35
GRASSES[b] (all late-growth bloom)	86–95	1.5–2.0	6–9	31–35	0.025–0.400	0.15–0.30

[a] Percent in feed dry matter.
[b] Grass hays and grasses are timothy, orchard grass, brome, Bermuda, fescue, reed canary grass, ryegrass, wheatgrass.

PURCHASING HAY

You can purchase hay by the bale at your local feed store, which is fine in a pinch but is the most expensive option. If you purchase your hay directly from the hay grower, you may be able to develop a regular account and place your order for the next year. You can also purchase your year's supply of hay at auction, where you will find a variety of hay types, bale sizes, quality, and prices. Although good-quality, barn-stored hay doesn't lose a great amount of its nutritive value when stored for a year, it is always best if you buy the current year's crop of hay, especially if it has been stored outdoors. Depending on how many horses you feed, it is usually most economical to buy the largest quantity of good hay that you can store. Instead of buying a pickup load at a time, if you buy a larger load, you'll probably not only get a better per-ton price but also the transportation costs will be greatly reduced.

Test hay quality

Because the nutritive quality of hay can vary so greatly, it is best to test hay before a large purchase, especially if it is to be used for young or lactating horses. Your Extension agent will instruct you on sampling techniques, and the test results will reveal crude protein, fiber, energy, and mineral content.

Top: **Buy the current year's hay rather than last year's hay.** *Bottom:* **Buying a semi-load is economical but can be risky. If the quality of the hay is poor, you will be stuck with a large amount of bad hay.**

WHICH BALE IS BEST FOR YOU?

Should you buy large round, large square, or small square bales?

- Large round bales can range from 4 feet wide by 4 feet in diameter to 8 feet wide by 6 feet in diameter and weigh from 500 to 2500 pounds.
- Large square bales come in sizes from 3 feet by 3 feet by 8 feet to 4 feet by 4 feet by 8 feet. The smaller bales weigh from 600 to 800 pounds.
- Small square bales are 3 feet by 1½ feet by 1¼ feet and weigh 40 to 70 pounds.

Large bales are usually less expensive per ton, and in some parts of the country large bales are more readily available than small square bales.

If you put a large bale out in a pasture to feed a group of horses, they tend to waste, trample, and foul the hay and then won't eat it. Hay lying at the base of the bale will be used for bedding, defecation, urination, or as breeding grounds for insects and rodents. If you try to feed portions from a large bale to individual horses, it is difficult and messy to move each ration from the bale to the feeders. And the open bale needs to be stored and protected while it is being used. Bales stored outdoors need to be protected from other animals eating them, such as deer and elk. Large bales require special equipment to transport to your farm and move around once on your farm.

Large round bales typically have a higher storage loss than small rectangular bales, especially when stored outdoors.

MINIMIZING LOSS WITH LARGE BALES

To minimize outdoor storage loss, do the following.

Choose dense bales. A dense bale sheds water best and sags least, putting less surface area in contact with the ground.

Choose bales covered with plastic twine, net wrap, or plastic wrap. Plastic twine spaced 6 to 10 inches apart holds the bale tight and resists damage from weather, insects, and rodents better than natural-fiber twine. Net wraps are porous materials designed to shed water and permit greater airflow at the bale surface. Solid plastic covers shed water and, if they are of ultraviolet (UV) light-stabilized plastic, result in the least storage loss.

Store bales on a high, dry location (or indoors if possible). A coarse gravel base will minimize bottom spoilage. Placing the bales across heavy poles or pallets provides air space between the bales and the soil to keep the bales dry.

Pack the bales tightly end to end in a long row. Stacking large round bales usually traps moisture, limits drying from sun and wind, and results in more loss. Consider covering the stack with a large tarp.

Hay auctions offer all types and qualities of hay: grass and alfalfa, large bales and small bales, premium hay and trash.

Summer Task: Caring for a Horse on Pasture

When turning out a horse to full-time pasture, let him get used to the radical change from dry feed to green feed over a period of 2 to 3 weeks. Some horses are prone to colic or laminitis when pastured, so it's best to turn out horses when the pasture grasses are mature. Give the horse a full feed of hay just before you turn him out. Limit his grazing to 30 minutes per day and gradually work up to being out on pasture full time. If a horse has been off pasture for a week or more, take a couple of days to reintroduce him to the green feed.

Horses actively seek the company of other horses and usually specific buddies. Horses will try to join or at least hang out with nearby horses even if a fence separates them. Not all horses get along. It is often necessary to separate mares from geldings and certain individuals from any other horses. If you have only one horse or a particular horse that is difficult to pair with others, a burro, ewe, goat, or cow might provide the companionship the horse needs.

If a horse is going to be turned out indefinitely, it's generally best to pull his shoes and trim his hooves. But if the horse is going back into work within a month or so, consider keeping the horse shod while on pasture. Although shod hooves are more damaging to pastures, and pastured horses are more likely to lose shoes than are stalled horses, the hooves often break at the site of the old nail holes when shoes are removed. It can take months for the horse to have a solid hoof to nail to again.

Whether a horse should be blanketed on pasture depends on the season, the horse, and your goals. For summer, a properly fitted PVC-coated mesh fly sheet and mask with ears will protect a horse from flies and gnats. He will be more comfortable, spend less time

Sassy and two foals are on a pasture that is dangerously lush; only by closely monitoring grazing time is a pasture like this safe.

Far left: **Sherlock's insecticide strip headband helps keep face flies at bay.** ***Left:*** **Sassy's intake is greatly slowed down with a grazing muzzle (by Best Friend).**

stomping and swishing his tail, and require very little if any fly spray, and his coat will be protected from UV rays. To prevent problems, be sure the sheet is the correct size and cut (prevents slippage or rubs), the leg straps and surcingles are properly adjusted (prevents tangling or loss), and the fly mask is the proper size and style for the horse (prevents eye irritation, mask loss, or rub sores). Fly collars or brow bands are also helpful for keeping face flies and gnats from bothering the horse's head.

During cold-weather turnout, many horses do well without a blanket as long as they have shelter. A horse's long winter coat provides insulation, and sebaceous secretions from the skin provide protection from moisture. Waterproof and breathable winter turnout blankets are ideal for old or thin horses and for horses in work that you want to keep clean. Long sides, neck extensions, and tail flaps provide added protection.

MONITORING A HORSE ON PASTURE

Many horses gain too much weight when put on pasture, as they pack away 6 pounds of grass per hour, 16 hours a day. Overweight horses are more prone to developing health problems such as colic, laminitis, and various lamenesses. Some horses (especially seniors) lose weight on pasture because they have dental problems or lack the initiative to eat enough to keep their weight. Weight can be monitored by a visual exam, a weight tape, or palpation of the ribs and tail head. Pasture horses that wear sheets should have a thorough weekly check that includes removing the sheets, grooming the horse, and brushing or washing the sheet.

Underweight horses can remain on pasture for the health benefits but be fed a grain and/or hay ration to supplement the pasture. Overweight horses can wear a grazing muzzle part time or full time to minimize the amount of feed ingested while still allowing the horse to have the exercise benefits of turnout.

Most regions have a few ideal pasture months: when the grass is growing, the temperatures are not extreme, and the insects are moderate. But sometimes being on pasture is unpleasant for horses. When precipitation is excessive, insects are relentless even with all control methods in place, the heat and sun are debilitating, or wind and cold temperatures make it too brutal for horses to be comfortable, the horses should be brought in from the pasture.

Sherlock is well trained and comes when called for his daily pasture check.

Pasture kit

- ☐ Compact halter and lead rope
- ☐ Fly spray, cream, wipes
- ☐ Gloves
- ☐ Mane detangler
- ☐ Treats or carrots
- ☐ Wound ointment
- ☐ Zinc oxide
- ☐ Extra fly mask
- ☐ Extra leg strap

I give Zinger a once-over on pasture with my kit nearby in case I need any supplies.

FEEDING A HORSE ON PASTURE

Horses need supplemental feed when the pasture starts waning and during winter turnout.

When feeding a single horse on pasture, the same rules and manners apply as would be required when feeding the horse in a stall or pen. No pushing, crowding, or aggressive behavior allowed. Decide how much personal space you need when carrying feed and placing it on the ground or in feeders. Then require that the horse stay out of that area until you give him the command *OK* as you leave the feed and walk away.

It is more difficult to safely feed a group of horses while you are on foot due to natural competition at feeding time. A pickup truck or a utility vehicle will help you carry the feed and can be pretty slick if one person drives while another throws out the feed. If you are feeding one or two horses and must transport the hay on foot, a homemade or commercial hay carrier will make the job easier and neater for you.

When putting out hay, make one or two more piles than there are horses so there are plenty of feeding stations, no matter what a horse's rank in the hierarchy.

When feeding grain to groups of horses, choose long troughs with plenty of room for twice as many

Daily pasture check

- ☐ Watch your horse as he approaches to discern any irregularity in his gait or if he is lagging behind the other horses.
- ☐ Check to see if your horse is alert and eager to get feed or treats.
- ☐ Start your visual inspection by looking carefully at the legs, the most common site of injury on pasture. If the horse is shod, be sure he has all his shoes.
- ☐ View each horse from both sides, the front, and the rear.
- ☐ Look for discharge from the eyes, head shaking, ears held at an odd angle, or an obvious blemish anywhere on the horse's body.
- ☐ Look for burrs or sticks in the mane and tail.
- ☐ Keep an eye out for damage from mutual grooming to the mane or sheet.
- ☐ Take care of those things that you can safely attend to out in the pasture. For anything that requires closer attention, lead the horse back to the barn.

horses to eat. For feeding grain individually, use broad, shallow pasture grain dishes that are difficult to tip over or roll and space them far apart.

Even if you are not taking feed out to the horses, a daily check (or twice-a-day check if a horse is wearing a blanket) is a good idea. If you make up a pasture kit to carry along, you'll have the items you'll need to take care of small issues.

Even if your horses are dead gentle and well trained, a halter and lead will come in handy if you need to stabilize a horse while you remove the wire around his leg or the bush he is dragging with his tail. Teaching your horses to come to a certain spot when you whistle is a help, and having a few wafers or carrots along to reinforce their behavior will pay off in the long run. Toss the treats on the ground in several spots and, as the horses search and nibble, you'll have plenty of time to give each of them a thorough visual examination.

WEEKLY PASTURE CHECK

Every few days or at least once a week, halter your horse and remove his blanket and give him a thorough hands-on check (see box on page 292).

If you lead your horse back to the barn for this weekly check, it will be a good review of in-hand manners and will give you the opportunity to perform routine health care tasks such as deworming, hoof care, mane and tail overhaul, application of fly cream, and change of sheet or fly mask. But don't plan to do everything in one session just to save time. In fact, try to schedule it so that you do only one large management task per week. That will give you a reason to gather up your horse once a week for the up-close-and-personal check.

Caution

It is dangerous to leave a halter on a pasture horse. He could catch the halter on fence posts, tree branches, or his own shoe when scratching his head with a hind leg. And a haltered horse is usually easier to steal. If it is essential that a horse wear a halter, be sure it is a safety, breakaway halter (as shown here) that has an element that will break if the horse gets caught on something.

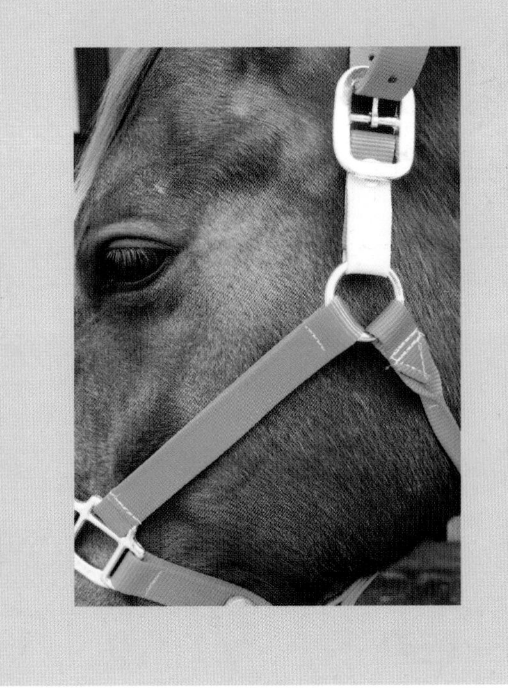

Weekly hands-on pasture check

- ☐ Remove blanket and mask.
- ☐ Look for any obvious problems such as wounds, scabs from fly bites, blanket or mask rub marks, and sores or lumps, which could indicate an abscess.
- ☐ Run your hands from throatlatch to shoulder down both sides of your horse's neck, paying special attention to the crest of the mane, where there might be signs of rubbing, and under the mane, where there might be skin problems or ticks.
- ☐ Do the same for the horse's entire body, paying special attention to the junction of the neck and chest and in front of the sheath or udder, two places where flies love to feast. Keep an eye out for rough bumps under the coat that could indicate rain rot. Also examine the inside of the hind legs to see if there have been any abrasions from leg straps. During winter, be sure to feel the ribs, as this is the main way to determine if your horse is carrying enough flesh. You should be barely able to feel the ribs.
- ☐ Check by palpation and visual exam every inch of the legs and hooves, making note of any nicks or bumps; roughness under the hair of the legs, which could indicate rain rot; scabs at the back of the pastern, which could be scratches; loose shoes; hoof chips or cracks or sticks or stones stuck in the frog. Take appropriate action.
- ☐ Give the tail a once-over, removing any hidden burrs or branches as you finger through the tail. Check the tail head, another favorite site for ticks. Apply a detangler or conditioner to the tail as needed.
- ☐ Finish with the head. Run your fingers inside the ears, noting if the horse is sensitive or if there are scabs or red spots inside from gnats. Look carefully around the eyes for small dings or rub marks from a mask. Be sure the eyes and nostrils are clear. Run your hand between the lower jawbones and over the fleshy portion of the throatlatch, two more favorite places for fly feasting. Finally, open the horse's mouth, looking for any evidence of awns or sticks lodged in his gums or lips.
- ☐ If your horse seems out of sorts, take his vital signs and appropriate action.

PASTURE MANNERS

For your own safety, ensure that all of your pastured horses retain their good manners. When you catch and turn your horse loose for the weekly checks, take the time to do things right.

Halter

It's safest for pasture horses to be without halters. If necessary, use a turnout halter; turnout halters have a breakaway mechanism that will ensure that if a horse does get hung up, the halter will release and free him. Just remember not to tie up the horse with a breakaway halter, or he could learn the bad habit of pulling to get free.

Catching

You need to be able to easily catch any horse anywhere for his own health and safety.

Horses should never turn away from you when you approach, and you can teach most horses to come when you call or whistle.

Horses are generally difficult to catch because of fear, resentment, or habit. Unhandled horses move away from humans due to unfamiliar movements, sound, and smell. Mistreated horses might evade humans from memory of bad treatment, hard work, ill-fitting gear, painful doctoring, or tactless training (for example, the frustrated person who finally catches a horse and then gives him a good swift slap with the halter rope).

No matter why a horse initially avoids being caught, even if the cause is removed, he may remain difficult to catch out of habit. If a horse avoids being caught, he needs to go back to the barn or round pen for groundwork before he can live out on pasture, because a horse you can't catch is basically useless.

Haltering

When haltering, approach the horse from the near (left) side and hold the unbuckled halter and rope in your left hand. With your right hand, scratch the horse on the withers and then move your right hand across the top of the neck to the right side. Pass the end of the lead rope under the horse's neck to your

Richard heads out to the lower forty to bring Sassy into the barn for her weekly detailed checkup.

right hand and make a loop around the horse's throatlatch. The loop is held with your right hand. If the horse starts moving, you can remind him with this loop to stay put while you halter. Hand the halter strap under the horse's neck to your right hand, which is holding the loop of rope. With your left hand, position the noseband of the halter on the horse's face and then bring the hands together to buckle the halter.

Turning Loose

If a horse jerks away, wheels, and kicks as you are turning him loose, you could get hurt. Make turning loose a lesson. If necessary, take a walk in the pasture for a minute or so. Take your time. Don't encourage your horse to anticipate turnout by letting him go as soon as you walk through the pasture gate.

Before you unhalter, drop a few treats on the ground in front of the horse. Then make a loop with the lead rope around the horse's neck, remove the halter, and release the loop. Walk away. The horse will drop his nose and stretch his back to look for the treats; he will soon associate turnout with reward and relaxation rather than throwing his head up and racing off.

19 Records

Good records will help you take better care of your horse and run a better horse operation. They are necessary if you will be claiming your horse operation as a business. In order for records to be useful, they need to be complete and accurate. Your record-keeping system should be designed with your personal habits in mind. If you make your forms easy to use and locate them in convenient places, you are more likely to make entries on a regular basis. Your record forms should be simple, but this does not mean using a single sheet of paper to record everything about your farm!

Categorize and organize the types of information you wish to record. You might want to use a recipe box with file cards, a spiral or ring binder specially designed for horse records, a computer software program, or a combination.

Management Records

Whether your horse venture is a hobby or business, you should keep accurate, complete records on health care, farrier work, training, and breeding.

Jotting everything on a calendar just doesn't work. Using a calendar to remind you of upcoming appointments and events is fine, but you will also need a place to record the various details related to horse ownership and care. A 5-inch by 8-inch card file works well. You can make dividers for each horse and subdivide these further into categories (for example, vet, training, farrier).

In the event of an emergency, you should always have a current set of feeding instructions that are easy for anyone to understand. An erasable board in the feed room works well for this.

In addition to daily records, you should keep a file folder on each horse to store documents such as a breeding certificate or bill of sale, registration papers, brand inspection, tattoo, freeze brands, permanent identification records (including photos and written description indicating height, weight, scars, brands), pedigree, and insurance policies.

Financial Records

If you are pursuing your horse endeavor as a business with profit intent, you should work closely with a tax counselor who has experience with farm accounts. You will have to keep formal income and expense records. You will need to keep your horse operation transactions separate from monies used for domestic matters, other farm ventures, hobbies, and your other work. One of the best ways to do this is to maintain a separate checking account for your horse venture. All equine expenditures and income and no others should be handled through this account. You can, of course, make cash withdrawals (hopefully profits) from the business account to use elsewhere.

Your horse business will need to show profit intent in order to retain your business or farm status. If that status is in question, the Internal Revenue Service (IRS) will use specific guidelines to determine if your horse operation is designed with profit intent. It will look at the manner in which you carry on your activity and the type of bank accounts,

It doesn't matter whether you make notes on a 3x5 card, in a notebook, or on your computer — just choose a method that you will use regularly.

records, and logs that you maintain. The IRS accepts the following types of documentation in an audit: receipts, checks, logs, third-person verification, and verbal description.

It will note if you have sought advice from experts, not only horse-related advisers but also marketing and financial counselors. The IRS will evaluate the time and effort you expend in your business: is it a weekend diversion, a full-time business, or something in between? It is advisable to keep a log of the hours you actually work in your business. You may be asked verbally or in writing if you have expectation and basis that your assets will appreciate. You should have a projected profit plan available, which is updated each year.

The IRS will also review your success in both similar and different types of businesses you have had. They will look at the previous history of profit or loss of your current horse business. They may look at your overall financial status and see how the horse business relates to your total income picture. It will do this to try to determine if you are legitimately profit-motivated or if you entered the horse business to use the deductions to offset your income from another area. Also, because many non-horse-owning people see only the pleasurable aspects of horse ownership, you may be asked to justify the business nature of your horse operation. You may know of the hard physical work, the early hours, and the many personal sacrifices that are required to run a successful horse operation, but an IRS representative may envision you only galloping across a meadow, heading for a tax shelter.

If you are in the horse business, just keep good records and find an experienced farm tax counselor. He or she will help you define and understand income and expenses in current terms.

INCOME

You must report all income related to your horse business. This includes horses bought for immediate resale, horses raised and sold, and assets acquired for the business and then sold (for example, a trailer, a saddle, or breeding stock). It also includes prizes and awards, show and track winnings, boarding fees, training fees, lesson fees, proceeds from hosting events and clinics, and income from the sale of farm-raised grain, hay, and straw.

If you engage in bartering, the services or goods you receive are considered income and must be recorded at the fair market value of the item or service. For your half of the trade in bartering, what you gave will be considered an expense if it would have been a deduction in the normal course of business.

EXPENSES

Horse business expenses generally fall into one of three categories: those operating expenses that are entirely horse-related and fully deducted in the year the expense is incurred; those that are shared expenses with domestic or other businesses and must be prorated and then the horse portion deducted in the year incurred; and those major expenses that are spread out (depreciated) over several years.

The operating expenses are fairly straightforward. Either your expenditures, such as feed and veterinary supplies, are "ordinary and necessary" for your business or they are not. Be sure you fully understand the definitions and requirements for an expense to qualify for a particular category. Some expenses are only partially deductible (such as business meals), and others are deductible only if your horse operation is showing a profit (such as an office in the home).

Expenses that need to be prorated include vehicles, taxes, and utilities. In each situation you must determine, to the best of your ability, what percentage of the expense is directly attributable to the horse operation. If a vehicle is shared between domestic and farm purposes, keep a logbook with the starting and ending mileage each time you use the vehicle for business-related activities, along with the business purpose of the trip.

Determine what percentage the business miles are of the total miles traveled by that vehicle during the year. Then your tax counselor will determine which is to your advantage — using a mileage allowance deduction (so many cents per mile) based on actual business miles traveled or taking a percentage of actual expenses (total of gas, oil, repairs, etc.) based on the percentage of business use. Either way, for your tax counselor to make this assessment, you will need to keep all receipts concerning vehicle operation and maintenance as well as a mileage log to determine the business percentage use. If you are using a leased vehicle, you are allowed to deduct only actual expenses.

Property taxes are usually determined by consulting the itemized breakdown on your property valuation and tax statement. The tax resulting from the assessed value of horse-related buildings is deductible, as is the tax assigned to the portion of the land you use for the horse operation. If you have 10 acres, you may be using 2 acres for house, yard, and other domestic uses and 8 acres for the horse operation. The tax related to the 8 acres is deductible

Prorated expenses that may be a little tougher to divide up are utilities. With electricity, for example, unless you have separate meters for your barn and your house, you will have to estimate what percentage is farm-related. If you are just adding horse facilities to your residential acreage, you can compare utility bills before and after the additions to help you calculate business-related kilowatt-hours and dollars. To further help you determine your utility business percentage, you may be able to get hourly rate estimates from your utility company on the kilowatts required to operate various electrical devices. Note that what you determine as a percentage for electricity may be totally different from the business percentage for gas, water, or other utilities. Most small horse operations located on a residence use an average of 20 to 30 percent of the total utilities.

Large assets purchased for the business (truck, trailer, buildings, tractor) and those converted to business use are often depreciated. Your tax counselor either will advise you to take a full expense in the year incurred for an item (depending on the cost of the item and the status of your profit or loss) or will determine what depreciation schedule each item will follow. Then each year of the depreciation schedule, a certain percentage of the cost of the tractor, for example, will be available to use as a deductible expense. It must be remembered, however, that during the year you sell a business asset, its expense potential will have to be recaptured and treated as income.

Whenever you pay $600 or more to an individual or a nonincorporated business in one year for services, you are required by law to fill out a Form 1099 and send it to the IRS and to the business or person for him or her to report as income. So, if your bills for veterinary work totaled $600 or more and your veterinarian's business is not incorporated, you should issue a 1099 to your veterinarian. The same goes for your farrier, farm laborers, and so on.

IF YOU FILE SCHEDULE F OR C

If your horse operation qualifies as a farm or ranch operation or if you operate a horse-related business such as training horses or giving lessons, then you will want to keep accurate records of your income and expenses. Definitions and requirements for your expenditures to qualify for tax-deductible business expenses vary from year to year. Work closely with your tax counselor for a current interpretation.

Recommended Reading

Anthony, Stan. *Farm and Ranch Safety Management.* Albany: Delmar, 1995.

Burch, Monte. *How to Build Small Barns and Outbuildings.* North Adams, MA: Storey, 1992.

———. *Monte Burch's Pole Building Projects.* North Adams, MA: Storey, 1993.

Campbell, Stu. *The Home Water Supply.* North Adams, MA: Storey, 1983.

Damerow, Gail. *Fences for Pasture and Garden.* North Adams, MA: Storey, 1992.

Engler, Nick. *Renovating Barns, Sheds and Outbuildings.* North Adams, MA: Storey, 2001.

Ewing, Rex A. *Beyond the Hay Days: A Refreshingly Simple Guide to Effective Horse Nutrition.* 2nd ed. LaSalle, CO: Pixyjack Press, 2003.

Fershtman, Julie I. *Equine Law and Horse Sense.* Franklin, MI: Horses and the Law Publishing, 1996.

———. *More Equine Law and Horse Sense.* Franklin, MI: Horses and the Law Publishing, 2000.

Greenwalt, Joni. *Homeowner Associations: A Nightmare or a Dream Come True?* Denver: Cassie Publications, 1998.

Hill, Cherry. *The Formative Years: Raising and Training the Young Horse from Birth to Two Years.* Ossining, NY: Breakthrough, 1988.

———. *Horse Care for Kids.* North Adams, MA: Storey, 2002.

———. *Horse for Sale: How to Buy a Horse or Sell the One You Have.* New York: Howell Book House, 1995.

———. *Horse Handling and Grooming.* North Adams, MA: Storey, 1990.

———. *Horse Health Care.* North Adams, MA: Storey, 1997.

———. *Stablekeeping: A Visual Guide to Safe and Healthy Horsekeeping.* North Adams, MA: Storey, 2000.

———. *Trailering Your Horse: A Visual Guide to Safe Training and Traveling.* North Adams, MA: Storey, 2000.

Hill, Cherry, and Richard Klimesh. *Maximum Hoof Power: A Horse Owner's Guide to Shoeing and Soundness.* North Pomfret, VT: Trafalgar Square, 2000.

Klimesh, Richard, and Cherry Hill. *Horse Housing: How to Plan, Build, and Remodel Barns and Sheds.* North Pomfret, VT: Trafalgar Square, 2002.

Knight, Dr. Tony. *A Guide to Plant Poisoning of Animals in North America.* Jackson, WY: Teton New Media, 2001.

Lewis, Lon. *Feeding and Care of the Horse.* 2nd ed. Media, PA: Williams & Wilkins, 1997.

Lodge, Ray, and Susan Shanks. *All-Weather Surfaces for Horses.* London: J. A. Allen, 1999.

Malmgren, Robert. *The Equine Arena Handbook: Developing a User-Friendly Facility.* Loveland, CO: Alpine Publications, 1999.

Martin, Deborah L., and Grace Gershuny. *The Rodale Book of Composting.* New York: Rodale Press, 1992.

McKenzie, Evan. *Privatopia: Homeowner Associations and the Rise of Residential Private Government.* New Haven, CT: Yale University Press, 1994.

Midwest Plan Service. *Horse Handbook: Housing and Equipment.* Ames: Iowa State University, 1994.

O'Keefe, John M. *Water-Conserving Gardens and Landscapes.* North Adams, MA: Storey, 1992.

Rynk, Robert, ed. *On-Farm Composting Handbook.* Ithaca, NY: Natural Resource, Agriculture & Engineering Service, 1992.

Seddon, Leigh. *Practical Pole Building Construction.* Charlotte, VT: Williamson Publishing, 1985.

United States Department of Agriculture. *Plants Poisonous to Livestock in the Western States.* Agriculture Information Bulletin 415. Washington, DC: U.S. Government Printing Office, 1980.

For a helpful resource guide and current information from Cherry Hill's Horse Information Roundup, visit Horsekeeping Books and Videos at www.horsekeeping.com.

Glossary

access panel. An opening in the wall or ceiling near a fixture for service access.

air change. The amount of fresh air required per hour to replace stale air in a given space.

ambient lighting. General illumination in a room, with no single, visible source of light.

arena. A fenced area for working horses; usually rectangular.

asphalt. A black mineral used in roof coverings and paving.

backhoe. Excavation machine that digs by pulling a boom-mounted bucket toward itself. Used to dig footings and trenches.

beam. Any large structural member used to support a load over an opening or from post to post.

bedding. Soft, absorbent material used to cover the floor of a stall.

berm. A small earthen embankment usually used for diverting runoff water.

board-and-batten siding. Wide vertical boards installed close together, with gaps between covered by narrow boards or strips (battens).

bots. The larvae of several species of bot fly *(Gasterophilus).*

bracing. Boards, metal rods, and strips used for strengthening a building or as temporary supports.

brand. A mark made on the hide of an animal by a hot or freezing iron to designate ownership.

breaker. A switchlike device that connects and disconnects power to a circuit.

breezeway. A center-aisle barn with a large door on each end of the aisle.

broom finish. The texture created when a stiff broom is pulled across the surface of a concrete slab before it has set.

building code. A set of safety standards that contains rules necessary for accepted safe building practices.

building permit. An authorized form that allows someone to legally construct or remodel a building.

building standards. Guidelines that detail the minimum design or the performance of a specific material.

cement. A powdered material composed mainly of limestone and sand. When mixed with water, it hardens and binds other materials, such as sand and gravel in concrete.

circuit. A continuous loop of current.

circuit breaker. A switchlike device that automatically "trips" and opens a circuit when the rated current is exceeded, as in the case of a short circuit. Not damaged by tripping and can be reset.

clear span. A building with an open area inside that contains no posts or columns.

clerestory. An outside wall with windows that is between two different roof levels.

code. Any published regulatory material.

colic. General term for abdominal pain in the horse.

column. A vertical structural component that supports a beam.

component stall. A stall made of preassembled walls that fit between support posts within the barn.

concrete. A mixture of water, cement, sand, and aggregate.

concrete block. A prefabricated structural component, typically 8 inches by 8 inches by 16 inches, made of concrete and used for walls.

condensation. The effect of warm, moist air contacting a cooler surface and depositing moisture (condensate) onto the cooler surface.

conduit. A plastic or metal tube for protecting electric cable.

contractor. An individual or company offering construction services.

convection. Heat transfer by the natural rising of heated air and the sinking of cooled air.

corral. A pen for keeping horses or cattle, usually devoid of grass.

coupling. A fitting that joins two pieces of pipe.

cross tie. A means of tying a horse across an alleyway. A chain or rope attached to each side of the aisle connects to the cheekpieces of the halter.

curing. The process in which mortar and concrete harden; best at 50 to 70°F; often requires 28 days.

d. "Penny." Refers to a specific nail size.

DC. Direct current, the type of electricity generated by batteries.

deck. Sheathing layer of a roof.

downspout. The pipe that leads water down from a gutter.

drip cap. A horizontal molding used to prevent water from running behind a panel, window, or other component.

drip edge. A continuous strip of material installed along the eaves and rakes to allow water to drip free without backing up under the roofing.

Dutch door. A door with two leaves or halves, one above the other.

easement. See *right-of-way.*

eaves. The bottom edges of a sloping roof; those portions of a roof that project beyond the outside walls of a building.

eaves trough. See *gutter.*

eutrophication. The process whereby excess nutrients in water cause excessive and unhealthy algae growth.

expansion joints. Lines cut into a concrete surface to allow for expansion and contraction. If the slab cracks, it will do so only along these lines.

farrier. See *horseshoer.*

fiberglass. Fine filaments of glass made into a fibrous insulating material or molded into a solid material for use as a translucent panel and other items.

finish grade. Surface elevation of lawn, driveway, or other improved surfaces after completion of grading operations.

firewall. A fire-resistant wall used to restrict or prevent the spread of fire between portions of a building.

fixture. Any permanently connected electrical or plumbing device.

flooring. Any material installed on the floor of a barn, such as rubber mats and wood plank.

floor plan. A drawing, a horizontal section, showing the arrangement of rooms, walls, partitions, doors, and windows and giving dimensions, names, and other information.

footing. The surface material on which a horse is worked or ridden; also, concrete pad used under a post or foundation wall to bear weight.

forced air. A mechanical warm-air system that uses a blower to increase the flow of heated air to specific areas.

foundation. Comprises the supporting building components that contact the soil, such as footing, piles, and caissons.

4WD. Four-wheel drive; a vehicle with power delivered to all four wheels; 4×4.

founder. See *laminitis.*

framing. The framework of a building, consisting of beams, columns, joists, rafters, etc.; the making of such a framework.

free span. See *clear span.*

French drain. A gravel-filled trench, sometimes containing a drainpipe at the bottom, used to disperse excess surface water into the ground.

frost line. The maximum depth to which the ground freezes in the winter.

FRP. Fiber-reinforced plastic.

fuse. A protective device containing a thin wire that melts when it exceeds a specified value to interrupt the flow of current. A fuse cannot be reset and must be replaced after use.

gable. The end wall of a building having a triangular shape formed by a single sloped roof on either side of a ridge.

gable roof. A simple triangular-shaped roof having two slopes.

galvanized. Coated with zinc to prevent corrosion.

gambrel roof. A roof having two slopes on each side of the ridge, with the lower slope steeper than the upper.

general contractor. An individual who undertakes or oversees a building project.

geotextile. See *landscape fabric.*

GFCI (GFI). Ground fault circuit interrupter; outlet or breaker that safeguards against shocks by shutting off electricity.

girder. A heavy beam of steel or wood that serves as a horizontal support and typically supports smaller members such as joists.

grade. Surface level of the ground.

grade of roofing. Grades A, B, and C are approved fire-rating roofs.

grading. The leveling and shaping of the ground.

gray water. Wastewater not containing toilet or food waste.

grill (grillwork). A steel grating used as a barrier to protect windows and as the top portion of stall walls and doors.

ground. A circuit wire that is connected to the earth, usually by means of a *ground rod.*

ground fault. See *short circuit.*

ground rod (grounding rod). A rod, usually 8 feet long and copper, driven completely into the earth and used to ground an electrical panel.

gutter (eaves gutter, eaves trough). A long, shallow trough installed under and parallel to the eaves in order to catch and direct water dripping from a roof.

halogen (quartz halogen). A bright, long-lasting light that uses a tungsten filament in a clear quartz tube filled with halogen gas.

horseshoer. One who shoes horses; *farrier.*

hot. The circuit wire that brings in the electrical current flow.

humus. The dark, uniform, finely textured, odorless product of the decomposition of organic matter; the end product of composting.

ID. Abbreviation for *inside diameter;* pipes are sized according to their inside diameter. Also abbreviation for *identification.*

insulation. Material that minimizes heat transfer through the walls, floors, and ceilings.

junction box (electrical box). A square, octagonal, or rectangular plastic or metal box open on one face that fastens to framing and houses wires, receptacles, and switches.

lameness. Pain or physical defect that interferes with normal movement, evidenced by varying degrees of limping.

laminated. Fabricated in layers that are glued or bonded together.

laminitis. Inflammation of the hoof causing severe lameness.

landscape fabric (geotextiles, weed barriers). Cloth made of meshed or tightly woven polypropylene and used to admit water yet keep soil or vegetation from passing through.

larvae. Insects in a very early stage of development. Fly larvae are called *maggots.*

leaching. Washing away, as to wash valuable minerals and other nutrients from the soil by means of rain and snow.

legume. Any of a group of plants of the pea family that store nitrogen in the soil; alfalfa and clover are examples of legume hays.

lime. Calcium hydroxide or slaked lime.

lintel. A beam over an opening.

load bearing. A wall or other similar structural component that supports more than its own weight; cannot be removed without providing an alternate means of support.

loafing shed. See *run-in shed.*

loose stocks. A chutelike enclosure that limits a horse's movement for grooming, veterinary care, or farrier care. Less confining than regular stocks.

manure. Animal dung.

mare motel. An open shelter used for protection from the sun and usually consisting of steel-pipe panels and a roof.

masonry. A wall or other structure made of individual concrete, clay, or stone units that are usually bonded together with mortar.

member. An individual component of a structure.

modular barn. A barn that is constructed from modules that are transported to the building site and assembled on a prepared foundation.

module. A complete factory-built unit of a building.

monitor roof (raised center aisle or RCA). A roof with a raised portion to admit light and/or ventilation.

muck. Manure and/or dirty bedding.

muzzle. The nose and lips of a horse; a device placed over a horse's mouth to prevent him from biting or eating.

neutral. The circuit wire that return the electricity to the source.

OC. On center; studs, joists, and other components are placed so many inches apart center to center.

OD. Outside diameter.

OEM. Original equipment manufacturer.

offset gable roof. Gable roof with the ridge off center of the building, so that one side is shorter than the other and often has a different slope.

ordinance. A local law enacted by a municipality such as a city or township.

OSB. Oriented strand board.

overhang. Commonly refers to the roof portion that extends past the exterior wall; it is often enclosed by a *soffit.*

overhead service. An electrical service where the conductors are brought to a building on poles overhead.

paddock. A grassy area of approximately ¼ to 1 acre that is designed to provide exercise as well as some grazing.

panel stall. A freestanding stall consisting of completed wall panels that bolt together.

parasite. A plant or animal that lives in (internal parasite) or on (external parasite) another. Internal parasites include bloodworms (strongyles), roundworms (ascarids), pinworms *(Oxyuris equi)*, and bots *(Gasterophilus)*. External parasites include lice and ticks.

partition wall. A wall that has no structural function and is installed to divide space.

pasture. A grassy area usually larger than 1 acre that is designed and maintained primarily to provide grazing.

pen. A non-grassy area that is designed primarily for outdoor living quarters for horses.

pitch. The ratio of the vertical rise of a roof to the total span, expressed as a fraction. For example, if the rise of a roof is 4 feet and the span is 24 feet, the roof has a pitch of 1/6 (4/24).

plot plan (site plan). A drawing showing a complete property, indicating the location of buildings, driveways, easements, and natural formations.

plumb. Vertically straight.

pole barn. A roof structure supported on wooden columns (poles).

post (column). A vertical structural component that is typically used to support a girder or other beam.

post and beam. Type of wall framing that uses posts that carry horizontal beams on which joists are supported; it allows for fewer bearing partitions and less material.

power. Watts, a measurement of the rate at which electrical energy is used; watts = volts × amps.

pressure-treated. Wood that has had chemical preservatives forced into it under pressure to protect it from insects and decay.

purlin. A horizontal member between trusses to which roofing is attached.

PVC. Polyvinyl chloride; a rigid white or cream-colored plastic.

rafter. A sloping roof member that extends from the ridge of the roof to the eaves and supports the roof covering.

RCA. Raised center aisle; a barn with a *monitor roof.*

ridge (peak). The top of a roof where two slopes meet; the highest point of a roof.

right-of-way. A strip of land granted by deed or easement for the construction and maintenance of utility lines, roadways, and driveways.

riparian. Waterside vegetation and land.

rise. The vertical distance between the top plate line of a barn and the ridge of the roof.

rolling. Natural horse behavior in which a horse lies down and rolls onto his back.

run. A long pen designed to encourage a horse to exercise. Also, the horizontal distance of a roof or stair step.

run-in shed (shed, loafing shed). A building, usually pole-framed, having only three sides and one large space that horses can enter and leave at will.

service. The conductors and equipment for delivering electricity from the electrical supply system to the main electrical panel of a building.

service panel (service distribution panel). The wiring and circuit breaker box (or fuse box) within a building from which utility outlet receptacle wiring originates. The main service panel in a building is fed from the outside power line.

setback distance. The legal distance a structure or building must be from a property line, a centerline of an adjacent road, or existing structures.

sheathing. The first layer of material applied over the framing and under the finish material to add support and attachment for the finish layer, such as on a roof and walls.

shed. See *run-in shed.*

shed roof (pitched flat roof). A roof having a single sloping surface.

shedrow. A type of barn consisting of a single row of stalls with no enclosed aisle.

shell. The basic minimum enclosure of a building, before installation of interior partitions, plumbing, wiring, and so on.

short circuit (ground fault). When current is interrupted by a hot conductor contacting a neutral or ground, it is an immediate fault to ground and should cause a breaker to trip or a fuse to blow.

slab. A level surface of concrete, commonly a floor, usually set on the ground.

slope. The incline of a roof as the ratio of vertical rise in inches to 12 inches of horizontal run, expressed as a fraction or as "X" in 12. For example, a roof that rises at the rate of 4 inches for each foot of run has a 4/12 or 4-in-12 slope.

soffit. The underside of a horizontal surface that projects beyond the wall line, such as an extended roof.

span. The horizontal distance between structural supports, such as of beams and of trusses; the width of a building.

splash board (skirt board). Pressure-treated boards or plywood placed at the bottom of a wall, below the siding, to protect the siding from ground moisture.

stocks. A chutelike enclosure that restrains a horse for veterinary care.

stucco. Cement plaster used as exterior wall covering.

stud. One of a series of slender wood or metal vertical structural members in walls. Wood studs are usually 2×4 or 2×6 boards spaced 16 or 24 inches on center.

subcontractor. A specialist (electrician, plumber, etc.) hired by a general contractor.

swale. A broad, shallow ditch or depression in the ground, either occurring naturally or excavated for the purpose of directing water runoff.

tamp. To pack down earth or other material firmly with a series of blows from a flat plate mounted onto a vertical handle.

tattoo. An indelible mark of identification on the inside of the upper lip of a horse, made by puncturing the skin and introducing some pigment into the punctures.

tongue and groove. A type of joint in which a tongue or spline along the edge of one board or panel fits into a groove of an adjoining board or panel.

treated wood. See *pressure-treated.*

truss. A framework of connected triangles used to support a roof without interior posts.

UL. Underwriters Laboratory, a private organization that evaluates and rates the safety of electrical equipment as either UL-listed or UL-recognized.

vapor barrier. Any waterproof material used to prevent passage of moisture, such as through walls and ceilings.

vent. An outlet or inlet for air.

ventilation. The process of supplying fresh air and removing stale air from a space.

volt. A unit of measure designating the amount of electrical pressure; volts multiplied by amps gives the wattage available in a circuit ($V \times A = W$).

watt. A unit of measure that designates the amount of electrical power consumed.

windrow. A mounded row of hay that has been raked together to dry before being baled; can also refer to a long manure pile.

wire size. American wire gauge (AWG) sizes range from 14 to 4/0; the larger the number size, the smaller the diameter.

Z trim. Sheet metal trim shaped like the letter *Z* with right angles; used as a drip cap and to cover the ends of steel panels.

zoning. The division of an area of land into smaller areas, the uses of which are restricted by law.

Index

Note: Numbers in *italics* indicate illustrations and photographs; numbers in **boldface** indicate charts.